THE
ENZYME MOLECULE

THE
ENZYME MOLECULE

W. FERDINAND

*Department of Biochemistry,
The University, Sheffield*

JOHN WILEY & SONS

London · New York · Sydney · Toronto

Library of Congress Cataloging in Publication Data

Ferdinand, W.
  The enzyme molecule.

  Includes bibliographical references.
  1. Enzymes. I. Title.
QP601.F378   574.1'925   76–7530
ISBN 0 471 018228 (Cloth)
ISBN 0 471 01821 X (Pbk)

Printed in Great Britain by J. W. Arrowsmith Ltd.,
Bristol, England

Dedication

To all my teachers, but in particular to
Dr G. T. Young, Dr L. A. Woodward and Dr R. Cecil
of the University of Oxford
and
Professor S. Moore and **Professor W. H. Stein**
of the Rockefeller University,
New York

Preface

The reasons for writing a book are no doubt various and subtle but in this case two can be identified as the major ones. The experience of teaching enzymology courses to second-year undergraduates for the last ten years at Sheffield University has convinced me of the need for a textbook that would cover the subject at this level. There are, to be sure, a number of excellent books on enzymes, and the standard textbooks of biochemistry devote large sections to aspects of enzymology. Nevertheless, there is no single source which is up-to-date and reasonably brief to which one can direct students and in which they can find all they need to know about the biological, structural and kinetic facets of the study of enzymes. That is a need which I have attempted to fill. The second reason for putting pen to paper stems from my conviction that the best reason for studying enzymes is to be found in their central role in biology. They are certainly of interest to chemists as molecules of great complexity and subtle properties—which is a good enough reason for a chemist to study them. That is how I first became interested in enzymes. But I have come to believe that enzymes should be studied primarily as biological entities rather than as chemical ones. The tools of the study are largely chemical but the object is a deeper understanding of biological processes mediated by enzymes.

Accordingly this book is directed at undergraduates who are studying enzymology as part of their course in biochemistry, botany, cell biology, genetics, microbiology, natural product chemistry, physiology or zoology. In addition postgraduate students coming to the study of enzymology for the first time should also find it valuable.

The first two chapters deal with the overall importance of enzymes in biology and with the basic tools of kinetics and thermodynamics needed to study them. Chapters 3 and 4 are concerned with the structure and catalytic mechanism of enzymes in chemical and three-dimensional molecular terms. This provides a framework for appreciating their biological function which is necessarily to be thought of in kinetic and regulatory terms. These latter topics are covered in the remaining chapters of the book. Wherever possible I have referred to real enzymes to illustrate different points. A total of thirty-seven enzymes, chosen from all the six major classes of enzymes, are used as examples. In this way I hope to give the reader some familiarity with the subject without bewildering him or her with too frequent references to the many thousands of known enzymes. Some enzymes, notably ribonuclease A and aspartate transcarbamoylase, are dealt with in some detail. At the end of each chapter I have included a list of key references both to books and, where appropriate, to the original literature to help the reader follow up the study of particular areas in greater detail if desired.

Throughout the book I have not hesitated to use mathematical formulations where they are needed. I do not believe the student of enzymology is well-served by shielding him or her from mathematical concepts and shorthand; the maths involved is at such a simple level conceptually that any student with a basis of algebra and an acquaintance with the differential calculus will easily take it in his stride. Nevertheless, since many students of biology have left their maths behind them at school, I have taken pains to develop the equations step by step in the early stages so that the process of following them should be painless for the reader.

To acknowledge all the teachers, colleagues and students who have enabled this book to be written would be an impossible task. Special thanks, however, are due to Professor Walter Bartley for his encouragement and for his many suggestions and constructive criticisms. Likewise they are due to my colleagues Dr Paul Engel, Dr Peter Banks, Dr Stanley Ainsworth and Dr Pauline M. Harrison. Lastly the book would not have been written at all without the patience and help of my wife Ann.

W. FERDINAND,
Totley Rise,
November 1975

Contents

List of symbols used xv

Chapter 1   THE ENZYME IN THE CELL 1

The Need for Enzymes in the Cell 1
The History of Enzymes 3
The Variety of Enzymes Found in the Cell 3
The Role of Enzymes in the Self-Regulation of Cellular Activities . 6
The Aims of Enzymology 8
Some Reasons for Studying Enzymes 10
   Enzymes in chemotherapy and medicine 10
   Enzymes in differentiation and development 11
   Enzymes and the chemical industry 12
Nomenclature and Classification of Enzymes 12

Chapter 2   BIOENERGETICS AND KINETICS . . . 14

Thermodynamic Considerations 14
   Enzymes act by increasing the rates of otherwise slow reactions . 14
   The enzyme as a biochemical rope and pulley . . . 16
   The measurement of free energy changes and their dependence upon
     concentrations 19
   Enthalpy and entropy in enzymology 20
Kinetic Considerations 21
   The measurement of reaction rates 22
   The catalytic effect of the enzyme 25
   The Michaelis–Menten hypothesis 26
   The Briggs–Haldane hypothesis 29
   The measurement of K_m and V_m 32
   Enzyme assays 36

Chapter 3 THE STRUCTURE AND PROPERTIES OF PROTEINS 41

The Amino Acids found in Proteins 41
Determination of the Molecular Weight of a Protein . . . 47
  Ultracentrifugal determination of molecular weight . . . 47
  Gel filtration 48
  SDS–gel electrophoresis 48
The Primary Structure of Proteins 50
The Spatial Organization of Protein Structure—Secondary and Tertiary Structure 61
  1. The hydrophobic interaction 62
  2. Hydrogen bonds 65
  3. Electrostatic interactions 72
  4. Electron delocalization 73
  5. Van der Waals forces 74
The Central Dogma 75
The Quaternary Structure of Proteins 76
X-Ray Diffraction Studies on the Three-Dimensional Structure of Proteins 81
Factors Affecting the Conformation of a Protein . . . 82

Chapter 4 ENZYME STRUCTURE AND FUNCTION . . . 85

Introduction 85
Ligand Binding Sites 85
  Enzyme cofactors 86
  Experimental detection and characterization of ligand binding sites 86
  The Scatchard plot 95
  Chemical modification studies 99
  Affinity labelling 101
  The induced-fit hypothesis 106
  Active sites and allosteric sites 109
  Specificity 110
Proenzymes, Isoenzymes and Mutations 112
Activation, Inactivation and Inhibition of Enzymes . . . 115
  1. Agents that alter enzyme activity by changing the covalent structure of the enzyme 116
  2. Agents that alter enzymic activity via changes in the enzymic environment 116
  3. Alteration of enzymic activity by the reversible, non-covalent binding of ligands 116
Catalysis 117
  The transition state 118
  Activation energy 119

Transition state analogues 121
Ways in which the enzyme can reduce the free energy of activation of
   a reaction 125
Entropic factors: Propinquity and orientation . . . 126
Enthalpic factors: Strain and general acid–base catalysis . . . 127
The Mechanism of Action of Bovine Pancreatic Ribonuclease A . . 131

Chapter 5 ENZYME KINETICS I: THE KINETICS OF
          INDEPENDENT ACTIVE SITES. 138

One-Substrate Kinetics 140
The effect of enzyme concentrations that approach those of substrate 140
The effect of product on the rate of enzyme-catalysed reaction . . 141
The Haldane relationships 144
The effect of isomerization of central complexes 145
The effect of enzyme isomerization 145
The effects of inhibitors and activators 146
The effect of pH on enzymic reaction rates 159
The effects of temperature on enzymic reaction rates . . . 161
The effects of ionic strength and dielectric constant on enzymic re-
   action rates 163
Kinetic Models 164
The structure of rate equations 164
Two-Substrate Enzymes 165
Random or branched mechanisms 165
Ordered or unbranched mechanisms 174
Enzymes with more than two Substrates or Products . . . 185

Chapter 6 ENZYME KINETICS II: THE KINETICS OF
          INTERACTING SITES 186

Homotropic and Heterotropic Interaction. 186
Ligand-Linked Conformational Change 187
Hill Plots and the Hill Coefficient 190
The Adair Hypothesis 192
The Monod, Wyman and Changeux (M.W.C) Model . . . 194
The binding curve for the M.W.C. model 196
The Sequential Model of Koshland 202
Aspartate Transcarbamoylase (ATCase). An Example of an Allosteric
   Enzyme 207
The catalytic subunit 212
The regulatory subunit 213
The ATCase oligomer 214

Chapter 7 ENZYMES AND THE CONTROL OF METABOLISM I: FINE CONTROL OF ENZYME ACTIVITY . . 217

Introduction 217
Steady State Fluxes 217
Flux Regulation by Feedback 220
Near-Equilibrium and Non-Equilibrium Reactions in Metabolic Pathways 224
Saturated or Substrate Independent Reactions: The Importance of V_m 227
The Pipe Analogy: Pacemakers and Bottlenecks . . . 228
The Theory of Kacser and Burns 228
Switch Mechanisms 233
The Role of Near-Equilibrium Reactions in Maintaining Metabolite Concentrations 236
Compartmentation as a Means of Control 237

Chapter 8 ENZYMES AND THE CONTROL OF METABOLISM II: COARSE CONTROL OF ENZYME ACTIVITY . 240

Coarse Control 240
The Operon Theory of Jacob and Monod 241
Operators and Repressors 242
What Controls the Rate of Enzyme Degradation? . . . 245
Factors Affecting Protein Turnover 246
Conclusion 247

Appendix 1 NOMENCLATURE AND CLASSIFICATION OF ENZYMES 248

Nomenclature 248
  1. Systematic names 248
  2. Recommended or trivial names 249
  3. Naming of multienzyme complexes 249
  4. Species and tissue differences 249
Classification 250
Key to Numbering and Classification of Enzymes . . 250
  1. Oxidoreductases 250
  2. Transferases 253
  3. Hydrolases 254
  4. Lyases 255
  5. Isomerases 255
  6. Ligases (synthetases) 256
Some Notes on the Rules of Nomenclature 256
Alphabetical List of Enzymes Referred to in the Text . . 258

Appendix 2 THE PURIFICATION OF PROTEINS . . . 261

Precipitation Methods 261
   1. Ammonium sulphate precipitation 261
   2. The use of other soluble salts for protein precipitation . . 265
   3. The use of organic solvents 265
Chromatographic Methods 265
   1. Molecular sieve chromatography or gel filtration . . . 265
   2. Ion-exchange chromatography 268
   3. Adsorption chromatography 270
   4. Affinity chromatography 270
Electrophoretic Methods 272
   1. Zone electrophoresis 272
   2. Isoelectric focusing 274
The Choice and Order of Methods of Purification . . . 275
Criteria of Purity 275

References 276

Author index 279

Subject index 281

List of Symbols Used

D	Diffusion coefficient
$\mathbf{D}$	Denominator of a rate equation
e	Base of natural logarithms
e_0	Total concentration of enzyme present in all forms
E	Enzyme
EX	Enzyme–X complex
$\mathbf{E}$	Energy
F	Free energy
ΔF	Free energy change
$\Delta F^{\ddagger}$	Free energy of activation
$\mathbf{F}$	Flux through a metabolic pathway
h	Hill coefficient
$\mathbf{h}$	Planck constant
H	Enthalpy or heat content
ΔH	Change in enthalpy or heat content
$\Delta H^{\ddagger}$	Enthalpy of activation
k	Rate constant
$\mathbf{k}$	Boltzmann constant
K or K_{eq}	Equilibrium constant
K_d	Dissociation constant
K_m	Michaelis constant
K_x	Dissociation constant for the dissociation of X from EX
$\mathbf{K}$	Controllability of an enzyme
L	A ligand
M	Molecular weight
n	Number of subunits in an oligomer
N	Avogadro number
$\overline{N}$	Average number of molecules of a ligand bound per molecule of protein
R	Universal gas constant
S	Entropy
ΔS	Change in entropy
$\Delta S^{\ddagger}$	Entropy of activation
$\mathbf{S}$	Sedimentation coefficient
T	Absolute temperature
v	Reaction velocity

v_0	Initial velocity of reaction
V_m	Maximum velocity of an enzyme-catalysed reaction
$\bar{V}$	Partial specific volume
$[X]$	Concentration of X
$\bar{Y}$	Fractional saturation of protein by ligand
Z	Sensitivity of an enzyme
Γ	Mass action ratio
ε	Elasticity of an enzyme
ε_r	Dielectric constant
μ	Ionic strength
ρ	Disequilibrium ratio

Chapter 1

The Enzyme in the Cell

Since the establishment of the cell theory by Schwann and Schleiden in the early nineteenth century, it has become clear that the cell is the fundamental unit of all living things. The study of cells is the proper subject of the growing discipline of Cell Biology and it is from this viewpoint that it is convenient to begin a discussion of the fundamental aspects of living things.

THE NEED FOR ENZYMES IN THE CELL

The most characteristic feature of any cell is that it is bounded by a membrane which contains a cell sap or cytoplasm. The type of membrane, its appearance, chemical composition and whether it is reinforced by a tough external capsule or not varies from one sort of cell to another, but in all cases it serves to define the limits of the cell. What lies inside the membrane may be thought of as 'alive', while on the outside of the membrane is the non-living, external environment.

Another feature of cells is that, for any particular type, the composition of the internal material is constant within closely defined limits and, in certain circumstances, it tends to increase in amount. That is to say, the cells can grow. This is possible only at the expense of material from the external environment.

It will be helpful in understanding the problems faced by cells in achieving growth, if we try to imagine how we would go about designing a simple unicellular organism. First of all one must be able to construct a membrane, maintain it against the ravages of the outside world and add to it as the cell grows. Membranes have been found to contain two main constituents, protein and lipid.

The external environment may or may not contain these substances; but even if it were to do so it is hardly likely that they would be the right kinds of protein and lipid to fit the needs of our hypothetical unicell. So the cell must be able to synthesize them from simpler substances available in the environment. As they are both made up of fairly complex organic components, amino acids in the case of proteins, or glycerol, fatty acids and various nitrogenous bases in the case of the lipids, a synthetic machinery is required first to construct these components, and then to put them together to make the correct kinds of protein and lipid.

But now another problem arises. In order to maintain the observed constancy of internal environment, the cell membrane must not be freely permeable to all the many kinds of chemical compound that may be encountered in the external world; and it must not allow valuable components required inside the cell to leak out. It must be a highly selective semipermeable membrane. In order to achieve such selectivity, mechanisms must operate within the membrane itself which recognize only those substances which should be allowed to cross it and which, moreover, actively encourage their transport in the proper direction. Clearly, the proteins and lipids present in the membrane, and their arrangement with respect to each other, must be of a highly specialized kind in order to achieve these special functions.

It is a feature of living cells that, although they may be presented with complex chemical structures such as proteins in the extracellular medium, nevertheless they usually break these down to simpler components before allowing them into the cell, and then build them up again to suit their own specifications. In most cases the degradative work is fairly easy, in fact if left alone for long enough many complex molecules such as proteins would be hydrolysed spontaneously. The cell, however, cannot wait for that to happen for, if it did, it would be likely to be broken down to its component parts itself. Some means of accelerating the process is required. Furthermore the reverse process, rebuilding the components into the complex structures of, for example, the membrane, does not take place spontaneously and the cell needs to be able to harness some source of energy in order to provide itself with the driving force to make its own complex structures. This can only come from outside, and all cells need to be able to utilize part of the material presented to them as food, in order to generate the energy needed to use the other part for maintaining and extending themselves. Plant cells can also utilize sunlight directly as their source of energy.

Already it can be seen that the simple requirements of a boundary—the membrane—and a constant internal environment—homeostasis—lead to a high degree of sophistication inside the cell. Complex machinery and organization are called for. In maintaining the constancy of the interior milieu the synthetic processes must be carried out time and time again with great fidelity. There is little room for error. In addition, however, these processes should not occur too slowly because the high degree of organization inside living cells leaves them vulnerable to the inevitable processes of decay which, though slow, are universally sure. The only way to avoid this is for the cell to be able to break down its food, obtain its energy and repair itself faster than destructive processes can degrade it. Unfortunately many of the chemical reactions that are observed inside cells in the normal course of their metabolic activity are inherently unlikely and slow processes. Except in living cells they hardly ever occur. This observation posed a serious problem to biologists of earlier generations and gave rise to the idea of a 'vital force' resident in cells which mysteriously directed their chemical activities.

The answer to the problem of speed lies, of course, in the existence of biological catalysts, enzymes, which pick out and speed up those specific chemical reactions needed for living processes. In order for life to exist there is no need for the enzymes to work particularly fast, but once life does exist competition will mean that those cells which contain enzymes that can work faster may be at an advantage. Conceptually the enzyme is the trump which allows the cell to exist. In this very general way the enzyme can be seen as fundamental to life, providing a way of keeping ahead of the universal tendency towards disorder, although the detailed energetics of how this is done will be dealt with later.

THE HISTORY OF ENZYMES

The discovery of enzymes came about gradually over a period that spans the latter half of the nineteenth century. The history of enzymes makes fascinating reading as described by Dixon.[1] The credit for discovery should probably go to Payen and Persoz who in 1833 published a report describing an extract of malt which was able to hydrolyse starch to glucose. The true significance of this discovery was not generally recognized and a long period of controversy followed until 1897. In that year E. Buchner reported an experiment that marked a turning point. In itself it was not a very remarkable experiment, more of an accident than anything else. An extract of living yeast cells had been made by grinding the yeast with an abrasive to break the cell walls. In order to preserve the resulting clear liquid for further study, sugar was added with the surprising result that fermentation occurred, just as if the living yeast cells had been present, though none could be demonstrated. In this way it became clear that one did not need the intact 'living' cell to carry out a typical metabolic process, but that some catalyst or catalysts present in yeast were capable of bringing about the overall process leading from sugar to alcohol. The work of many investigators, among them, not unnaturally, representatives of the brewers, was needed before the nature of enzymes lay revealed. Perhaps this early association between enzymes and alcohol is what has made enzymology so attractive to so many biochemists. Each enzyme in the cell catalyses a particular chemical reaction or group of closely related reactions. This specificity, together with the truly enormous increase in reaction rates brought about by enzymes, is what distinguishes them from inorganic catalysts and gives them their importance in the cell. Enzymes behave differently from inorganic catalysts in other ways as well, particularly in respect of their fragility, as we shall see later. In one respect, however, they are similar: the rate of the catalysed reaction is always directly proportional to the concentration of catalyst present. This is not always clearly understood and deserves special emphasis. Twice the amount of catalyst in a given volume gives twice the reaction rate.

THE VARIETY OF ENZYMES FOUND IN THE CELL

Many hundreds of different enzymes have now been isolated and all have been found to be proteins, macromolecules made up of amino acids polymerized together to form long chains. The cellular synthesis of enzyme proteins from their constituent amino acids, and in many cells the synthesis of those amino acids from even simpler precursors, can be seen to require the intervention of other enzymes to catalyse the processes. Already one can begin to discern something of the complexity of the metabolic activity occurring in cells. All the processes of living cells are interrelated and interdependent. Returning to the consideration of a unicellular organism, degradative enzymes must be exported from the cell to break down complex molecules to their building blocks. Enzymes are needed in the membrane to direct and catalyse the transport of small molecules into and out of the cell. Others catalyse the oxidation of some of these molecules in order to provide energy with which still others will catalyse the assembly of macromolecules including all the enzymes. Each enzyme acts upon one particular kind of molecule, its substrate, and catalyses its conversion to a product, which may then serve as substrate for another enzyme and so on. Such chains of enzymically catalysed reactions, each chain leading from an initial substrate via a number of intermediates to a final product, constitute the metabolic pathways of living matter.

The problem of organizing all these activities so that they all remain in step is made worse by the fact that the external environment upon which the unicell depends is not constant. The types of raw material upon which the cell must operate vary from moment to moment and, of course, are altered by the very activity of the cell itself. In this situation of potential confusion, some kind of overall organization is clearly required. A means of stipulating which enzymes are made at a particular time is needed, so as to match the metabolic capacity of the cell with its food supply. Some mechanism for determining how rapidly a particular enzyme shall operate is also necessary to ensure that no one activity gets out of step with the others, for that would lead to waste and cells cannot afford to be uneconomical. In other words the whole cellular organization needs to be regulated. Unless we are prepared to revert to the ideas of the nineteenth century vitalists, it is also clear that, since nothing else is available to regulate cellular activity, it must regulate itself.

Before considering the problems of self-regulation it is necessary to enlarge on another difficulty facing our hypothetical unicellular organism, and that is the need for a central store of information. This need arises because of the complexity of the macromolecular structures of the cell, particularly its proteins. As each protein is made up of a long string of amino acids, covalently joined together, and as there are twenty different types of amino acids, it is easy to see that a very large number of different proteins is possible. This variety is extremely useful to the cell, but it does present a problem of information. Each enzyme and each structural protein, for example, in the cell membrane, most probably has its own unique structure, its proper sequence of amino acids. To make more of that particular protein, something in the cell has to 'know' the correct order in which to join together its constituent amino acids. The information needed for this resides in another class of macromolecules found in all cells, the nucleic acids, in particular, deoxyribonucleic acid or DNA. The DNA present in a cell can be thought of as the central library where all the necessary information about the chemical structure of the enzymic and structural machinery of the cell is stored. Whenever a new molecule of some particular protein is needed, either to replace one that has become degraded, or to allow the cell to grow, or to meet some new contingency because of a change in food supply, there in the DNA the information can be found to make it correctly. In principle, provided the information stored in the DNA could be maintained intact, and provided self-regulatory mechanisms could be superimposed upon the cellular activities, there would be nothing to prevent a unicellular organism from maintaining itself indefinitely. When the external supply of food was abundant it would enlarge itself and store some reserves against harder times. In adverse circumstances it would regulate itself so as to restrict its activities to repairing such damage as was inevitable and wait for external abundance before resuming its growth.

There is, however, a natural limit to the size of a cell. This is dictated by the fact that all the chemical reactions occurring inside it, the movement into it of the small molecules it requires, and the excretion of its waste products are limited by molecular diffusion. As the cell becomes bigger diffusion limits restrict its metabolism and would eventually prevent further growth.

By dividing into two the cell maintains its size within the optimal range and growth of the cells can proceed. But in order for the two cells to continue as before each must have a central library, which raises the need for duplication of the DNA. The DNA library must contain within itself not only the information on how to make all the proteins of the cell, but also information on how to replicate itself. If it did not, the need would arise for a second store of information on how to make more DNA and for yet a third store concerned with the replication of the second, and so on: the cell would find itself in the philosopher's dilemma of the infinite regression.

The structure of DNA is well suited to self-replication since it is composed of two polynucleotide strands each the complement of the other. The strands have only to be separated, fresh complementary nucleotides paired off against the nucleotides of each strand, and all that remains is to join together these new nucleotides in order to complete the replication. This is not the place to describe the details of DNA structure and replication which have been brilliantly set out elsewhere.[2] The purpose of introducing the subject is only to illustrate the fact that here again, in an activity that is absolutely central to the continuity of life, enzymes are essential. They are needed to catalyse the synthesis of nucleotides and the joining together of the correct nucleotides in order to make the daughter strands. Other enzymes are needed to preserve the integrity of DNA in between the acts of replication that occur before cell division.

Nevertheless, despite the elaboration of enzymes to direct and catalyse the precise replication of DNA and to preserve the information in the completed DNA molecules, mistakes do occur. When this happens there are three possible results. If the new sequence of nucleotides leads to the synthesis of proteins that are hopelessly inadequate for their tasks in the cell, that cell dies. On the other hand the mistake or mutation might not be very serious, in that the altered protein resulting from the new information might be functionally indistinguishable from its predecessors, or at least able to perform its task, albeit less efficiently. The third possibility is that the new protein may actually perform its task in the cell better than the original, in which case the particular cell in which the mutation occurred, and all its daughters containing the mutation, will have an advantage over the other cells. These three types of mistake, lethal, neutral and advantageous mutations, provide the wherewithal for evolution. It is paradoxical that, in order for the future of a species to be assured in a given environmental niche, where the cellular surroundings lie within a certain range of conditions, the library of information in the DNA of the species must be faithfully reproduced from generation to generation. The paradox arises because, if the environment changes, survival can then only come about by adaptation, which in the last analysis can only occur by mistakes being made in the transmission of the genetic information of DNA. So, to ensure survival whatever happens, the genetic and enzymatic machinery for maintaining and replicating DNA must be almost, but not quite, perfect.

THE ROLE OF ENZYMES IN THE SELF-REGULATION OF CELLULAR ACTIVITIES

The stark outlines of the picture of a functioning cell are now almost complete and it only remains to return to the problem of self-regulation. A mechanical engineer faced with the problem of designing a machine that will regulate itself, proceeds as follows. If the speed at which the machine runs has to be maintained within narrow limits, the speed itself is made to operate a switch that cuts off the supply of fuel when that desired speed is reached. The device that accomplishes this is called a governor.

In a similar way an electronic engineer will design a self-regulating amplifier circuit using the feedback principle. The idea is exactly similar in that the output of the device, either current or voltage, can be used to shut off the input, when a desired level is reached.

Cells utilize the governor or feedback principle in regulating their own metabolism. To achieve homeostasis the concentrations of the component substances inside the cell must be held within close limits. It is clear that the rate at which a given component is made or used up must be regulated by the concentration of that component. If its concentration begins to climb above the optimal level, it may be because it is being produced too rapidly; and the increased concentration must somehow or other damp down the processes which synthesize it. This is called feedback inhibition, or negative feedback. Alternatively the concentration may be rising because the substance is not being used fast enough by processes which utilize it. In this case the increased concentration must speed up the latter processes by feedforward activation, or positive feedforward. The concept of metabolite concentration regulating the rate of its own synthesis or utilization is absolutely central to the homeostatic mechanism; it will be necessary to go into the details of such mechanisms more thoroughly later. For the moment, however, the question of how metabolic rates could possibly be affected by metabolite concentrations must be asked.

Earlier in this chapter it was seen how the chemical reactions typical of living cells are inherently slow reactions that require enzyme catalysis to make them proceed at the necessary rates. One must be clear that the reaction rate of any chemical process depends upon the concentrations of the reactants. Enzyme catalysed reactions are no exception to this rule, though the dependence of reaction rate on substrate concentration may take some unusual forms (see Chapters 5 and 6). On the whole it is true that if the concentration of a substrate is increased from a low value the rate of enzyme catalysed reaction will increase. Consider for a moment the metabolic pathway by which a molecule of A is converted to one of Z. The substance A could be a raw material entering the cell, and Z a molecule needed in the synthesis of a macromolecular structure. The pathway can be depicted as:

$$A \rightarrow B \rightarrow C \rightarrow D \cdots \rightarrow \cdots X \rightarrow Y \rightarrow Z$$

each arrow representing a specific chemical reaction catalysed by an enzyme. The rate of the first reaction in the pathway, $A \rightarrow B$, will be increased if the concentration of A rises from virtually zero to some new value. The increased concentration of B that results will likewise cause an increase in the rate of the second step, and so on down the pathway. This is fine provided that the cell needs the substance Z. But if it already has enough Z for its immediate needs the feedback principle must operate by Z slowing down the initial reaction $A \rightarrow B$, otherwise a pile-up of unwanted intermediates B, C, D, etc., will ensue. Clearly there are two quite distinct problems to be understood in connection with the regulation of metabolic rates by metabolite concentrations. The first is how rate depends upon the reactant concentrations. In our example this problem would be concerned with the dependence of the rate of the process $A \rightarrow B$, upon the concentration of A. It must be added that, because any process can also go backwards, in principle at any rate, we may have to consider the dependence of rate on the concentration of B also in this category.

The other problem is how the rate of the process $A \rightarrow B$ depends upon the concentration of substances other than the substrate and product of the reaction, substances such as Z the end-product of the whole pathway. This is not so easy to envisage until we recall that it is an enzyme catalysed reaction. This being the case it will be clear that the primary target for control of metabolic rates is the enzyme itself. Remove the catalytic factor and the rate must fall; replace it and the rate must increase. In attempting to understand how metabolic regulation occurs we must seek ways in which metabolite concentrations can affect the catalytic power of enzymes.

One way in which this could happen is by changing the amount of enzyme. As with all catalysts, the amount of enzyme present determines the rate of the catalysed reaction. The amount of a given enzyme present in a cell, or rather its cellular concentration, is determined by two factors. Enzymes are inherently unstable protein molecules and are breaking down all the time in the cell. Indeed there is good evidence that certain enzymes exist in the cell solely for the purpose of catalysing the destruction of others. So one factor determining the concentration of a given enzyme is its rate of destruction. The other is of course its rate of synthesis. When these two are equal, the enzyme concentration remains constant. Let us assume for the moment that degradation of enzymes proceeds at a more or less constant rate all the time. In that case, the way to influence the concentration of an enzyme is to affect its rate of synthesis. This kind of control certainly occurs and is called coarse control.

Having said that, however, one appears to be no further towards an understanding of metabolic regulation. All we have said is that, in order to control a particular reaction rate one must control another reaction rate, that is the rate of synthesis of a particular enzyme. There must be some other way in which metabolites can influence the catalytic power of enzymes. In fact this involves an alteration not in the amount or concentration of an enzyme, but in the catalytic efficiency of the enzyme. Instead of removing some of the enzyme or making more of it, one modifies the existing enzyme, making it a poorer or better catalyst. This is called fine control. These terms, fine and coarse control, are not particularly good descriptions of what actually happens, but in the absence of better ones they will be used throughout this book to describe control by modification of existing catalytic power on the one hand, fine control, and alteration in the amount of catalyst on the other hand, coarse control.

A discussion of fine control will have to be deferred until Chapter 7 because it is impossible to describe how a metabolite can, as it were, tune the enzyme molecule to perform its catalytic task more or less efficiently until something of the structure and function of the enzyme has been dealt with. Suffice it to say that the metabolite combines with the enzyme molecule and alters it subtly, either inhibiting or activating it.

THE AIMS OF ENZYMOLOGY

So far, this discussion of the cell and the place occupied in it by the enzyme has proceeded at a conceptual rather than a practical level. Many aspects of the living cell and its metabolism have been touched upon in order to try to illustrate two important aspects of living matter. The first is the organizational interdependence of its many complex systems. The second is the ubiquitous and indeed essential role played by enzymes. It is not too much to claim that the enzyme lies at the heart of the secret of how cells live. Enzymology seeks to understand the role of enzymes in cellular activity.

In 1932, in the Introduction to his book *The Work, Wealth and Happiness of Mankind*, H. G. Wells wrote,

Suppose, to borrow an idea from Mr Bernard Shaw, some young man or young woman, instead of being born in the usual fashion, were to be hatched out of an egg at the age of twenty, alertly intelligent but unformed and uninformed. He or she would blink at our busy world and demand, 'What are they all up to? Why are some so active and some so inactive? Why are some toiling so industriously to produce things and some, it seems, doing nothing but consume? Why is this? What is going on?' [reproduced by permission of The Estate of H. G. Wells]

If we transfer the analogy from the world to the microcosm of the cell and observe the enzymes in it instead of the people, the same questions would be asked and enzymology would be the attempt to answer. In this book our aims cannot be so lofty and we shall try only to depict some of the more essential roles of enzymes, their structure, and an idea of the mechanisms by which they achieve their functions in the cell.

Before beginning this task it is first necessary to correct some misunderstandings that may have arisen so far, out of an eagerness to present all the activities of enzymes in cells in as broad a frame as possible. The first thing to point out is that cells are not mere membranous bags full of enzymes. Although that impression may have been given, it is far from the truth. Cells come in all sorts of sizes and shapes and with a wide variety of internal structures. Excellent introductions to the structures of cells and tissues can be found elsewhere.[3] The unicellular bacteria have little discernible internal structure even when observed under the electron microscope, although the DNA is all gathered up in one part of the cell. But even here there is good reason to believe that certain enzyme systems are located in quite definite places inside the cell, in some cases attached to, or associated with, the inner face of the membrane. More complex cells have a number of structures or organelles inside them together with complex internal membranous surfaces ramifying throughout them. Once more certain metabolic activities and the enzymes needed for them are precisely located on, or in, these structures. The idea of compartmentation arises, that is the sealing off of one part of the cell from another by membrane barriers so that activities can be carried out in one place under different constraints from those applying in another part of the cell.

Another false impression that may have been created is that the enzyme is a kind of all-purpose magic molecule of limitless potential, able to catalyse reactions with remarkable speed and specificity, direct the events of the cell with unerring accuracy and in turn to be regulated by events inside the cell. The concept of the enzyme does incorporate all of these features, but not all at once. Individual enzyme molecules are chemical entities of a quite specific structure that is well understood in an increasing number of cases. The structure is beautifully tailored to the function of the particular enzyme. The astronomical number of different enzyme proteins that are possible, because of the number of ways of arranging twenty different amino acids in long chains, is what really makes possible the variety of enzyme functions. The folding up of the long chains into a roughly spherical but high specialized three-dimensional shape and the possibilities for aggregation of these chains, give rise to all the observed properties and functions of the enzymes. These structural aspects and their bearing on the role of the enzymes will be discussed in Chapters 3 and 4.

SOME REASONS FOR STUDYING ENZYMES

The discerning student may be wondering what practical advantages are conferred by the study of enzymes. It should hardly be necessary to stress that, as enzymes are involved in so many of life's processes, an understanding of biology to some extent at least demands an acquaintance with enzymes. But there are more direct and palpable reasons for studying them. The following sections pick out one or two areas of current interest. The list is by no means exhaustive, but it illustrates the diversity of ways in which enzymology is important.

Enzymes in chemotherapy and medicine

Clearly the whole organization of the cell is such a finely balanced and self-regulating system that it might be thrown easily out of order. A well directed 'biochemical spanner', aimed at a particular enzyme, can rapidly destroy the smooth functioning of a cell. This fact is potentially both hazardous and useful to mankind.

In the case of disease caused by pathogenic micro-organisms for example, it has been possible to develop compounds which specifically hit the microbe while missing the man. In many cases this occurs because the drug in question inactivates a microbial enzyme, but for one reason or another hardly affects the equivalent enzyme in the human body; in other cases there is no equivalent human enzyme and no other enzymic system is affected significantly by the drug. Up to now much of the research leading to new antimicrobial drugs has proceeded on a hit-or-miss basis in which a large number of compounds are tested. When one is found that is effective against the microbe it is tested on animals and finally on volunteers. Increasingly, however, the hope is that by studying the particular enzyme systems of pathogenic organisms one may find ways of inactivating them. This logical approach requires a great deal of biochemical understanding of the metabolic pathways of micro-organisms, the particular microbial enzymes that catalyse them and the ways in which they differ from mammalian systems.

The goal of chemotherapy is to be able to design subtle reagents that will find their way through the microbial defences to specific target enzymes while at the same time leaving the human body unaffected. In this quest the pharmacologist has been helped, perhaps surprisingly, by microbes themselves. Many micro-organisms, particularly the actinomycetes, synthesize and excrete compounds which have antibacterial activity. They probably do this to protect themselves. These antibiotics as they have come to be called, can be synthesized and even improved upon by the drug manufacturers. The way in which some of them act is now beginning to be understood; it appears that they exert their effect by blocking specific enzyme catalysed processes in the bacteria against which they are directed. Drugs have other uses besides killing pathogens. Sometimes, for hereditary or environmental reasons, the mechanisms of the body fail to function correctly. Drugs can often help in rectifying the malfunction, but here again a more complete understanding of the specific ways in which they act, the systems they interact with and the enzymes involved in these systems will one day enable the physician to develop more powerful tools for controlling the disorders.

These are some of the ways in which one can tinker with cellular metabolism either to hamper or help it. Viruses, on the other hand, act in quite a different way. They invade the cell and make use of certain cellular processes, perverting them to the synthesis of more viruses. No general way of controlling virus infection has yet emerged but intense study of the viral takeover process is under way. The nature of the new viral enzymes made by the host cell under instruction from viral DNA or RNA is being investigated and it may transpire that a way of acting against the viruses will emerge from these studies.

Another serious problem in which the study of enzymes has a significant part to play is that of cancer. In this case some of the body cells appear to lose part of their essential control mechanisms and proliferate uncontrollably. One might typify cancer as a failure of metabolic regulation in a small number of cells, perhaps only one to start with. What initially causes this imbalance is not yet known, but it is probable that any one of many things may initiate cancer. Mutation induced by radiation, carcinogenic chemicals, or a failure of the DNA repair systems may be responsible; or it may be that a virus starts the process by turning off some control mechanism essential for the balance of body cells. Studies of the enzymes of cancerous tissues have shown that many of them appear to be affected, some are missing and others are altered, or the control processes normally regulating their activity are affected. Whatever the causes of cancer and however it is ultimately understood and brought under control, the study of enzymes will have a contribution to make.

Enzymes in differentiation and development

From a biological rather than a medical point of view, one of the most fascinating aspects of modern biological research is that which concerns itself with differentiation. By this term one means the process whereby one type of cell divides and produces daughter cells which are not identical in form and function. This is the process by which a seed grows to produce not only root and stem but leaf and flower tissue in plants. It has been shown that a single cell from the root of the mature plant has within it all the genetic information to generate cells typical of any other part of the plant. Clearly some of the information is not expressed in root cells, while a different part of the information is dormant in stem or leaf cells. This same process underlies embryonic development in animals. There is clearly a kind of programmed control over the biochemical activities of the cell and its daughters. This problem is immensely complex but a start has been made with·simple organisms such as the cellular slime moulds which can assume one form in one set of circumstances or another form in another. Differentiation during the metamorphoses of insects is also a subject of intensive study at the present time. Study of the metabolic processes occurring during differentiation shows that some enzymes are simply not made until either an external stimulus or a built-in programme causes the unmasking of the information needed for their synthesis, whereupon they are produced while other enzymes cease to be made. The change in enzyme pattern necessarily produces a different intracellular organization in the differentiated cells.

Enzymes and the chemical industry

Lastly, in an entirely non-biological context, the study of enzymes promises to revolutionize the ability of the organic chemical industry to produce complex substances. Many reactions which are extremely difficult to carry out by the traditional methods of organic chemistry can be brought about with ease if the right enzyme is available. The isolation of such enzymes and the study of how best to use them in an organic chemical context is extremely important. A recent development in the industrial use of enzymes arises from the finding that their attachment to solid mesh structures at the same time stabilizes the delicate enzymes against denaturation and provides the chemist with a convenient catalyst. He has only to cause a flow of suitable reagents past or through the mesh at neutral pH and room temperature and the desired product emerges at the other side. Needless to say this production line approach to chemistry has great attraction for the chemical industry and in particular the brewers. Instead of having to brew by the traditional batch process in which yeast is added to an infusion of malt and fermentation allowed to proceed to completion, the hope is that the individual enzymes involved in fermentation could be harnessed in sequence on a series of support meshes. The wort would be allowed to flow through the train of meshes and beer would emerge at the other end.

We have come full circle, back to alcohol where the study of enzymes originated. It is now necessary to turn to the discussion of what enzymes are and how they function in detail.

NOMENCLATURE AND CLASSIFICATION OF ENZYMES

In the course of this discussion it will be necessary to refer to many different enzymes by name. To avoid confusion it should be stressed that enzymes are named according to the reaction they catalyse, by adding the suffix '-ase' to either the substrate or the reaction. For example, the hydrolysis of urea to ammonia and carbon dioxide:

$$NH_2 - CO - NH_2 + H_2O \rightarrow 2NH_3 + CO_2$$

is catalysed by an enzyme called urease. This is a convenient working name or *trivial name*. The enzyme also has a formal name and a classification number, which describe explicitly the nature of the reaction according to a set of well-defined rules. In this case the full name is urea amidohydrolase and the number is 3.5.1.5. The rules for nomenclature and for assigning the classification number have been laid down by the Commission on Enzymes of the International

Union of Biochemistry and are described in Appendix 1 at the end of the book, together with a summary of the various types of enzymes classified according to the rules of the Commission.

In addition, a list of all the enzymes referred to in the text, together with the reactions they catalyse, will also be found in Appendix 1.

Chapter 2

Bioenergetics and Kinetics

In studying the biological activities of enzymes we are always looking at their effect upon chemical reactions. In many cases it appears that an enzyme actually brings about a chemical reaction that otherwise would not occur and bio-chemists may sometimes catch themselves referring to the enzyme which 'causes' a certain reaction.

One must be careful to distinguish between two possible modes of action of the enzyme. It may act simply by speeding up a normally slow reaction, so slow perhaps that its rate would be undetectable, but nevertheless a reaction that was theoretically possible. On the other hand, it might actually make possible a chemical reaction that was otherwise impossible.

THERMODYNAMIC CONSIDERATIONS

Enzymes act by increasing the rates of otherwise slow reactions

Let us examine, first of all, the proposition that the enzyme can cause a reaction to occur that is impossible in its absence. There are two types of 'impossibility' here. Clearly it is beyond the power of enzymes to bring about reactions that contravene basic laws of chemistry, the law of conservation of mass for example, or which result in transmutation of elements. On the other hand, one can visualize chemical reactions in which the atoms of the reactants are rearranged to give new compounds, in a manner that is not normally found to occur spontaneously, and ask whether an enzyme could make it occur. As an example, consider the conversion of carbon dioxide and water to glucose and oxygen:

$$6CO_2 + 6H_2O \rightarrow C_6H_{12}O_6 + 6O_2 \tag{1}$$

At room temperature and atmospheric pressure this reaction is never found to proceed to any detectable extent in aqueous solution. Could the addition of a suitable enzyme bring it about? The organic chemist would say that the reaction is thermodynamically impossible for the following reason. If one started with the elements carbon, hydrogen and oxygen and first of all allowed them to combine to make carbon dioxide and water, a certain amount of free energy would become available, that is energy which could be used to perform

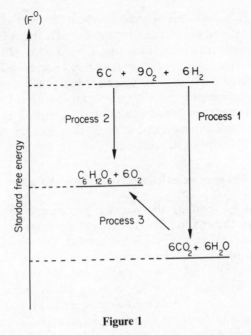

Figure 1

work (Process 1 in Figure 1). On the other hand, if the same elements were now allowed to combine to form glucose and oxygen, less free energy would be produced; that is to say, Process 2 would be able to perform less work as shown in Figure 1. It is an experimental law of thermodynamics that chemical reactions only proceed when free energy is liberated. In other words, only processes corresponding to downward pointing arrows in Figure 1 occur spontaneously. It can be seen quite clearly from Figure 1 that free energy would be released if $C_6H_{12}O_6 + 6 O_2$ were to react to produce $6 CO_2 + 6 H_2O$, but not in the reverse direction, i.e. Equation 1, shown as Process 3 in Figure 1. That is why the organic chemist could say with such certainty that the reaction would not take place in the direction shown in Equation 1 and Process 3. He could also add, with equal certainty, that the presence of any enzyme would not induce the reaction to take place. The reason for this would be that, although there may be more than one way of getting from reactants to products in a chemical reaction, nevertheless the free energy change will always be the same whatever the route taken, because the free energy of the reactants and that of the products are both fixed constants under a particular set of physical conditions, so the change in free energy must always be the same. Thermodynamically then, enzymes change nothing. If a reaction is energetically impossible, an enzyme will not make it possible.

This is not to say that the reaction shown as Process 3 (i.e. Equation 1) cannot be *made* to proceed by somehow providing it with the necessary free energy. Indeed in plant cells it apparently does occur by a series of chemical steps, each catalysed by an enzyme. In order to understand how this happens it is necessary to go a little deeper into biochemical thermodynamics or bio-energetics. Before doing so one must say immediately that what occurs in plant cells is not the simple reaction of Equation 1, but a series of reactions in which CO_2 and water figure, among other reactants such as ATP and NADPH. Glucose is produced together with other products such as ADP and NADP, which are then reconverted to ATP and NADPH with the release of oxygen. The absorption of light provides the necessary free energy. The role of enzymes in these processes is fundamental, but in no case does an enzyme promote a thermodynamically impossible reaction.

The enzyme as a biochemical rope and pulley

One must begin by recalling that certain processes of change occur spontaneously if given the chance. The hackneyed examples are that water flows downhill, if it can, and weights fall if unimpeded. The reverse processes never occur spontaneously. Water can be made to flow uphill and weights can be lifted, but only by performing work upon them. We say that a source of free energy must be harnessed in order to bring about these non-spontaneous processes. If we look more closely into what is meant by a source of free energy it always turns out to be some spontaneous process. For example, a weight can be made to rise by coupling it, via a rope and pulley, to another weight which is then allowed to fall. The only proviso is that the second weight must be a heavier one. In other words the spontaneous process that is harnessed must be more spontaneous than the one that is being driven backwards against its natural tendency.

It must be clear that if one wishes to drive a non-spontaneous process one must first find out how non-spontaneous it is and then look for a spontaneous process whose spontaneity is numerically greater than that of the first one. A measure of the spontaneity of processes is required. The required measure is the free energy change of the process, in other words its ability to perform work at constant temperature and pressure. The free energy change is denoted as ΔF, or sometimes ΔG.

Spontaneous processes are characterized by a negative free energy change, that is to say when they occur the free energy of the components decreases. Non-spontaneous processes, on the other hand, although they cannot actually occur can be thought of as having a positive free energy change. The former are said to be exergonic, and the latter endergonic.

It must be stressed that the free energy change of a process tells us nothing about how fast it occurs. One might think that a highly spontaneous process was one that occurred rapidly, but that is not what is meant by this particular use of the word spontaneous. The free energy change measures only the amount of work that can be done. Many highly spontaneous processes exist which proceed only extremely slowly at room temperature, such as the chemical reaction between dry oxygen and hydrogen to produce water. On the other hand, processes with a small free energy change can occur exceedingly fast, like the ionization of water to hydroxide and hydronium ions. There is no connection between the thermodynamic spontaneity or free energy change of a process and its kinetics or rate.

Armed with the knowledge that the free energy change is a measure of the spontaneity of a chemical reaction, and bearing in mind that if it is positive the reaction will not occur, no matter whether an enzyme is present or not, it is now possible to return to the question of using exergonic or spontaneous reactions to drive endergonic ones. The free energy change, ΔF, of a chemical reaction depends upon the concentrations of reactants and products that are present. The standard free energy change, ΔF^0, is defined as the free energy change when all reactants and products are present at unit concentration. Consider the reaction:

$$\text{acetyl CoA} + CO_2 \xrightarrow{25\,^{\circ}C} \text{malonyl CoA} \tag{2}$$

The standard free energy change for this reaction is about $\Delta F^0 = +4\cdot5$ kcal/mol, so it tends to be non-spontaneous. Yet acetyl CoA *is* converted to malonyl CoA as the first step in the synthesis of fatty acids. This endergonic reaction is driven by the exergonic hydrolysis of ATP:

$$ATP^{4-} + H_2O \xrightarrow{25\,^{\circ}C} ADP^{3-} + \text{orthophosphate ion} + H^+ \tag{3}$$

$$\Delta F^0 = -8\cdot9 \text{ kcal/mol}$$

Since the free energy change of ATP hydrolysis is larger than that required to drive malonyl CoA synthesis, the two could clearly be coupled together provided a suitable mechanism was available, in just the same way as a heavy weight falling could be coupled to lift a lighter one. Merely allowing ATP hydrolysis to proceed in the same cell as the acetyl CoA and CO_2 would not bring about malonyl CoA synthesis, of course. That would be as absurd as dropping a large weight on one side of the room and expecting a lighter one to rise into the air on the other side of the room. A biochemical rope and pulley are required, to couple the endergonic and exergonic reactions together.

It must be clear that the actual process by which a falling weight lifts a lighter one is not the same as the sum of the two separate processes involving the uncoupled falling of a heavy weight and the (imaginary) rising of a light weight, although the net result is the same as far as the weights are concerned. In the same way the actual process of malonyl CoA synthesis can be expected to be different from the two processes depicted in Equations 4 and 5, although the net result must be the same:

$$\text{acetyl CoA} + CO_2 \rightarrow \text{malonyl CoA}; \Delta F^0 = +4\cdot5 \text{ kcal/mol} \tag{4}$$

$$ATP^{4-} + H_2O \rightarrow ADP^{3-} + HPO_3^{2-} + H^+; \Delta F^0 = -8\cdot9 \text{ kcal/mol} \tag{5}$$

$$\text{Net: acetyl CoA} + CO_2 + ATP^{4-} + H_2O$$

$$\rightarrow \text{malonyl CoA} + ADP^{3-} + HPO_3^{2-} + H^+;$$

$$\Delta F^0 = -4\cdot4 \text{ kcal/mol} \tag{6}$$

What is required is a new set of processes, each of which is exergonic and has the same net result as Equation 6. This is provided by the enzyme acetyl CoA carboxylase which thus acts as a biochemical rope and pulley for coupling together reactions 4 and 5. The enzyme, denoted as E in the following equations, contains biotin as a bound cofactor. The E–biotin complex provides a mechanism whereby ATP, acetyl CoA and CO_2 can be brought together to produce malonyl CoA, ADP and inorganic phosphate:

$$\text{E–biotin} + CO_2 + ATP^{4-} + H_2O$$

$$\rightarrow \text{E–biotin} \cdot CO_2 + ADP^{3-} + HPO_3^{2-} + H^+;$$

$$\Delta F^0 = -4 \cdot 2 \text{ kcal/mol} \qquad (7)$$

$$\text{E-biotin} \cdot CO_2 + \text{acetyl CoA} \rightarrow \text{E–biotin} + \text{malonyl CoA};$$

$$\Delta F^0 = -0 \cdot 2 \text{ kcal/mol}$$

$$\text{Net: acetyl CoA} + Co_2 + ATP^{4-} + H_2O$$

$$\rightarrow \text{malonyl CoA} + ADP^{3-} + HPO_3^{2-} + H^+;$$

$$\Delta F^0 = -4 \cdot 4 \text{ kcal/mol, as before} \qquad (6)$$

By selecting the thermodynamically possible reactions depicted in Equation 7 and catalysing them, the enzyme brings about the synthesis of malonyl CoA even though in the same cell, at the same time, the malonyl CoA is spontaneously decarboxylating to acetyl CoA (Equation 4 in reverse), and ATP is spontaneously hydrolysing to ADP and phosphate (Equation 5). Although many of the enzymes present in a living cell are involved in coupling exergonic and endergonic reactions, by no means all of them act in this way. Many enzymes act simply by accelerating exergonic reactions of value to the cellular economy.

It must be apparent by now that enzymes exert their effect kinetically. That is, they function only by speeding up selected reactions. They can only act within the constraints imposed by the thermodynamic limitations. In that sense the enzyme resembles a conjuror, rather than a magician. A magician could make impossible things happen, but a conjuror, by sleight of hand, works so rapidly that one almost imagines that something impossible has occurred. If he does it for us slowly we can see how easy it really is, but at the same time we have to admire his skill. Enzymology concerns itself with understanding how the conjuror does it. There are two major parts of this study. The first task is to measure reaction rates and discover how they depend upon metabolite concentrations. This part is the study of enzyme kinetics. The second task is to appreciate the mechanism of enzyme action, which requires a knowledge of both enzyme kinetics and the chemical structure of enzymes.

One can see how, in principle, the existence of specialized enzymes in plant cells allows them to harness the free energy available to them from the absorption of sunlight, in order to drive the endergonic synthesis of glucose and oxygen from carbon dioxide and water.

The measurement of free energy changes and their dependence
upon concentrations

One may ask how to measure the spontaneity, the free energy change, of a chemical reaction such as A + B → P + Q occurring in solution. Imagine that we have a certain concentration of A, denoted by [A], and a concentration of B denoted by [B], in solution together with concentrations [P] and [Q] of P and Q respectively. We wish to know whether A and B will be converted to P and Q. In that case the reaction would be spontaneous or exergonic and the free energy change would be negative. Alternatively P and Q might be converted to A and B under these conditions, so the reaction A + B → P + Q would be non-spontaneous, or endergonic, and the hypothetical reaction would be associated with a positive free energy change.

Whichever of these alternatives turned out to be the correct one, the reaction would proceed in one direction or the other until equilibrium was reached, after which no further change would occur, and the new concentrations $[A]_{eq}$, $[B]_{eq}$, $[P]_{eq}$ and $[Q]_{eq}$ would remain constant. An equilibrium constant K_{eq} can be defined for the reaction where

$$K_{eq} = \frac{[P]_{eq} \cdot [Q]_{eq}}{[A]_{eq} \cdot [B]_{eq}}$$

A prior knowledge of K_{eq} would be of some help in deciding whether the reaction would proceed from left to right. A large value of K_{eq} tells us that the reaction is inherently very likely to proceed, whereas a value of K_{eq} much less than unity tells us that it is inherently unlikely. Without, however, knowing the actual concentrations [A], [B], [P] and [Q] that are present in solution at the start of reaction, we still should not be able to say whether the reaction would be spontaneous. Both the equilibrium constant and the concentrations [A], [B], [P] and [Q] are needed before the free energy change of the reaction can be found. In fact the true measure of spontaneity, the free energy change ΔF, is given by the equation:

$$\Delta F = -RT \ln K_{eq} + RT \ln \frac{[P][Q]}{[A][B]} \tag{8}$$

where R is the universal gas constant ($1 \cdot 987$ cal deg^{-1} mol^{-1}). T is the absolute temperature and ln denotes the natural, or Naperian, logarithm. By measuring K_{eq} and the concentrations of reactants and products and inserting them into Equation 8, one can find the value of ΔF for any reaction.

The equation can be tested quite easily by putting in some values of K_{eq}, [A], [B], [P], and [Q], and working out the value of ΔF. It is best first to rearrange the equation to the form

$$\Delta F = RT \left(\ln \frac{[P][Q]}{[A][B]} - \ln K_{eq} \right) = RT \ln \frac{[P][Q]}{[A][B]K_{eq}}$$

By playing with the values inserted into the equation one can find out how the sign and size of ΔF depend upon the equilibrium constant and the concentrations.

There are several points to note about Equation 8. The first is that if $[A] = [A]_{eq}$, $[B] = [B]_{eq}$, $[P] = [P]_{eq}$, and $[Q] = [Q]_{eq}$, that is if the reaction actually starts at equilibrium, then $\Delta F = 0$. This merely tells us that at equilibrium the free energy change of any reaction is neither positive nor negative but zero. The second is that if $[B] = [A] = [P] = [Q] = 1$ then, since $\ln 1 = 0$, $\Delta F = -RT \ln K_{eq}$. This particular value of the free energy change, the value when all reactants and products are present at unit concentration (i.e. 1 M), is denoted ΔF^0, the standard free energy change. It is nothing more than the equilibrium constant in disguise! Values of ΔF^0 are commonly quoted in the biochemical literature and are sometimes used as a measure of the spontaneity of particular reactions but, like the equilibrium constant, they are only partial measures of spontaneity. Without knowing the concentrations of reactants and products as well as ΔF^0, one cannot say whether a reaction will be exergonic or endergonic since

$$\Delta F = \Delta F^0 + RT \ln \frac{[P][Q]}{[A][B]}$$

Further insight into the intricacies of biochemical thermodynamics and the derivations of the equations given here can be found in standard texts on the subject.[4]

Enthalpy and entropy in enzymology

However, before leaving the topic for the moment, it is necessary to point out that there are two components of the free energy change ΔF. Wherever a process of change occurs heat may either be absorbed or given out, or there may be no alteration in the heat content of the system undergoing change. If heat is given out the system must have moved to a state of lower heat content, or enthalpy, denoted by the symbol H. In that case the enthalpy change accompanying the process, ΔH, is said to be negative. It is easy to see that when heat energy is lost, the system must tend to become more stable, so processes with a negative value of ΔH tend to be spontaneous ones. On the other hand, many spontaneous processes are known where ΔH is positive, so some other driving force must be present to offset the unfavourable increase in enthalpy. This driving force is the entropy change ΔS. The entropy, S, is a measure of the disorder of a system. Whereas enthalpy tends to decrease, entropy tends to increase, that is to say when change occurs spontaneously, there is a tendency for the system to become more disordered. If this tendency happens to be strong enough it can even compensate for an increase in heat content. The overall spontaneity of a process, ΔF, is compounded of these two driving forces as expressed in the relationship $\Delta F = \Delta H - T\Delta S$, where the absolute temperature T, also enters into the equation. Thus if ΔH is large and negative, ΔF will be negative even though ΔS may be negative (i.e. a decrease in the disorder or increase in the order of the system). On the other hand, in cases where $T\Delta S$ is large and positive, ΔF can still be negative although ΔH may be positive. These two tendencies, the one towards lower enthalpy and the other towards greater disorder underlie all processes of change and equilibrium, not only chemical and biochemical ones. Often, of course, ΔH is negative and ΔS positive so that there is no conflict and ΔF is unequivocally negative. In the study of enzymes, however, and particularly in the study of the forces that underlie their structure and stability, one frequently finds that there is a very fine balance between enthalpic and entropic tendencies, each pulling opposite ways, so that small changes in the conditions, especially the temperature, can lead from stability to instability and vice versa. Further consideration of these topics will be left until Chapter 3.

KINETIC CONSIDERATIONS

Before dealing with enzyme kinetics, a few basic aspects of general chemical kinetics must be established. Until now chemical reactions such as $A \rightarrow P$ have been discussed as if they resembled other natural processes like falling weights. In one respect, however, they have to be treated differently because when one studies them, the reactions of individual molecules cannot be studied in the same way as one can observe a single falling weight. It is always a collection of molecules that is studied, some of which will possess more energy than others. Thermodynamically this does not matter. Chemical thermodynamics concerns itself with whether a collection of molecules of A will become converted to a collection of molecules of P under certain circumstances, or what proportion of them will be converted, and where the equilibrium between A and P lies. In kinetic studies, on the other hand, one always has to bear in mind that, whatever the general trend, some molecules of A will be in the process of conversion to P while, at the same time, the reverse process will be occurring with other molecules. Even at equilibrium the interchange of A and P is going on, although now there is no net change of A to P or vice versa. It is a *dynamic* equilibrium. In short, kinetics always has to recognize that chemical reactions are reversible. In principle they are all reversible, though in practice some are effectively irreversible because the equilibrium favours one side of the reaction so strongly, but one always writes the reaction with double arrows: $A \rightleftharpoons P$. In case this should cause confusion after what has been said about non-spontaneous reactions *never* occurring unless driven, it is worth reiterating that thermodynamics deals with the overall trend of a reaction. In this way one can consider the spontaneous reaction $A \rightarrow P$, where at given concentrations of A and P the overall reaction proceeds with the formation of P, secure in the knowledge that under these conditions there will be no *net* conversion of P to A. At the same time, one recognizes that some molecules of P will be converted to A, in any given time span, but during that time a greater number of molecules of A will be converted to P.

The measurement of reaction rates

A prerequisite for experimental kinetic studies is a method of measuring the rate at which a chemical reaction proceeds. This can be done by measuring either the rate at which a product of the reaction appears or the rate at which a reactant disappears. For the simple uncatalysed reaction $A \rightleftharpoons P$ the concentration of A will fall as time passes and that of P will increase, as shown in Figure 2. The rate or velocity, v, of the reaction at any time, t, is defined as the slope of either line at that time. In the language of differential calculus:

$$v_t = -\left(\frac{d[A]}{dt}\right)_t = \left(\frac{d[P]}{dt}\right)_t \tag{9}$$

where [A] and [P] are the concentrations of A and P, and the subscript t means 'at time t'. The two slopes $(d[A]/dt)_t$ and $(d[P]/dt)_t$ are shown on the figure. As implied in Equation 9, the two slopes must be numerically equal but of opposite sign, simply because the rate at which A disappears *must* be equal to the rate at which P appears, since there is nowhere else for A to go and nowhere else for P to come from. This implies that the reaction occurs without any hidden intermediates between A and P. The figure also shows that the rate v_t decreases with time and finally tends to zero as equilibrium is approached.

One merely has to measure the concentration of the reactant or product at various times, plot a graph as in Figure 2 and measure the slope of the graph at

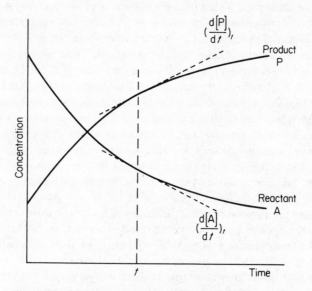

Figure 2 Graph showing how the concentrations of A and P change with time for a simple uncatalysed reaction, $A \rightleftharpoons P$

any desired point to obtain the rate of reaction at that point. The most convenient way of measuring the concentration of a substance in solution is by spectrophotometry or fluorimetry, since the concentration can be monitored continuously without disturbing the reaction. If for any reason this is impracticable, samples may have to be withdrawn from the reaction mixture at intervals and the concentration measured by some other assay, after stopping the reaction of course.

In the example shown in Figure 2 the product P is already present at the start of reaction, so the situation is now part of the way towards equilibrium. If one had started with only A in solution at a known concentration, $[A]_0$, the initial rate v_0, or v_i as it is sometimes denoted, would have been given by $-(d[A]/dt)_0$ and it can be shown experimentally that in the absence of P this is always proportional to the initial concentration $[A]_0$. In other words:

$$v_0 = -\left(\frac{d[A]}{dt}\right)_0 = k[A]_0$$

where k is the rate constant for the reaction step $A \longrightarrow P$. If one had chosen instead to start with only P in solution, and no A present, the rate of disappearance of P would have been given by $v_0 = -(d[P]/dt)_0$ and this is found to be proportional to $[P]_0$, the initial concentration of P:

$$v_0 = k'[P]_0$$

where k' is a new rate constant, now referring to the reverse step $A \longleftarrow P$.

When both A and P are present at concentrations $[A]$ and $[P]$, each tends to convert to the other. The rate at which $A \longrightarrow P$ is given by $k[A]$, and the rate at which $A \longleftarrow P$ is given by $k'[P]$, so the net rate of conversion of A to P is given by

$$v = k[A] - k'[P]$$

This equation holds for any concentrations of A and P because k and k' are true constants under given conditions. Going back to Figure 2 where A and P are both present throughout, it is now quite easy to predict the value of v_t provided one knows or can measure the concentrations $[A]_t$ and $[P]_t$ and so long as one knows k and k'. On the other hand, if one measures v at a number of values of $[A]$ and $[P]$, it is possible to work out the values of k and k' from a series of equations such as:

$$v_1 = k[A]_1 - k'[P]_1$$

$$v_2 = k[A]_2 - k'[P]_2$$

In principle, since there are only two unknowns, k and k', these two equations should be sufficient, but one would usually make a series of rate measurements v_1, v_2, v_3 etc., at values $[A]_1$, $[P]_1$, then $[A]_2$, $[P]_2$, then $[A]_3$, $[P]_3$, etc., so as to avoid experimental error.

The fundamental point to grasp in kinetics is that each step in a reaction, such as $A \longrightarrow P$, or $A \longleftarrow P$, has assigned to it a unique rate constant, k, such that the rate of that step is given by the product of k and the concentration of the substance that is reacting. In a more complex process such as $A + B \longrightarrow P + Q$, there is still a rate constant k, but now the rate of the process is given by $k\,[A][B]$, the product of the rate constant and both reactant concentrations. The reverse process $A + B \longleftarrow P + Q$ would have another rate constant k' and its rate would be given by $k'[P][Q]$. Processes or reaction steps in which only one type of molecule is involved as reactant, such as $A \longrightarrow$ are called first order processes and the constant is a first order rate constant. Where two reactants are concerned, for example in the process $A + B \longrightarrow$, it is a second order process with a second order rate constant. Note that a reaction such as $A \rightleftharpoons P + Q$ consists of a first order process in one direction and a second order process in the other.

The rate of a reaction is measured as a rate of change of concentration, usually as moles/litre/second or M/s. This is sometimes more conveniently written as $\mathrm{mol}\,l^{-1}\,s^{-1}$ or $M\,s^{-1}$. For a first order process where $v = k \times a$ concentration, the dimensions of k are v/concentration that is s^{-1}. For a second order process the rate is given by:

$$v = k \times \text{one concentration} \times \text{another concentration}$$

$$\therefore \quad v = k \times (\text{concentration})^2$$

$$\therefore \quad k = \frac{v}{(\text{concentration})^2}$$

so the dimensions of a second order rate constant are $M^{-1}\,s^{-1}$ or litre moles^{-1} seconds^{-1}. The fact that first and second order rate constants are expressed in different units, need cause no concern; they are both proportionality constants and their different dimensions merely reflect the fact that they imply proportionality between rate and, on the one hand, a single concentration (first order) and, on the other hand, the product of two concentrations (second order).

In passing, it is worth noting that when a reaction is at equilibrium, and no further net change is occurring, the rate of the forward step is equal to that of the reverse step. The ratio of the concentrations then gives the equilibrium constant of the reaction. For the reaction $A \rightleftharpoons P$

$$K_{eq} = \frac{[P]_{eq}}{[A]_{eq}} \tag{10}$$

But we can also write:

$$v_{eq} = 0 = k[A]_{eq} - k'[P]_{eq}$$

$$\therefore \quad k[A]_{eq} = k'[P]_{eq}$$

$$\therefore \quad \frac{k}{k'} = \frac{[P]_{eq}}{[A]_{eq}} \tag{11}$$

From Equations 10 and 11 we now find that a relationship exists between K_{eq}, a thermodynamic constant, and the two rate constants k and k' of a reaction: $K_{eq} = k/k'$. The kinetic properties of a chemical reaction are in this way constrained or limited by the thermodynamic constants that apply to it. The limit, however, is not very restrictive; in principle k and k' can assume any values between 0 and ∞ provided only that their ratio k/k' is equal to K_{eq}. In practice the upper limits for rate constants are set by various physical considerations, as discussed by Gutfreund.[5] The upper limit for first order rate constants is about $10^{13}\,\mathrm{s}^{-1}$ and for second order rate constants about $10^{11}\,\mathrm{M}^{-1}\,\mathrm{s}^{-1}$.

The catalytic effect of the enzyme

Having looked at the kinetics of the uncatalysed reaction $A \rightleftharpoons P$, now the effect of adding an enzyme must be investigated. Starting with a mixture of A and P as in Figure 2, the time course of the reaction will appear very similar to that figure except that everything will occur much more rapidly. The time scale will be much contracted and equilibrium will be reached more rapidly. The rate of reaction could be measured just as before if one were quick enough. Alternatively the concentrations of A and P can be reduced to low values where the uncatalysed rate is exceedingly slow, whereupon the addition of enzyme will give rise to measurable rates. An investigation of the dependence of rate upon concentration of A in the presence of a fixed concentration of enzyme, now reveals that the simple equation for the initial rate in the absence of product:

$$v_0 = k[A]_0$$

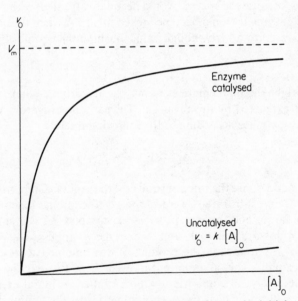

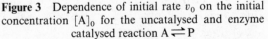

Figure 3 Dependence of initial rate v_0 on the initial concentration $[A]_0$ for the uncatalysed and enzyme catalysed reaction $A \rightleftharpoons P$

now no longer holds. A graph of v_0 against $[A]_0$, instead of being a straight line of slope k, is now a curve, as shown in Figure 3. As the initial concentration $[A]_0$ is increased the rate at first increases in a more or less linear fashion, but then falls off to a limiting value, V_m or V_{max}, at very high values of $[A]_0$. Analysis of the shape of this curve reveals that for many enzymes it is part of a rectangular hyperbola governed by the equation:

$$v_0 = \frac{V_m[A]_0}{K_m + [A]_0} \tag{12}$$

where K_m is a constant, the Michaelis constant. When $[A]_0$ becomes very large, approaching infinity, Equation 12 simplifies to $v_0 = V_m$, since K_m becomes negligible in the denominator and $[A]_0$ cancels out. It was also found that if the substrate concentration $[A]_0$ was held constant in a series of experiments in which the amount of enzyme was increased, the rate was proportional to enzyme concentration.

The Michaelis–Menten hypothesis

These observations led several of the early enzymologists such as Brown in 1902, Henri in 1903 and Michaelis and Menten in 1913, to make the suggestion that the enzyme E actually combines reversibly with the reactant or substrate A to form a complex EA, which then breaks down to liberate the product P together with unchanged enzyme E. This proposal explained why the initial rate reached a limiting value when the substrate concentration was increased to high values, because the enzyme would become 'saturated' with A. Essentially all of the free enzyme E would become locked up in EA when the concentration of A was increased to a high enough value, pushing the equilibrium

$$E + A \rightleftharpoons EA$$

over to the right. In mathematical terms, the hyperbolic shape of the curve could also be explained by this proposal. The early investigators assumed that the reaction took place according to the following model:

$$E + A \underset{k_{-1}}{\overset{k_1}{\rightleftharpoons}} EA \overset{k_2}{\rightarrow} E + P \tag{13}$$

where k_1, k_{-1} and k_2 are the rate constants for the separate steps in the reaction. The enzyme E and substrate A were supposed to combine together extremely rapidly (k_1 = large), and the enzyme substrate complex EA was supposed to be able to break down again to E and A extremely rapidly (k_{-1} = large). The breakdown of EA to E and the product P was assumed to occur relatively slowly (k_2 = small). This assumption allowed them to treat the system as if E and A were always in equilibrium with EA, after an initial extremely short period of time during which EA was building up to the equilibrium concentration. Thereafter, and for all practical purposes, it was assumed that the relatively slow 'leakage' of EA to E + P did not perturb the equilibrium. This assumption is now known as the Rapid Equilibrium, or R.E., Assumption.

In any sequence of events, such as depicted in Equation 13, the overall process, the conversion of A to P in this case, cannot occur faster than the slowest step in the sequence. That step, known as the rate limiting step, is clearly the conversion of EA to E + P in this case, and its rate defines the overall rate of reaction. The slow drain on EA to produce P will be rapidly compensated by the net conversion of E + A to EA, so the substrate A actually disappears at the same rate as P appears.

Under these circumstances then, the measured initial rate will be given by:

$$v_0 = k_2[EA]_{eq} \tag{14}$$

where $[EA]_{eq}$ is the equilibrium concentration of the EA complex. The experimental value of $[EA]_{eq}$ at a given initial concentration of A will not be known, but it is defined by an equilibrium constant and the concentrations of E and A thus:

$$EA \rightleftharpoons E + A: \qquad K_d = \frac{[E]_{eq}[A]_{eq}}{[EA]_{eq}}$$

$$\therefore [EA]_{eq} = \frac{[E]_{eq}[A]_{eq}}{K_d} \tag{15}$$

where K_d is the dissociation constant of the EA complex. It is the reciprocal of the association constant K_a for the reaction $E + A \rightleftharpoons EA$, but in enzyme kinetics it is usual to use dissociation constants rather than association constants.

The values of $[E]_{eq}$ and $[A]_{eq}$ are no more accessible to us experimentally than was $[EA]_{eq}$, but progress can now be made towards expressing $[EA]_{eq}$ in terms of constants and the known value of $[A]_0$. The concentration of enzyme is usually very small in comparison with that of its substrate so the assumption can be made that, at equilibrium, only a negligible proportion of A will have been converted to EA. This allows one to say

$$[A]_{eq} = [A]_0$$

Further, although neither [E] nor [EA] is known at any point during reaction, their sum must always equal the original concentration of enzyme present at the start, e_0:

$$\therefore e_0 = [E] + [EA] \tag{16}$$

This is known as the Conservation Equation and is always true since enzyme is not destroyed during the catalytic reaction; it must be present either as E or as EA. From Equation 16,

$$[E]_{eq} = e_0 - [EA]_{eq}$$

substituting for $[E]_{eq}$ in Equation 15 and replacing $[A]_{eq}$ by $[A]_0$, we obtain:

$$[EA]_{eq} = \frac{(e_0 - [EA]_{eq})[A]_0}{K_d}$$

$$\therefore \quad K_d[EA]_{eq} = e_0[A]_0 - [EA]_{eq}[A]_0$$

$$\therefore \quad [EA]_{eq}(K_d + [A]_0) = e_0[A]_0$$

$$\therefore \quad [EA]_{eq} = \frac{e_0[A]_0}{(K_d + [A]_0)}$$

and substituting this value for $[EA]_{eq}$ into Equation 14, we now obtain an expression for the rate of reaction in terms of constants and the known value of $[A]_0$:

$$v_0 = \frac{k_2 e_0[A]_0}{(K_d + [A]_0)}$$

Comparison of this Equation with the experimentally observed Equation 12, i.e. $v_0 = (V_m[A]_0)/(K_m + [A]_0)$, shows that $V_m = k_2 e_0$ and $K_m = K_d$. The first of these equalities is just what would be expected from the kinetic model used to derive the rate equation, because the maximum rate of reaction, V_m, must be achieved when all the added enzyme, i.e. e_0, is present as EA. The second equality tells us that, according to this kinetic model, the constant K_m is nothing more than the dissociation constant of the EA complex.

The assumptions made in this model and the mathematical derivation of the rate equation applying to it are as follows:

Rapid Equilibrium is achieved and maintained at all times between the enzyme and its substrate;

the rate limiting step is the breakdown of the EA complex to product;

a negligible proportion of the initial concentration of substrate is used up in achieving the equilibrium;

enzyme is not 'used up', and can only exist as free E, or as the complex EA.

A further implicit assumption in this treatment is that the reverse reaction from P to A does not occur, because no step depicting it is present in Equation 13. Although this is a serious omission, because all reactions should be considered as theoretically reversible, nevertheless when considering only initial rates of reaction in the complete absence of product, it does not matter. Even if the model had been expressed in a sounder way as $E + A \rightleftharpoons EA \rightleftharpoons E + P$, in the absence of P no back reaction from $E + P$ to EA can occur anyway. It does, however, raise a more profound objection to the Rapid Equilibrium Assumption. If the fully reversible reaction were to be considered, why should the left-hand equilibrium $E + A \rightleftharpoons EA$ be assumed to occur so much more rapidly than the other? It is quite impossible to consider the enzyme catalysed conversion of P to A in the absence of A by means of the Rapid Equilibrium Assumption, once we have made that assumption for the forward direction.

Another objection that could be raised against the Rapid Equilibrium Assumption is that we have no *a priori* reason for supposing that k_2 will be much less than k_1 and k_{-1} for all enzymes; after all the enzyme is actually *catalysing* the conversion of A to P and, in some cases, the breakdown of EA to E + P is likely to be very fast, perhaps as fast as the binding of A to E, and probably as fast as its breakdown to E + A again. For all these reasons the original derivation of the rate equation by the Rapid Equilibrium Assumption, the Michaelis and Menten approach to the problem, is less useful than it might be. It provides a possible explanation of the experimentally observed kinetic behaviour, but not a very convincing one.

The Briggs–Haldane hypothesis

A more fundamental approach to the problem must be adopted. Consider what happens to a solution of the substrate A from the instant when enzyme E is first added to it. Let the total concentration of added enzyme be e_0, as before, and the initial concentration of substrate be $[A]_0$. It is reasonable to assume, as before, that enzyme and substrate reversibly form an enzyme–substrate complex EA according to the model:

$$E + A \underset{k_{-1}}{\overset{k_1}{\rightleftharpoons}} EA \underset{k_{-2}}{\overset{k_2}{\rightleftharpoons}} E + P \tag{17}$$

At first the rate of disappearance of A is given by $-(d[A])/dt = k_1 e_0 [A]_0$, the minus sign merely indicating disappearance, but very quickly a finite concentration of EA builds up and starts to contribute to the re-formation of A and E, so that the net rate of disappearance of A now becomes:

$$-\frac{d[A]}{dt} = k_1(e_0 - [EA])([A]_0 - [EA]) - k_{-1}[EA] \tag{18}$$

The quantities in the brackets are the new instantaneous concentrations of free enzyme and substrate, taking account of the fact that some of the initial concentration of each has been removed to form the instantaneous concentration, $[EA]$, of EA (cf. the conservation equation used previously).

At the same time, as soon as EA appears, the second leg of the sequence can come into operation, generating P at a rate given by:

$$\frac{d[P]}{dt} = k_2[EA] \tag{19}$$

In fact EA is now being generated from E and A at a rate

$$k_1(e_0 - [EA])([A]_0 - [EA])$$

and is disappearing in two directions, to re-form E and A at a rate $k_{-1}[EA]$ and to form P at the rate $k_2[EA]$. The net rate of formation of EA is thus:

$$\frac{d[EA]}{dt} = k_1(e_0 - [EA])([A]_0 - [EA]) - k_{-1}[EA] - k_2[EA]$$

$$= k_1(e_0 - [EA])([A]_0 - [EA]) - [EA](k_{-1} + k_2) \tag{20}$$

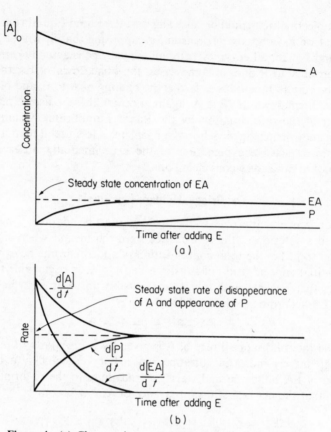

Figure 4 (a) Changes in concentration in the first few milliseconds after adding enzyme E to substrate A in the absence of
product P. (b) Changes in rate during the first milliseconds after
adding enzyme E to substrate A in the absence of product P

As more and more EA is formed, the rate of its formation from E and A decreases, because of the decrease of the quantities in the first two brackets, while
at the same time the rate of its breakdown, $[EA](k_{-1} + k_2)$, increases. The
result is that as EA increases in concentration, its net rate of formation
(Equation 20), slows down. Eventually a point will be reached where its net
rate of formation is zero, in other words it is appearing from E and A just as
fast as it is disappearing to re-form E and A, and to form E and P. At this time
its concentration will stabilize and $[EA]$ = constant. As soon as this happens
Equation 18 shows that the rate of disappearance of A becomes constant,
while Equation 19 shows that the rate of formation of P also becomes constant.
A hidden assumption underlying these statements is, of course, that EA reaches
a constant concentration before a finite concentration of P has had time to
build up. Otherwise the second bracket in Equation 18 would have to be
modified to $([A]_0 - [EA] - [P])$ to include the fact that some P had appeared,

and an additional term would have to be added to Equation 20 to take account of the rate of re-formation of EA from E and P (i.e. $k_{-2}[\mathrm{EP}]$).

The changes occurring in the extremely short period after adding enzyme to A, and leading to the so-called 'steady state', where [EA] is constant, are shown in Figure 4. The figure also shows that in the steady state the rate of disappearance of A is equal to the rate of appearance of P, which must be the case because the concentration of the only other component that can contain A, that is EA, is constant by definition in the steady state.

The steady state is achieved very rapidly after mixing the enzyme with substrate. Fast reaction techniques have shown that in most cases studied, it is achieved within a few milliseconds, and before any finite concentration of product P has appeared. This means that when one measures the 'initial' rate of reaction by the usual experimental methods, one is actually measuring the steady state rate of reaction, because the conventional kinetic techniques are too slow to catch the rate in the first few milliseconds.

So we can now write some equations relating v_0 to $[\mathrm{A}]_0$ in terms of kinetic constants and e_0:

$$v_0 = k_2[\mathrm{EA}] \tag{21}$$

$$\frac{d[\mathrm{EA}]}{dt} = k_1(e_0 - [\mathrm{EA}])([\mathrm{A}]_0 - [\mathrm{EA}]) - [\mathrm{EA}](k_{-1} + k_2) = 0 \tag{22}$$

where [EA] denotes the steady state concentration of EA. Although [EA] is unknown in any particular experiment, it can be eliminated from these two equations. One usually makes the additional assumption that $([\mathrm{A}]_0 - [\mathrm{EA}]) \simeq [\mathrm{A}]_0$. This assumption is generally valid because e_0 is so much smaller than $[\mathrm{A}]_0$ that even if the whole of the enzyme were converted to EA it would not make an appreciable difference in the concentration of free A.

From Equation 22:

$$k_1(e_0 - [\mathrm{EA}])[\mathrm{A}]_0 = [\mathrm{EA}](k_{-1} + k_2)$$

$$\therefore \quad k_1 e_0[\mathrm{A}]_0 - k_1[\mathrm{EA}][\mathrm{A}]_0 = [\mathrm{EA}](k_{-1} + k_2)$$

$$\therefore \quad [\mathrm{EA}](k_{-1} + k_2 + k_1[\mathrm{A}]_0) = k_1 e_0[\mathrm{A}]_0$$

$$\therefore \quad [\mathrm{EA}] = \frac{k_1 e_0[\mathrm{A}]_0}{(k_{-1} + k_2 + k_1[\mathrm{A}]_0)}$$

dividing numerator and denominator by k_1:

$$\therefore \quad [\mathrm{EA}] = \frac{e_0[\mathrm{A}]_0}{[(k_{-1} + k_2)/k_1] + [\mathrm{A}]_0}$$

substituting this value of [EA] into Equation (21):

$$v_0 = \frac{k_2 e_0[\mathrm{A}]_0}{[(k_{-1} + k_2)/k_1] + [\mathrm{A}]_0} \tag{23}$$

Comparison of Equation 23 with the experimental Equation 12 now shows that $V_m = k_2 e_0$ as expected, but now $K_m = (k_{-1} + k_2)/k_1$ instead of K_d as predicted by the rapid equilibrium model. The dissociation constant K_d is, of course, equal to k_{-1}/k_1, so the two approaches to explaining Equation 12 actually provide quite similar descriptions of the Michaelis constant K_m. In fact the rapid equilibrium treatment is a special case of the more general steady state treatment. In the rapid equilibrium case the extra assumption has been made that $k_2 \ll k_{-1}$, so that the rate limiting step is the breakdown of EA to product. Considering the value of K_m predicted by the steady state treatment:

$$K_m = \frac{(k_{-1} + k_2)}{k_1}$$

it is obvious that when $k_2 \ll k_{-1}$, K_m becomes equal to k_{-1}/k_1, that is, equal to K_d. Note that the steady state treatment makes no assumptions about which step is rate limiting and as such it is a more valid approach to the general description of enzyme kinetics. It was first used by Briggs and Haldane in 1925. Another point in its favour is that in the living cell, where substrate is supplied continuously and product is removed continuously, steady state conditions are actually maintained. This point will be returned to later. In the meantime, however, we are faced with the possibility that, in the case of any particular enzyme that we might be studying, the rapid equilibrium description might actually be the correct one. How would one distinguish? So far it has been shown that both steady state and rapid equilibrium models will lead to the same experimental rate law, the only difference being in the meaning given to K_m, the Michaelis constant. In the special case of the rapid equilibrium model, K_m will be equal to K_d the dissociation constant of the enzyme for its substrate. In the more general steady state model, K_m will be greater than K_d. In some cases, though not usually for an enzyme with only one substrate, it is possible to measure K_d by independent methods. In those cases, measurement of K_m would decide the question. In any case, the value of K_m is one of the fundamental kinetic properties of an enzyme and one must be able to measure it. The other fundamental measurable is V_m, the maximum rate of the enzyme or, more properly, V_m/e_0, the maximum rate per unit concentration of the enzyme; and one must be able to evaluate it also.

The measurement of K_m and V_m

Figure 5 shows a so-called Michaelis curve, in which initial rate v_0 is plotted against $[A]_0$ according to Equation 12, $v_0 = (V_m[A]_0)/(K_m + [A]_0)$. One can see that whereas a rough estimate of V_m can be obtained by extrapolation of the curve towards $[A]_0 = \infty$, nevertheless an accurate estimate is difficult to obtain from this type of curve. If an accurate value of V_m was available, inspection of Equation 12 shows that K_m is equal to the substrate concentration, $[A]$, at

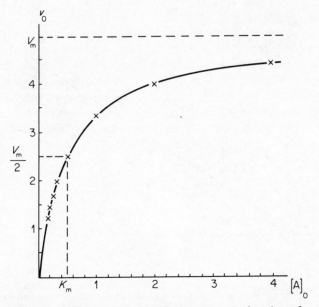

Figure 5 Michaelis plot of initial rate v_0 as a function of initial substrate concentration $[A]_0$ for an enzyme obeying the rate equation:

$$v_0 = \frac{V_m[A]_0}{K_m + [A]_0} \qquad (12)$$

where $V_m = 5$ and $K_m = 0{\cdot}5$.
The crosses represent experimental values of v_0 obtained at increasing values of $[A]_0$ in a series of kinetic experiments

which v_0 is half V_m:

i.e. $$\frac{V_m}{2} = \frac{V_m[A]}{K_m + [A]}$$

$$\therefore \quad K_m + [A] = 2[A]$$

$$\therefore \quad K_m = [A]$$

This value is shown in Figure 5.

Without resorting to computer-aided methods, the best way to obtain an accurate estimate of V_m and K_m from the experimental data is to replot them in a different manner. A very common procedure is that due to Lineweaver and Burk in which $1/v_0$ is plotted against $1/[A]_0$. From Equation 12 by simple inversion:

$$\frac{1}{v_0} = \frac{K_m + [A]_0}{V_m[A]_0} = \frac{K_m}{V_m} \cdot \frac{1}{[A]_0} + \frac{1}{V_m} \qquad (24)$$

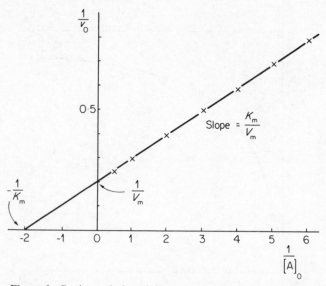

Figure 6 Reciprocal plot of the same experimental data shown in Figure 5 according to Lineweaver and Burk

so the so-called reciprocal plot will be a straight line of slope K_m/V_m and intercept on the $1/v$ axis equal to $1/V_m$, as shown in Figure 6. It can easily be shown that when $1/v_0 = 0$, $1/[A]_0 = -1/K_m$, so that if the line through the experimental points is extrapolated back until it cuts the abscissa, the value of $-1/K_m$ is obtained without further ado. Since it is a simple matter to take reciprocals of the raw data before plotting, and since it is much easier to extrapolate a straight line than a hyperbola, this is the preferred method of plotting data when V_m and K_m are required. It does, however, have one disadvantage because it puts great weight on those points furthest from the ordinate, which are those obtained at the lowest values of $[A]_0$ and v_0. These may not be the most accurate experimental points so one may be misled into estimating incorrect values for V_m and K_m. A method that avoids this danger involves plotting $v_0/[A]_0$ against v_0. The Michaelis equation (Equation 12) can be rearranged as follows:

$$v_0 = \frac{V_m[A]_0}{K_m + [A]_0} \tag{12}$$

$$= \frac{V_m}{(1 + K_m/[A]_0)}$$

$$\therefore \quad v_0 + \frac{v_0 K_m}{[A]_0} = V_m$$

$$\therefore \quad v_0 = V_m - \frac{v_0}{[A]_0}K_m$$

A plot of v_0 against $v_0/[A]_0$ will therefore have a slope of $-K_m$ and an intercept when $v_0/[A]_0 = 0$, of V_m as shown in Figure 7. Another method of plotting the experimental data so as to obtain a straight line is to plot $[A]_0/v_0$ against

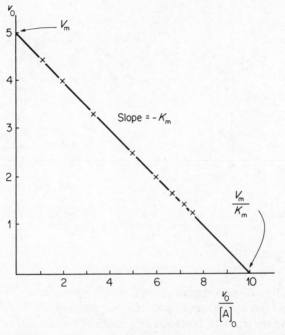

Figure 7 Eadie or Hofstee plot of the experimental data of Figure 5

$[A]_0$. The Michaelis equation (Equation 12) can be transformed as follows:

$$v_0 = \frac{V_m[A]_0}{K_m + [A]_0}$$

$$\therefore \quad \frac{v_0}{[A]_0} = \frac{V_m}{K_m + [A]_0}$$

$$\therefore \quad \frac{[A]_0}{v_0} = \frac{K_m + [A]_0}{V_m} = \frac{K_m}{V_m} + \frac{[A]_0}{V_m}$$

A plot of $[A]_0/v_0$ against $[A]_0$ (Figure 8) therefore gives a straight line of slope $1/V_m$, vertical intercept (when $[A]_0 = 0$) equal to K_m/V_m, and horizontal intercept (when $[A]_0/v_0 = 0$) equal to $-K_m$. Although this method is not widely used, it is in many respects the best of the available procedures for obtaining V_m and K_m from linear plots of experimental data.

36

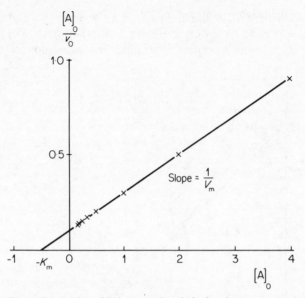

Figure 8 A plot of $[A]_0/v_0$ against $[A]_0$ for the same experimental data as shown in Figure 5

Enzyme assays

In order to begin any experimental work involving the study of any enzyme, one needs to be able to measure its amount or concentration. Whether one is interested in the enzyme as a biologist or a medical biochemist, or for whatever other reason, the first requirement is for an enzyme assay. From the foregoing theoretical considerations, it will be clear that the best way to detect the enzyme is to allow it to catalyse its specific chemical reaction under fixed experimental conditions, where the reaction rate will be proportional to the enzyme concentration. In fact the amount of enzyme is defined in terms of reaction rate under such fixed conditions. The International Unit (U) of enzyme is that amount which will catalyse the conversion of 1 μmol of its substrate in 1 minute under defined conditions of temperature and pH that vary from one enzyme to another. Enzyme concentrations are expressed in units/millilitre (U/ml) or units/litre (U/l).

The first experimental requirement for making an enzyme assay is a method of following the reaction, preferably continuously, so that the initial rate of reaction in the presence of a fixed concentration of substrate can be measured. The initial rate is measured so as to avoid the complications due to the presence of product, and because the enzyme may become inactivated during the course of the assay procedure. In this way the measured rate will be directly proportional to enzyme concentration in the assay solution. It is usual, however, to employ a substrate concentration at least five times greater than the K_m of the enzyme, so that the enzyme will be working 'on the Michaelis plateau', that is in the region of the Michaelis curve (see Figure 5), where the rate hardly changes with variations in substrate concentration. This minimizes the possibility of error due to slight variations of the substrate concentration from one assay to another.

The other experimental conditions that must be fixed are the temperature because, like all chemical reactions, enzymic rates vary with temperature, the pH of the assay solution because enzymes are susceptible to changes in pH, and the concentrations of all solutes such as buffer ions. The latter restriction applies because enzymes are susceptible to changes in ionic strength and may interact specifically with solute molecules in a way which alters their catalytic activity.

A typical enzyme assay is described below for the enzyme fumarase which catalyses the reaction:

$$fumarate + water \rightleftharpoons malate$$

The procedure is due to Massey[6] and relies on the fact that fumarate absorbs light in the near ultraviolet whereas malate does not. Into a spectrophotometer cell is placed a solution of sodium fumarate in a phosphate buffer of pH 7·3. The cell is placed in the thermostatically controlled cell carrier of a recording spectrophotometer and allowed to reach the pre-set temperature of 20 °C. The wavelength of the illuminating beam is set to 300 nm and the absorbance is recorded over a short period of time to ensure that it is not changing. A small amount of enzyme solution, say 0·1 ml, is then added to the cell contents and mixed rapidly, after which the absorbance is again recorded over a period of several minutes. An experimental record of the type shown in Figure 9 is obtained. Matters are always arranged so that the concentration of sodium fumarate in the cell immediately after adding the enzyme is 0·017 M, and the concentration of the phosphate buffer is 0·033 M, in a final volume of 3·0 ml.

The initial slope of the experimental record is measured, let us say x absorbance units/minute, and divided by the molar extinction coefficient of fumarate under these conditions, say y l/mol, to give the initial rate of change of fumarate concentration x/y mol/l/min, which is equivalent to $x . 10^3/y$ μmol/ ml/min. Now the cell contained 3 ml of solution, so in it $3x . 10^3/y$ μmol of substrate were lost per minute. From the definition of the enzyme unit, this means there were $3x . 10^3/y$ units of enzyme in the cell. This amount of enzyme was added to the cell in an original volume of 0·1 ml, so the enzyme concentration in the added solution was $3x . 10^4/y$ U/ml.

This standard assay would be used under all conditions where the amount or concentration of fumarase was to be measured. By varying the substrate concentration, on the other hand, with a fixed amount of enzyme, it was used by Massey[7] to measure the K_m and V_m of fumarase as shown in Figure 10. The individual points are values of $1/v_0$ at different values of $1/[A]_0$, i.e. the reciprocal of fumarate concentration, in a series of otherwise identical assays.

In many instances the reaction catalysed by the enzyme of interest is more complex than that of fumarase. In the clinical evaluation of patients suffering

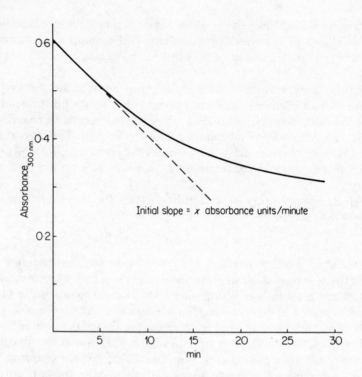

Figure 9 Typical recording spectrophotometer trace obtained in the assay of fumarase (redrawn, with permission, from Reference 7)

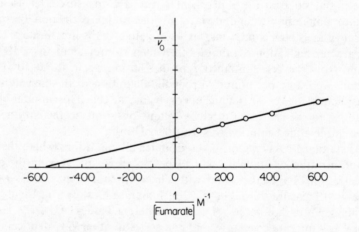

Figure 10 Lineweaver–Burk plot for the enzyme fumarase (redrawn, with permission, from Reference 7)

from conditions resulting from damaged heart muscle, it is necessary to monitor the level of an enzyme released from the heart and present in the serum. The enzyme is lactate dehydrogenase (LDH) which catalyses the reduction of pyruvate by NADH to form lactate and NAD. Five types of lactate dehydrogenase may be present in serum, but the types released into the blood after damage to the heart muscles are found to catalyse the reduction of α-ketobutyrate to α-hydroxybutyrate much faster than the other types. The compound α-ketobutyrate resembles the natural substrate of the enzymes, that is pyruvate, and as such is called a substrate analogue. The heart-type LDH may be assayed by use of this analogue in the following reaction:

$$\text{α-ketobutyrate} + \text{NADH} + \text{H}^+ \rightleftharpoons \text{α-hydroxybutyrate} + \text{NAD}^+$$

The disappearance of NADH is followed spectrophotometrically by the decrease in absorbance at its absorption maximum, 340 nm. As before, constant α-ketobutyrate concentration, temperature and pH are employed, but this time constant NADH concentration must be maintained as well in all assays. The procedure is in all other respects similar to the fumarase assay.

In other cases neither the substrate(s) nor the product(s) absorbs light at a practically useful wavelength, so it is common practice to couple the generation of product by the enzyme in question to another enzymic reaction that does involve an absorbing species. An example is in the assay of the enzyme UDP-galactose-4-epimerase which catalyses the reaction:

$$\text{UDP-galactose} \overset{\text{eperimase}}{\rightleftharpoons} \text{UDP-glucose}$$

These substances do not differ in their absorption at wavelengths in the useful range of laboratory spectrophotometers, so the enzyme UDP-glucose dehydrogenase is employed in the assay to oxidize the UDP-glucose formed. The second reaction involves the reduction of NAD$^+$ to NADH:

$$\text{UDP-glucose} + 2\,\text{NAD}^+ \overset{\text{dehydrogenase}}{\rightleftharpoons} \text{UDP-glucuronic acid} + 2\,\text{NADH} + 2\,\text{H}^+$$

By this means each molecule of UDP-galactose epimerized in the first reaction, gives rise to two molecules of NADH which absorb light at 340 nm.

A high activity of the second enzyme must be present in the spectrophotometer cell to ensure that UDP-glucose is whisked away and oxidized before its concentration can build up. Nevertheless, a disadvantage of this type of assay is that a small concentration of product UDP-glucose *must* be present before the second enzyme can work. This may lead to a lag period while the product builds up and has the theoretical disadvantage that some product is present during the assay of the epimerase. For the purposes of enzyme detection and measurement, this type of assay is widely used and is very satisfactory; but care must be taken when it is used in formal studies of enzyme kinetics.

In those cases where it is impossible or undesirable to couple one reaction to another so as to be able to carry out the assay in a recording spectrophotometer, it may still be possible to use a continuous monitoring procedure. For example, reactions in which H^+ is produced or used up can be followed with a pH meter, while those involving the use or production of oxygen can be followed with an oxygen electrode or in a Warburg manometer. Failing these, one has to fall back upon sampling techniques in which portions of the reaction mixture are removed at intervals, the reaction is stopped, by acidifying the solution, for example, or by heating it, and then chemical techniques are used to separate and estimate the substrate or product. Such methods are detailed by Dixon and Webb.[8]

In conclusion it is worth remembering that in some cases where an enzyme can be assayed by more than one method, different answers are obtained. This is not always so, but the answer may depend on the method used.

Chapter 3

The Structure and Properties of Proteins

The serious examination of individual enzymes can only be attempted when they have been isolated in a pure or relatively pure condition. The fact that enzymes are often very sensitive, especially to heat and extremes of pH, rendered this task all the more difficult for the early investigators. After about 1920, however, a number of successes were achieved. Wilstatter and his co-workers carried out several purifications of enzymes between 1922 and 1928, but Sumner was the first to obtain a crystalline enzyme, urease, in the year 1926. Even then the crystalline urease was quite impure. There followed a series of classical isolations, by Northrop and his colleagues, of crystalline proteolytic enzymes. Northrop in addition established that the enzymes he had isolated were proteins, a fact which had not been clear until then.

A very large number of enzymes has now been purified and every one has proved to be a protein. Sometimes, it is true, the protein requires non-protein cofactors in order for it to function as an enzyme, but there is now no doubt that enzymes are proteins. The methods used to purify them are described in Appendix 2. It has become clear that if we wish to understand the way in which enzymes function, we must first learn as much as possible about the structure and properties of proteins.

THE AMINO ACIDS FOUND IN PROTEINS

As we saw earlier, proteins are polymers of high molecular weight where the individual monomers are amino acids. Twenty amino acids occur in proteins, although some of them can appear in more than one guise. The fundamental twenty are listed in Table 1, from which it can be seen that all of them are L-α-amino carboxylic acids with the structure

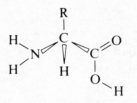

where only the side chain R differs from one to another; in proline, the side chain turns back onto the nitrogen atom. This is only strictly true for the amino acids that occur in proteins. There is a large number of other naturally occurring amino acids, some of which are not α-amino acids and some of which have the D-configuration, but these are never found incorporated in proteins.

In Table 1 the amino acids are grouped according to the type of side chain they possess. The first group of two, aspartic and glutamic acids, comprises those with a carboxyl group in the side chain which can ionize to give a negatively charged carboxylate ion. The second group contains asparagine and glutamine with an amide group in the side chain, threonine and serine with hydroxylic side chains, and glycine with only a hydrogen atom as its side chain. Glycine is the only one of the 20 protein amino acids without optical activity. Although the side chains of amino acids in this second group are uncharged, nevertheless they are able to interact favourably with water, like those of the first group. All the amino acids in the first two groups are consequently said to be hydrophilic (i.e. water-loving).

Table 1
The 20 amino acids that occur in proteins

Name	Structure	Name of residue	Residue abbreviation 3-letter	1-letter
1. L-Aspartic acid	$CH_2 \cdot COOH$ $\vert$ $H_2N \cdot CH \cdot COOH$	Aspartyl	Asp	D
2. L-Glutamic acid	$CH_2 \cdot CH_2 \cdot COOH$ $\vert$ $H_2N \cdot CH \cdot COOH$	Glutamyl	Glu	E
3. L-Asparagine	$CH_2 \cdot CONH_2$ $\vert$ $H_2N \cdot CH \cdot COOH$	Asparaginyl	Asn	N
4. L-Glutamine	$CH_2 \cdot CH_2 \cdot CONH_2$ $\vert$ $H_2N \cdot CH \cdot COOH$	Glutaminyl	Gln	Q
5. L-Threonine	$CH_3 \cdot CH \cdot OH$ $\vert$ $H_2N \cdot CH \cdot COOH$	Threonyl	Thr	T
6. L-Serine	CH_2OH $\vert$ $H_2N \cdot CH \cdot COOH$	Seryl	Ser	S
7. Glycine	H $\vert$ $H_2N \cdot CH \cdot COOH$	Glycyl	Gly	G
8. L-Cysteine	CH_2SH $\vert$ $H_2N \cdot CH \cdot COOH$	Cysteinyl	Cys	C
9. L-Methionine	$CH_2 \cdot CH_2 \cdot S \cdot CH_3$ $\vert$ $H_2N \cdot CH \cdot COOH$	Methionyl	Met	M

Table 1 *continued*

Name	Structure	Name of residue	Residue abbreviation	
			3-letter	1-letter
10. L-Proline	CH_2——CH_2 CH_2 $CH \cdot COOH$ NH	Prolyl	Pro	P
11. L-Alanine	CH_3 $H_2N \cdot CH \cdot COOH$	Alanyl	Ala	A
12. L-Valine	$CH_3 \cdot CH \cdot CH_3$ $H_2N \cdot CH \cdot COOH$	Valyl	Val	V
13. L-Leucine	CH_3 $CH_2 \cdot CH$ CH_3 $H_2N \cdot CH \cdot COOH$	Leucyl	Leu	L
14. L-Isoleucine	$CH_3 \cdot CH \cdot CH_2 \cdot CH_3$ $H_2N \cdot CH \cdot COOH$	Isoleucyl	Ile	I
15. L-Tyrosine	CH_2 ⬡—OH $H_2N \cdot CH \cdot COOH$	Tyrosyl	Tyr	Y
16. L-Phenylalanine	CH_2 ⬡ $H_2N \cdot CH \cdot COOH$	Phenylalanyl	Phe	F
17. L-Tryptophan	CH_2 $H_2N \cdot CH \cdot COOH$ NH	Tryptophyl	Trp	W
18. L-Lysine	$CH_2 \cdot CH_2 \cdot CH_2 \cdot CH_2 \cdot NH_2$ $H_2N \cdot CH \cdot COOH$	Lysyl	Lys	K
19. L-Histidine	CH_2 $H_2N \cdot CH \cdot COOH$ HN═N	Histidyl	His	H
20. L-Arginine	NH $CH_2 \cdot CH_2 \cdot CH_2 \cdot NH$ $H_2N \cdot CH \cdot COOH$ NH_2	Arginyl	Arg	R

The two sulphur containing amino acids cysteine and methionine comprise the third group in Table 1. Here, once the amino acid has become incorporated into a protein by the biological machinery of protein synthesis, the side chains can become modified by oxidation. The same reactions also occur with free cysteine and methionine. In the case of cysteine the simplest oxidation product is the disulphide cystine, in which the –SH, or sulphydryl group, of one cysteine molecule becomes joined to that of another:

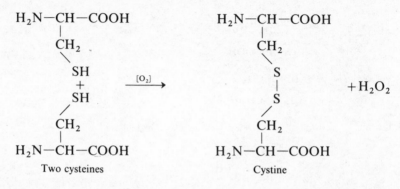

    Two cysteines   Cystine

At pH values above 7 this reaction occurs readily in the presence of dissolved oxygen and is catalysed by traces of metal ions such as Cu^{2+}. The existence of the disulphide cystine in proteins arises from the incorporation of cysteine followed by oxidation within the protein. That is why only cysteine is listed among the 20 amino acids occurring in proteins. The –SH group of cysteine dissociates at alkaline pH to give a mercaptide ion:

$$R-SH \rightleftharpoons RS^- + H^+$$

Methionine can also undergo air oxidation but this is a much less important process leading to the rather unstable methionine sulphoxide,

$$\begin{array}{c} CH_2 \cdot SO \cdot CH_3 \\ | \\ H_2N \cdot CH \cdot COOH \end{array}$$

If it occurs in a protein, however, it may have consequences for the structure, as we shall see.

The fourth group of amino acids in Table 1 contains those having aliphatic side chains without chemically reactive substituents. These side chains are said to be non-polar or apolar because they contain no ionizable or polarizable groups. They are also called hydrophobic, literally water-hating, because they are not very soluble in water, in much the same way as aliphatic hydrocarbons. Methionine in the previous group also has a hydrophobic side chain. Proline is the only amino acid without a primary amino group; its side chain turns back onto the nitrogen to form a secondary amine or imine, so that proline is the only imino acid that occurs in proteins. After proline has been incorporated into some proteins it can become hydroxylated at the 3 or 4 position of the ring to give hydroxyproline. The only well-characterized occurrences of this are in the structural proteins collagen and elastin which are not enzymes and so will not concern us.

The aromatic amino acids in the next group consist of phenylalanine and tryptophan with hydrophobic side chains, together with tyrosine whose side chain is partly hydrophobic like that of phenylalanine but partly hydrophilic on account of the hydroxyl group.

Lastly the three basic amino acids lysine, histidine and arginine each have hydrophilic side chains in which there is a nitrogenous base which readily accepts a proton to form a positively charged group.

Before beginning to study the chemical structure of a protein it is valuable to know which of the 20 amino acids it contains, and how many residues of each. This fundamental information, the amino acid composition of the protein, is found by first hydrolysing the protein to its constituent free amino acids, then determining the amounts of these using an automatic amino acid analyser.

The hydrolytic step is usually carried out by heating a sample of the protein (about half a milligram) with 6 N-HCl in a sealed and evacuated glass tube. The hydrolysis is carried out at 110 °C for at least 18 h and even then some amino acids may not be completely released; others, however, decompose slowly during acid hydrolysis while tryptophan is completely destroyed, and the amides asparagine and glutamine are completely converted to aspartic and glutamic acids respectively. Alkaline hydrolysis with 5 N-sodium hydroxide in the presence of hydrolysed starch preserves the tryptophan, but causes losses of the sulphur amino acids and asparagine and glutamine. The only way of preserving all the amino acids intact is to hydrolyse with a mixture of proteolytic enzymes. This has not been a popular method because not all the necessary enzymes have been commercially available in stable form, and because of doubts about whether complete release of all the amino acids occurs, although that can be checked for the stable amino acids by comparison with an acid hydrolysate. Many enzymes are now available chemically attached to solid supports in the form of gels or beads to make them more stable. The existence of proteolytic enzymes stabilized in this way may make enzymatic hydrolysis the method of choice in future.

The mixture of free amino acids produced by hydrolysis is next applied to a cation-exchange column in an automatic amino-acid-analyser. The ion-exchange resin is a sulphonated polystyrene/divinyl benzene copolymer which at a pH of about 2 binds all the amino acids tightly to the top few millimetres of the column. A wide range of column sizes is in use, some of them being up to 150 cm in length, but all perform the basic task of separating the amino acids one from another when a series of buffers of increasing pH and/or ionic strength is pumped through. In this way the amino acids emerge in sequence from the bottom of the column in the flowing stream of buffer, which is then mixed with ninhydrin reagent in order to measure their concentrations. Ninhydrin, or triketohydrindene hydrate, reacts at 100 °C with most amino acids to give a purple product, Ruheman's purple. With proline it gives a yellow product. The reaction is complex but for most amino acids is as follows:

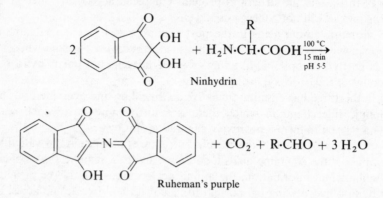

After reaction has taken place the colour intensity in the flowing stream is measured photometrically and recorded as absorbance on a moving strip chart. The photometer monitors the absorbance at two wavelengths, 570 nm for the normal purple colour and 440 nm for the yellow colour given by proline. The appearance of the resulting record is shown in Figure 11. A series of peaks of

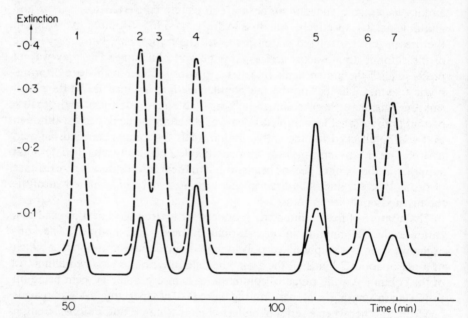

Figure 11 Reproduction of part of a typical chromatogram obtained from an automatic amino acid analyser in the author's laboratory. The pecked trace records extinction at 570 nm, the solid trace at 440 nm. The seven peaks shown represent: 1. Aspartic acid; 2. Threonine; 3. Serine; 4. Glutamic acid; 5. Proline; 6. Glycine; 7. Alanine. Note that the 440 nm trace is much lower than the 570 nm trace for all amino acids except proline where the reverse is true

absorbance occurs in the effluent stream of buffer, each peak representing the emergence of an amino acid from the ion-exchange column. The position of the peak defines which amino acid it is and the area under the peak is proportional to the amount present.

The first amino acid analyser was constructed about 1958 by Professors Stanford Moore and William H. Stein at the Rockefeller Institute, as it then was, in New York. This event proved to be a breakthrough that opened the way to the successful determinations of many protein structures. The first of these, that of the enzyme ribonuclease A, was also worked out by Moore and Stein and their co-workers, for which they were awarded the Nobel Prize in 1972.

DETERMINATION OF THE MOLECULAR WEIGHT OF A PROTEIN

Before commencing structural studies on a protein, it is essential to know its molecular weight. There are two commonly used experimental approaches to this problem. The first is to see how rapidly it sediments in a centrifugal field, and the second is to see how readily it will penetrate a gel of calibrated pore size. There are two variants of the latter approach, the second involving electrophoresis.

Ultracentrifugal determination of molecular weight

Probably the most common method until quite recently has been to measure the sedimentation coefficient S, and the diffusion coefficient D for the protein in a buffer of neutral pH, and to calculate the molecular weight, M, from the Svedberg equation $M = RTS/(D(1 - \overline{V}p)$ where R and T are the universal gas constant and the absolute temperature, $\overline{V}$ is the partial specific volume of the protein or the reciprocal of the density of the protein molecule, and p the density of the buffer used as solvent. The sedimentation coefficient, S, is a measure of how fast the protein sediments in a centrifugal field of unit force and can be measured directly with the aid of a modern analytical ultracentrifuge.[9] The diffusion coefficient, D, can also be measured in an ultracentrifuge run at very low speeds. Ideally the gravitational force should be zero for this measurement, but at very low speeds it is negligible and use is made of the optical measuring systems of the analytical ultracentrifuge to measure the rate of spreading of an artificial boundary between the protein solution and pure solvent. From this rate D can be calculated.[9] The density p of the solvent is easily measured. Finally $\overline{V}$, the partial specific volume of the protein can be calculated from the known amino acid composition, or measured experimentally, but as it lies between 0·70 and 0·75 for most proteins, a value of 0·725 is often taken without further ado. This is a dangerous procedure because small errors in $\overline{V}$ can lead to large errors in M so, for accurate estimates of M, $\overline{V}$ should be measured carefully.

An alternative method of finding the molecular weight is the equilibrium method which avoids the need to measure D. Instead of measuring how fast the protein sediments through a high centrifugal field, one employs a lower field, centrifuging at a lower speed until the protein is spread throughout the whole observation chamber of the centrifuge. It takes several days to achieve equilibrium, where the tendency to sediment is everywhere exactly balanced by diffusion. If the protein is sufficiently stable to last long enough for equilibrium to be established, this is a good method. The distribution of protein throughout the field is measured and the molecular weight calculated directly.[9] There is one drawback to this method in that it reveals nothing about the purity of the protein, whereas the sedimentation velocity method reveals impurities if they have different S values. Neither method can yield the molecular weight if the protein is impure, but the sedimentation velocity method is more likely to reveal any impurity.

Gel filtration

In principle this method depends on the fact that large protein molecules cannot penetrate the pores of a granular gel whereas small proteins can. In practice a column of the gel granules such as Sephadex, is packed and washed with a suitable buffer. A mixture of proteins of known and widely different molecular weights is applied to the top of the column and eluted with the buffer. The largest of these marker proteins emerges first because it has had to pass between the granules of gel, being unable to penetrate them. Smaller proteins partially penetrate the granules and emerge later, the smallest emerging from the bottom of the column last of all because it has completely penetrated the gel granules. A calibration graph is plotted, relating molecular weight to elution volume as shown in Figure 12. The protein of unknown molecular weight is now passed through the column and its elution volume is measured. Reference to the calibration graph then gives an estimate of its molecular weight.

This method can give misleading results if the protein molecule in question is markedly non-spherical. Rod-shaped or elongated molecules are retarded by the gel granules more than spherical ones of the same molecular weight.

SDS–gel electrophoresis

A method which overcomes the problem of shape employs the anionic detergent sodium dodecyl sulphate (SDS; $CH_3(CH_2)_{11}SO_3Na$) or sodium lauryl sulphate to give it another name. This detergent destroys the native shape of the protein molecule and forms a complex in which the protein serves the same sort of function as a pincushion. Molecules of SDS become inserted like pins into the protein cushion, where the outward-pointing head of the pin represents the negative charge of the sulphonate group, the stem of the pin represents the dodecyl group. In this situation the charge on the complex is almost entirely due to the negative sulphonate groups, which far outweigh the intrinsic charge on the protein. For this reason the charge to mass ratio is always constant no matter how big or small the protein involved in the complex. A big protein

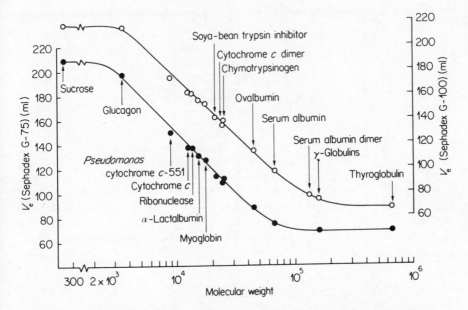

Figure 12 A calibration curve for the estimation of molecular weight by gel filtration on Sephadex G-75 (●) and G-100 (○) columns. V_e is the elution volume. Column dimensions 2·4 × 50 cm (reproduced, with permission, from P. Andrews in *Biochemical Journal*, **91**, 225 (1964))

complexes more SDS and consequently carries a bigger negative charge. The protein–SDS complex is now placed on the top of a continuous gel of poly-acrylamide and subjected to an electric field which moves the complex down through the gel. Electrophoretic mobility, the rate of free movement in aqueous solution under the influence of unit electric field, is related to the charge to mass ratio, so all protein–SDS complexes would have the same electrophoretic mobility in free solution. The bigger complexes, however, cannot move through the gel as rapidly as small ones because of the restrictions of pore size. Consequently the observed rate of movement through the gel depends only on the size, that is the molecular weight, of the original protein. Shape is not a consideration because the shape is the same for all the protein–SDS complexes. Once again a calibration curve is constructed with known proteins and, in addition, marker proteins are usually electrophoresed along with the unknown to correct for differences between gels and electrophoretic conditions from one occasion to the next (see Figure 13). The only drawback to this method is that if the native protein consists of a number of polypeptide subunits loosely held together, the action of SDS will break down the aggregate and only the subunit molecular weight will be measured. For proteins with molecular weights lying between about 10,000 and about 200,000 the method works well, but it is unreliable with smaller or larger proteins.

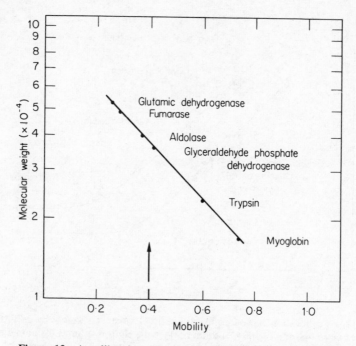

Figure 13 A calibration graph for the estimation of molecular weight by SDS–gel electrophoresis. All proteins were electrophoresed through polyacrylamide gels under standard conditions. The arrow represents the mobility of a protein of unknown molecular weight which was estimated as 37,000 from the graph (reproduced, with permission, from K. Weber and M. Osborn in *Journal of Biological Chemistry*, **244**, 4409 (1969))

THE PRIMARY STRUCTURE OF PROTEINS

In the polymerized molecule of protein, amino acids are joined in long unbranched polypeptide chains. The carboxyl group of one amino acid is condensed with the amino group of the next to form a peptide bond by the elimination of water:

$$H_2N-\underset{\underset{R_1}{|}}{CH}-\underset{\underset{O}{\|}}{C}-OH + H_2N-\underset{\underset{R_2}{|}}{CH}-\underset{\underset{O}{\|}}{C}-OH + H_2N-\underset{\underset{R_3}{|}}{CH}-\underset{\underset{O}{\|}}{C}-OH + \cdots + H_2N-\underset{\underset{R_n}{|}}{CH}-\underset{\underset{O}{\|}}{C}-OH$$

$$\searrow (n-1)H_2O$$

$$H_2N-\underset{\underset{R_1}{|}}{CH}-\underset{\underset{O}{\|}}{C}-NH-\underset{\underset{R_2}{|}}{CH}-\underset{\underset{O}{\|}}{C}-NH-\underset{\underset{R_3}{|}}{CH}-\underset{\underset{O}{\|}}{C}-\cdots-NH-\underset{\underset{R_n}{|}}{CH}-\underset{\underset{O}{\|}}{C}-OH$$

In this diagrammatic representation, the peptide bonds that link the amino acid residues together are shown in boxes. The diagram also shows that each polypeptide chain has one α-NH_2 group remaining at one end and an α-COOH group remaining at the other. The amino-terminus is, by convention, always written at the left-hand end of the chain and the carboxyl-terminus at the right. The sequence of amino acid residues is known as the *primary structure* of the protein and, in a typical protein, it may be between 100 and 1000 residues long.

Primary structures of this sort of length clearly present a considerable problem of nomenclature. The individual amino acid residues in a protein are named from the parent amino acid by replacing the -ic or -ine ending by the suffix -yl, as shown in Table 1. Thus aspartic acid becomes the aspartyl residue and glycine the glycyl residue in a polypeptide. Even the primary structure of the enzyme bovine pancreatic ribonuclease A, one of the smaller proteins, consists of 124 residues and would take up a great deal of space if written out in full. A system of three-letter abbreviations for the amino acid residues is now universally accepted and these are given in Table 1. Using them, the primary structure of ribonuclease A is shown in Figure 14. The residues are numbered from the N-terminus, so that one can, for example, distinguish between the histidyl residues at positions 12 and 119, merely by referring to them as HIS 12 and HIS 119. It is worth mentioning that the three letter abbreviations refer strictly only to the amino acid residues in a polypeptide, but one sometimes finds them also used to refer to free amino acids. In order to compress the naming of primary sequences even further, a system of one letter abbreviations has been recommended recently, and although not yet widely used, is also listed in Table 1.

The experimental approach to determining the primary sequence of a protein is simple in essence though often difficult in practice. In principle, a method of removing amino acid residues one by one from only one end of the chain is all that is needed. Identification of the amino acids sequentially removed in this way would automatically reveal the primary sequence. This approach is essentially based upon end-group determination and could work equally well with either N-terminal residues or C-terminal ones. Although methods exist for identifying both, in peptides or proteins, most of them destroy the rest of the peptide in the process making them unsuitable for sequence analysis. These methods, which will be described later, are nevertheless of great value in that they allow one to ascertain whether the protein being studied consists of a single polypeptide chain, or more than one such chain.

Some proteins, like ribonuclease A (Figure 14), consist of only one chain. Others contain two or more which may be the same or different in primary structure, held together either relatively loosely by non-covalent forces, or by disulphide bond formation between cysteinyl residues in the different chains. Before commencing sequence analysis it is essential to be sure that only one polypeptide chain is present, otherwise the sequential removal of residues from more than one type of chain would make it impossible to reconstruct the sequence in any of them. End group analysis, preferably at both N- and C-termini, is one way of establishing how many chains are present. If more than one is found, they must be separated and their sequences determined separately. It

52

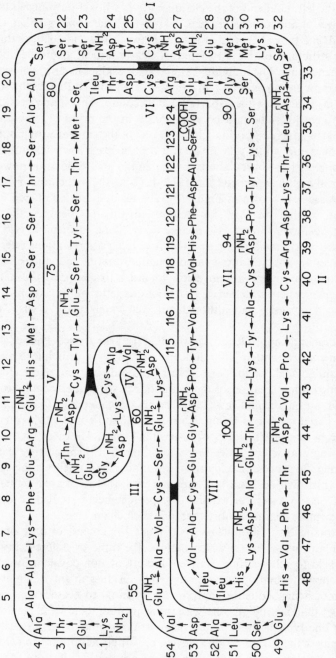
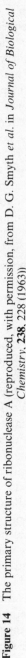

Figure 14 The primary structure of ribonuclease A (reproduced, with permission, from D. G. Smyth *et al.* in *Journal of Biological Chemistry*, **238**, 228 (1963))

is usual to first break any disulphide bonds present in the protein because they could be holding polypeptide chains together. Even if they are involved only in intra-chain bonds, as with ribonuclease A (Figure 14), they will make sequence analysis impossible if left intact. One method employs reduction to the –SH groups, followed by alkylation with iodoacetic acid to prevent re-formation of disulphide upon exposure to air:

$$R\text{–}SS\text{–}R^1 \xrightarrow{[2H]} RSH + R^1SH \xrightarrow{I \cdot CH_2 \cdot COOH} RS \cdot CH_2 \cdot COOH + R^1 \cdot S \cdot CH_2 \cdot COOH$$

peptide with di-   cysteinyl   carboxymethylcysteinyl
sulphide bridge   peptides   peptides

Another method involves oxidation with performic acid, yielding the stable sulphonic acid:

$$R\text{–}SS\text{–}R^1 \xrightarrow{H \cdot \overset{\overset{O}{\|}}{C}OOH} R \cdot SO_3H + R^1SO_3H$$

peptide with di-   peptides containing the sulphonic acid
sulphide bridge   derivative of cysteine, cysteic acid

Having cleaved any disulphide bonds and separated the polypeptide chains if more than one was present, the way is now open to determine the primary sequence of each chain. Only one end group method preserves the remaining peptide after a single terminal residue has been removed, and that is the Edman degradation which works at the N-terminus. The polypeptide is made to react with phenylisothiocyanate (PITC) at a slightly alkaline pH, when the free α-NH$_2$ group reacts to form a phenylthiocarbamyl (PTC) peptide.

The PTC peptide is subjected to mild acid conditions that result in a cyclization and cleavage process, releasing the anilinothiazolinone (ATZ) derivative of the original N-terminal amino acid, together with the rest of the peptide or protein, now with a new N-terminal residue. The ATZ derivative is readily removed and converted to a phenylthiohydantoin (PTH) which is identified by chromatographic procedures.

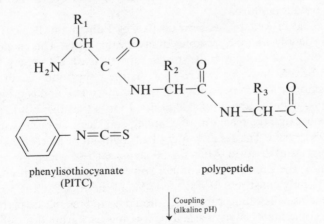

phenylisothiocyanate   polypeptide
(PITC)

Coupling
(alkaline pH)

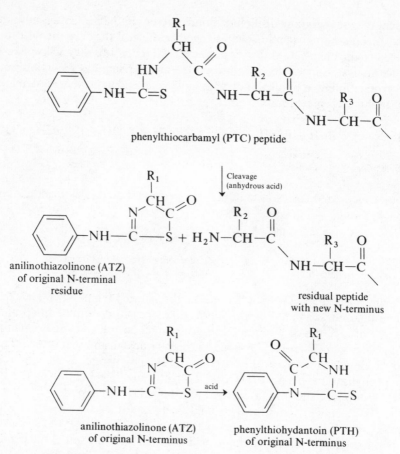

phenylthiocarbamyl (PTC) peptide

Cleavage
(anhydrous acid)

anilinothiazolinone (ATZ)
of original N-terminal
residue

residual peptide
with new N-terminus

anilinothiazolinone (ATZ)
of original N-terminus

phenylthiohydantoin (PTH)
of original N-terminus

The remaining peptide chain is put through the procedure once more, thus identifying the second residue in the sequence, and so on to the third and subsequent residues in turn.

An automated procedure has been devised by Edman and Begg, which can sequentially identify up to 70 residues in favourable cases. This method relies on the insolubility of proteins in the special solvents used. Ultimately, however, this method breaks down for two reasons. As the polypeptide chain becomes shorter and shorter it becomes soluble in the reagents used and is progressively removed along with the anilinothiazolinones. Even if this does not happen, the method fails because the reactions of coupling and cleavage occur with at best only 97%–98% yield. The result is a gradual build-up of heterogeneity in the chains present at each stage. After say 50 steps, each of which proceeds with 98% efficiency, the product will contain some chains of the expected length but these will only constitute $(0.98)^{50} \times 100\%$ of the total, that is 36.3%. The remaining chains will be longer than they should be and the release of the correct ATZ will become increasingly masked by a rising background of other ATZs from the contaminating chains. In spite of these drawbacks the use of automated sequencing instruments promises to be most useful.

In the meantime, one has to use alternative approaches to discover the complete sequences of enzymes and other proteins. The first task is to break the polypeptide chain into shorter peptides by cleaving specific peptide bonds. This can be achieved either by the use of enzymes or by chemical procedures. In the former case the enzymes used are called proteolytic enzymes or peptidases because they hydrolyse peptide bonds. A wide variety of such enzymes occurs naturally, some of which are so voracious that they would catalyse the hydrolysis of nearly all the peptide bonds in the polypeptide, leaving only free amino acids or very short peptides. Clearly these are to be avoided in sequence studies. Other peptidases are so selective as to cleave only the peptide bonds following or preceding specific amino acids. Trypsin is the most specific of these, causing cleavage only after lysyl or arginyl bonds. Chymotrypsin is only a little less specific, leading to cleavage at the bonds following the aromatic phenylalanyl, tyrosyl or tryptophyl residues, and also, but more slowly, at bonds following methionyl, leucyl, glutaminyl or asparaginyl residues. Other peptidases such as papain, thermolysin, pepsin and subtilisin can also be used but they are normally much less specific about the bonds they cleave. Nevertheless it is sometimes possible to achieve a high degree of specificity in well controlled circumstances. For example, subtilisin will cleave only a single bond in the ribonuclease A structure (Figure 14) if used at 0 °C for only a few minutes, in which case the bond between Ala 20 and Ser 21 is broken very efficiently even though there are many other susceptible bonds in the RNase A structure which can be cleaved by subtilisin under other conditions.

A wide variety of chemical procedures is available for bringing about specific hydrolysis of peptide bonds, but nearly all suffer from the disadvantage that they are not very efficient. The best and most widely used is the method in which cyanogen bromide (CNBr) is used to effect cleavage at methionyl residues:

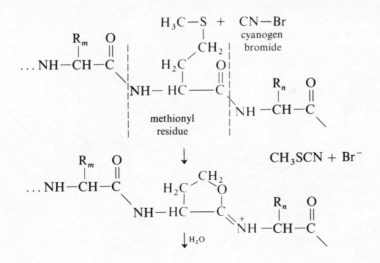

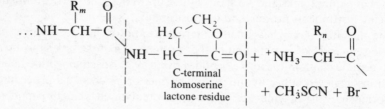

The result of the reaction is to produce one peptide, which was the N-terminal portion of the original peptide as far as the methionyl residue, in which there is now a C-terminal homoserine lactone residue. The other peptide produced represents what was the C-terminal portion of the original chain.

By one means or another the primary structure of the protein is broken into smaller pieces which must then be separated from each other and purified by chromatographic methods. These fragments may be amenable to automatic sequence determination under certain circumstances. Those which are not are broken down still further until one arrives at peptides about 6–8 residues long at which stage the manual Edman procedure can be applied to determine the sequence of each one. A slightly different variant of the Edman reaction is used for this manual procedure, as compared with the automated one described earlier but the principles are the same.

Some workers, instead of identifying each amino acid as it is released in turn, in the form of its phenylthiohydantoin, prefer to adopt either the sub-tractive Edman procedure or the dansyl–Edman procedure. In the former case the amino acids present in the peptide are analysed on small portions of the material before and after the Edman reaction. The difference in amino acid composition reveals the N-terminus removed. The dansyl–Edman procedure, on the other hand, involves taking a small portion of the original peptide and reacting it with 1-dimethylaminonaphthalene-5-sulphonyl chloride (dansyl chloride) which combines with the amino groups present:

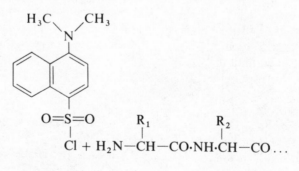

dansyl chloride   peptide

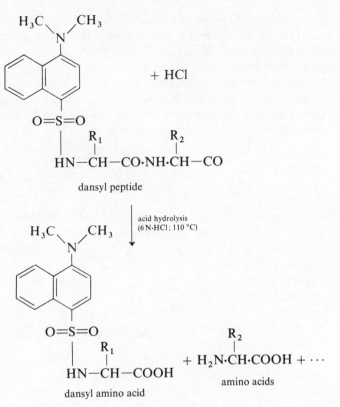

dansyl peptide

acid hydrolysis
(6 N·HCl; 110 °C)

dansyl amino acid

+ H$_2$N·CH·COOH + ⋯
amino acids

The resulting dansyl peptide is vigorously hydrolysed to break all the peptide bonds, releasing free amino acids together with the dansyl derivative of the original N-terminal amino acid. Dansyl amino acids are highly fluorescent and can be identified very easily in tiny quantities after chromatographic separation. In this way the N-terminal amino acid is *identified* in the small portion taken for dansylation, while it is *removed* by the Edman reaction from the undansylated portion. After removal the new N-terminus is once again identified by dansylation of a portion and removed by Edman degradation of the rest, and so on until the sequence of the peptide is determined. The reason for doing it this way is that dansyl amino acids are easier to separate than phenylthiohydantoins in some cases and, because of their fluorescence, can be identified using only a small quantity.

Up to now two methods of identifying the N-terminal residues have been mentioned, the Edman method which preserves the rest of the peptide and the dansyl method which does not. Other methods are available for use where either of these two is unsatisfactory. For example, when it is necessary to make a quantitative measurement of the amount of an N-terminal residue the cyanate method would be used. This method is particularly suitable for establishing the purity of a polypeptide, or finding out how many polypeptide chains there are in a protein. The reagent is potassium cyanate which generates the cyanic acid needed to form an *N*-carbamoyl peptide. Cyclization yields a hydantoin together with free amino acids which can be removed by an ion-exchange column leaving the purified hydantoin. This is hydrolysed to yield the original N-terminal amino acid which is estimated by amino acid analysis. As with the dansyl method, the peptide is lost once the N-terminus has been identified and the method cannot be used sequentially.

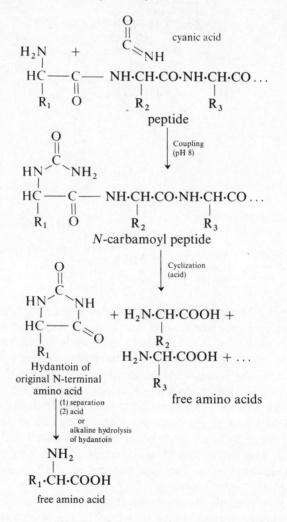

Certain proteolytic enzymes, the amino peptidases, remove the N-terminal amino acid from a peptide chain, without breaking bonds within the chain. They would be ideal for sequence work if they could be induced to stop once they had removed just one residue. Unfortunately they cannot and, since different amino acid residues are removed at different rates, the interpretation of results obtained with amino peptidases can be fraught with uncertainty.

Sometimes analysis of C-terminal residues is of great value in sequence studies. For example, we have not yet considered the problem of how to put in order the various peptides derived by limited cleavage of the original protein. This is illustrated by the following example. Suppose we have a peptide whose N-terminal residue, A, and amino acid composition are known. Suppose we have been able to cleave it into three pieces with chymotrypsin, and further suppose that the amino acid sequences of all three pieces have been determined. With luck only one of the pieces will have A as the N-terminal residue, so this clearly was the N-terminal part of the original peptide. But we will have no idea which of the other peptides came next in the original sequence.

$$A \ldots \ldots \ldots \ldots \ldots \ldots \ldots Z$$
$$\downarrow \text{chymotrypsin}$$
$$A \ldots B + C \ldots D + E \ldots Z$$

C-terminal analysis of the original peptide and the three daughter peptides would establish Z as the last residue in the sequence and would allow us to put the peptide E...Z in its proper place. This would be sufficient information to establish the whole sequence since clearly the peptide C...D would have to be the middle one. C-terminal analysis can be carried out with carboxypeptidases, enzymes that specifically remove only the C-terminal residue from a peptide. As with the amino peptidases there can be difficulties in interpreting the results of such experiments, but in the example shown in Figure 15 not only

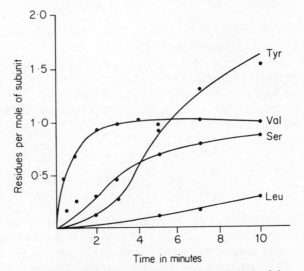

Figure 15 Analysis of the C-terminal sequence of the protein glutamine synthetase using carboxypeptidase A (reproduced, with permission, from H. S. Kingdon *et al.* in *Journal of Biological Chemistry*, **247**, 7925 (1972))

was the C-terminal residue established, but some information as to sequence was also obtained. In this case a protein was treated with the enzyme carboxypeptidase A and the release of free amino acids was monitored by amino acid analysis of the solution. The first one to appear was valine, after which serine started to appear, then after a lag of a few minutes tyrosine gradually built up in concentration, followed much later by leucine. From the results of the experiment it was concluded that the sequence of the protein at its C-terminal end was ... Leu–Tyr–Tyr–Ser–Val. The presence of two tyrosyl residues was deduced from the much greater release of tyrosine than of any of the other amino acids, and they were placed next to each other in the sequence because the release of tyrosine was a smooth process which was effectively completed before leucine started to appear.

The only other satisfactory experimental method of C-terminal analysis is a chemical one in which the peptide is treated with anhydrous hydrazine. All the peptide bonds are cleaved and amino-acid hydrazides are liberated from all the residues except the C-terminal one which appears as a free amino acid:

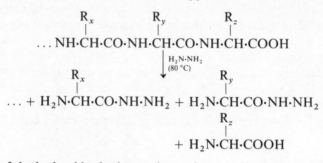

Removal of the hydrazides by ion exchange leaves only the C-terminal amino acid which is estimated by amino acid analysis.

Although in the example quoted earlier, N- and C-terminal analysis allowed us to put three daughter peptides into their correct order in the parent, this would not have been possible had there been four daughters. In that case the first and last peptides would have been placed, but the order of the remaining two would have been left in doubt.

$$A .... Z$$

$$\downarrow \text{chymotrypsin}$$

$$\begin{matrix} (1) & (2) \text{ or } (3)? & (3) \text{ or } (2)? & (4) \\ A ... B & + \; CDEFGH & + \; JKLMN & + \; O ... Z \end{matrix}$$

This is the classical dilemma of sequence work and can only be resolved by going back to the parent peptide and re-cleaving it into new daughters by a different method of specific cleavage. For example, whereas in the example given chymotrypsin was used, one might repeat the cleavage with trypsin or cyanogen bromide. This would generate a new set of peptides one of which would contain a stretch of sequence common to two of the peptides from the first set, thus providing the necessary 'overlaps'. For example, if a peptide with the sequence FGHJKL was found it would be clear that the original set of peptides was correctly ordered as shown, whereas if LMNCDE were found the reverse order for the middle two peptides would be indicated.

Further details of the problems of sequence determination and the methods of solving them can be found in the many excellent review articles and books on this subject.[10]

Assuming that the primary structure of a protein is known, what does this tell us about its overall three-dimensional structure? The protein is not merely a rigid rod consisting of amino acid residues joined up in line astern: the chain twists back on itself to form a convoluted and involuted structure.

THE SPATIAL ORGANIZATION OF PROTEIN STRUCTURE—SECONDARY AND TERTIARY STRUCTURE

Apart from the peptide bond which is strictly planar and rigid (Figure 16), the backbone of repeating NH·CH·CO units allows considerable flexibility in the three-dimensional structure of a polypeptide. It is not surprising to find that the side chains, R, interact with each other and with the solvent in which the

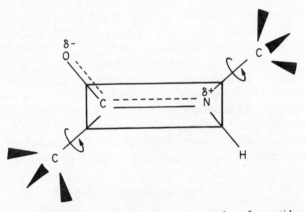

Figure 16 Three-dimensional representation of a peptide bond. The peptide bond is shown in the box as a partial double bond due to resonance between two canonical forms:

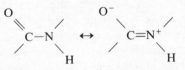

This means that all the atoms shown in the diagram must lie in the same plane. The peptide bond is rigid: rotations about the C—C bond and the N—C bond are possible as indicated but no rotation is possible about the peptide bond itself

polypeptide is dissolved, to give rise to a certain three-dimensional arrangement that is more favoured than others. The forces between side chains are individually weak. Collectively, on the other hand, they can produce a coil or sheet or apparent tangle which is nevertheless a well-defined molecular conformation with substantial stability. The particular 3-D structure assumed by a given polypeptide will depend on the amino acid residues it contains, and their order in the chain. If this were all, life would be a lot simpler, especially for the enzymologist, but the most stable three-dimensional structure, or conformation, is also markedly dependent upon the solvent environment of the protein. Small changes in the temperature, pH, or ionic strength, may cause radical shifts in the balance of forces between side chains, leading to the stabilization of a quite different conformation. The same effect may result from the addition of solutes to the solution, if they can either interact with the side chains of the protein or affect the structure of the solvent which, of course, is usually water.

This is the fundamental reason why proteins can be so 'delicate' and why, in general, they must be treated carefully. Small changes in their environment may produce radical changes in their structure which, if taken far enough, may completely destroy their biological activity. When this happens the protein is said to be denatured. Its native, or biologically active, conformation has been destroyed. Denaturation may be reversible, but is often irreversible, leading to the production of an insoluble mass, as when egg white is cooked.

On the other hand, conformational alteration may not proceed as far as to denature the protein. Small changes in structure resulting from the specific binding of solutes, or changes in the environment, can lead to modifications of biological activity. This is the basis of the flexibility of proteins in response to changes in cellular conditions and in particular underlies the phenomenon of regulatory enzymes. For this reason a more detailed study of the ways in which side chains may interact with each other and with the aqueous environment is needed. Protein chemists group these types of interaction into five classes.

1. The hydrophobic interaction

The first and probably the most important is the so-called hydrophobic interaction. Put at its simplest, this is due to the tendency of side chains that are not water soluble to get away from the aqueous medium surrounding the protein. They, therefore, tend to cluster together in the 'core' of the folded polypeptide chain, leaving the hydrophilic side chains at the surface to interact happily with the water surrounding them. The polypeptide will tend to fold itself up in such a way that the maximum number of hydrophobic side chains is able to escape from the aqueous phase into a non-aqueous core. In such circumstances hydrophobic groups or side chains are forced into proximity, not out of any attraction for each other but out of a need to avoid water. There is no such thing as a 'hydrophobic bond' although it is often referred to! The nearest approach to a hydrophobic *bond* may arise inside the non-aqueous core of a protein when side chains forced into proximity will tend to arrange themselves according to the very weak van der Waals forces, discussed later. The primary force behind the hydrophobic interaction is not the van der Waals force but the inability of water to accommodate non-polar residues within its structure.

Water has a highly ordered structure in which each water molecule forms up to four hydrogen bonds with neighbouring water molecules, two by donating its hydrogen atoms to the oxygen atoms of other water molecules and two by receiving hydrogen atoms from other water molecules, as shown in Figure 17.

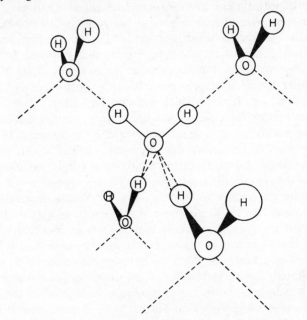

Figure 17 A tetra-hydrogen-bonded water molecule
at centre, surrounded by four other water molecules.
The dotted lines denote hydrogen bonds

This tendency depends upon the polar nature of the water molecules. The electronegative oxygen atom pulls electrons away from the hydrogen atoms thus generating a dipole with a partial negative charge on oxygen and a partial positive charge spread over the hydrogens. Other water molecules in the vicinity will tend to be aligned so that their dipoles are orientated with their positive (hydrogen) ends towards the negative (oxygen) end of the one in question and vice versa. The formation of a hydrogen bond enhances the negative charge on the oxygen of the molecule acting as hydrogen donor, thereby encouraging it to accept hydrogen from another molecule, and vice versa, so that the process of hydrogen bond formation is a *cooperative* one. In other words when it has happened once it tends to happen twice more. Two things prevent the water molecules from forming a completely regular hydrogen-bonded lattice structure. The first is the fact that the individual molecules possess energy of vibration, rotation and translation, which disrupts or distorts the lattice; and the second is the fact that an ordered lattice structure in which each molecule formed four tetrahedrally directed hydrogen bonds would possess a very low *entropy*. This thermodynamic function is a measure of the *disorder* of a system and it always tends to be as high as possible. So one model for the structure of water suggests that it may contain lattice structures with tetra-hydrogen-bonded water in the centre and tri-, di- or mono-hydrogen-bonded molecules at the edges. In between these 'floating lattices' may be water molecules without any hydrogen bonds at all. Alternatively all the water molecules may be tetra-hydrogen bonded but some of the bonds, instead of being tetrahedrally directed, may be distorted rather than broken. The molecules at the edges of regular lattices would have more than four nearest neighbours with which they would interact by van der Waals forces. It should be added that the regular lattice structures must be very short-lived, shifting and re-forming elsewhere very rapidly, whichever model is correct.

The reason hydrophobic substances are hydrophobic is not hard to find. When a non-polar molecule is dissolved in water it has no means of interacting with the water molecules, other than by weak van der Waals forces. If the non-polar molecule finds itself in the middle of a regular tetrahedral lattice, the energy of the tetra-hydrogen-bonded molecules around it will be reduced slightly because they can now undergo van der Waals interactions with it, whereas there was only an empty space in its absence. If, on the other hand, it finds itself next to a water molecule that is forming less than four regular hydrogen bonds the energy of the lattice will be increased. This is because the non-polar molecule has replaced a polar water molecule which would have been able to interact more favourably. These alterations in the energy levels of the variously hydrogen-bonded water molecules bring in train certain changes in the overall structure of the water. The Maxwell–Boltzmann distribution law tells us that, at a given temperature, the number of molecules n, which have energy E, is proportional to $e^{-E/kT}$, where k is the Boltzmann constant. If E is increased, n must decrease and vice versa. Thus at a given temperature, the number of regularly tetra-hydrogen-bonded water molecules must increase and the number of all other forms must decrease in the vicinity of a dissolved non-polar molecule. Energy will be liberated as heat in this process, but order will be generated. In other words, although the system goes to a situation of lower enthalpy (H), which tends to make it stable and thus favours the solution of non-polar molecules, on the other hand the entropy (S) decreases which works against their solution. The deciding factor in cases like this, is the change in free energy ΔF, where:

$$\Delta F = \Delta H - T\Delta S$$

If ΔF is negative the process will occur, if it is positive the process will not occur. Now for the solution of non-polar molecules such as hydrocarbons, it has been found experimentally that ΔH is negative. It is effectively the reduction in internal energy brought about by the shift into regular tetra-hydrogen-bonded states; ΔS on the other hand is also negative and at temperatures around

20–40 °C the product $T\Delta S$ is bigger than ΔH, so ΔF is positive. As a result non-polar substances are effectively insoluble in water.

By extending these experimental results on hydrocarbons to non-polar side chains of proteins, the driving force behind the hydrophobic interaction is thought to be entropic. By forming a non-polar enclave, the hydrophobic side chains of a protein avoid the increased order and decreased entropy that would follow from their intrusion into the water structure. It also follows from the above considerations that the hydrophobic interaction should be stronger at higher temperatures, because the term $T\Delta S$ becomes more and more unfavourable towards the dissolution of non-polar groups, as T is increased. This only operates up to a point, usually around 50–60 °C, because above this temperature the ordered structure of water progressively breaks down and the intrusion of non-polar residues then becomes favourable on account of the enthalpic term. Nevertheless for the temperatures at which living systems usually operate, the theory of hydrophobic interactions predicts that they will be stronger at higher temperatures. Hydrophobic interactions are also stabilized by the addition of ionic solutes to the protein solution, but are weakened by the addition of non-ionic solutes such as alcohols or dioxane, and by the presence of high concentrations of urea or guanidinium salts.

2. Hydrogen bonds

Hydrogen bonds can be formed between –NH or –OH groups on the one hand acting as H donors, and the oxygen atoms of hydroxyl, carbonyl or carboxyl groups on the other hand, acting as acceptors. We have already seen how water molecules can form hydrogen bonds with each other, but they can also form hydrogen bonds with groups in the protein, such as the carboxyl, carboxamido, hydroxyl and amino groups of various side chains. In addition they can form hydrogen bonds with the C=O or N—H groups of the peptide bonds in the polypeptide backbone.

On the other hand, there are important hydrogen bonds that do not involve water molecules at all. One of the most important of these is the hydrogen bond formed between the C=O group of one peptide bond and the N—H of another. These inter-peptide hydrogen bonds have been found to be particularly important in stabilizing the conformations of fibrous proteins such as fibroin in silk and keratin in hair and feathers, where two fundamental types of repeating structure are based upon them. These two, the α-helix and the β-sheet structures (Figure 18), are also found to a lesser extent in the non-fibrous globular proteins, including most of the enzymes. The individual inter-peptide hydrogen bond:

$$
\begin{array}{c}
\qquad\qquad\quad | \\
\qquad\qquad\quad C{=}O \\
\qquad\qquad\quad | \\
\quad | \qquad\qquad\quad \\
\quad C{=}O \ldots H{-}N \\
\quad | \qquad\qquad | \\
H{-}N \\
\quad |
\end{array}
$$

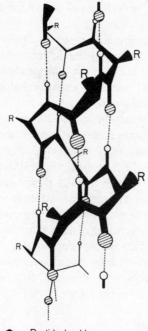

Key:

⌇ Peptide backbone
○ Hydrogen atom of peptide NH group
⊖ Oxygen atom of peptide CO group
⋮ Hydrogen bond
R Side chain group

Figure 18(a) The α-helix structure for a polypeptide chain. This is a picture of a right-handed α-helix. There are 3·6 amino acid residues per turn of the helix. The C=O group of each peptide bond points almost vertically downwards and is hydrogen-bonded to the upward pointing N—H group of the third peptide bond along the chain from it towards the carboxyl terminus

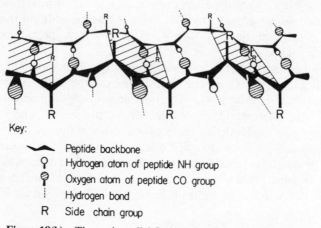

Key:

⌇ Peptide backbone
○ Hydrogen atom of peptide NH group
⊖ Oxygen atom of peptide CO group
⋮ Hydrogen bond
R Side chain group

Figure 18(b) The antiparallel β-pleated sheet structure. Two stretches of polypeptide chain, running in opposite directions, are cross-linked by hydrogen bonds between a C=O group of one chain and an N—H group of the other. The side chains point up or down alternately, above or below the plane of the sheet

is not particularly strong when compared with the alternative peptide–water hydrogen bonds that are possible for a polypeptide dissolved in water. Here again, there is an interplay of enthalpic and entropic effects to be considered. Much work has been carried out on the theoretical interpretation of the so-called 'helix–coil transition' which serves as a model for what happens when an ordered native protein structure becomes denatured. The 'helix' used in the model is the α-helix, while the 'coil' to which the α-helix breaks down is the random coil, an idealized concept in which the polypeptide chain is completely opened out and all peptide bonds made available for interaction with water. Some synthetic polypeptides are found to undergo just such a helix–coil transition in well-defined circumstances, so the model is not entirely devoid of reality, though as we shall see most enzymes do not exist entirely as the α-helix in the native form, nor are they entirely random coils in the denatured form.

If we consider the random coil all the peptide carbonyl oxygen and amide N—H groups will be involved in H bonds with water molecules. Also there will be very little conformational restraint on the orientations of neighbouring amino acid residues. In other words, the conformational entropy will be high. The formation of peptide–peptide hydrogen bonds with the release of water is accompanied by a reduction in the enthalpy of the system, i.e. ΔH for their formation is about $-1\cdot5$ kcal/mol

$$
\begin{array}{ccccc}
 & & & C{=}O & & C{=}O \\
 & & & | & & | \\
C{=}O\ldots H_2O & H_2O\ldots H{-}N & & C{=}O\ldots H{-}N & \\
| & & \rightarrow & | & | & + 2\,H_2O \\
H{-}N & + & CH & H{-}N & CH \\
| & & | & | & | \\
CH & & & & \\
| & & & & \\
\end{array}
$$

Their formation is accompanied, on the other hand, by the introduction of more order into the system because the two peptide units are now locked together and not free to rotate, so the conformational entropy decreases and ΔS is about -4 entropy units per mole. Thus at 25 °C (298 K) the free energy change ΔF is about -300 cal/mol

$$\Delta F = \Delta H - T\Delta S$$

$$\therefore \quad \Delta F = -1500 + 298 \times 4 = -308 \text{ cal/mol}$$

This calculation ignores the fact that other amino acid residues besides the ones directly linked by the hydrogen bond, will also be immobilized to a greater or lesser degree. So the entropy change will actually be more negative than -4 entropy units per bond formed, making the value of ΔF positive unless these other residues can also form inter-peptide hydrogen bonds, thereby gaining the $1\cdot5$ kcal/mol of stabilizing enthalpy. In fact the α-helix is one way in which multiple inter-peptide hydrogen bonds *can* form, thus offsetting the unfavourable entropy decrease. The β-pleated sheet structure is another. In both cases, however, the structure is more stable the more residues are involved, and only becomes stable at all in the case of the α-helix when more than approximately 16 residues are involved. As there are 3·6 residues per turn of the α-helix, this means that helices of less than about four turns are not expected to be stable in water, unless other compensatory forces can be called upon to offset the entropic factor. Also, as the temperature is increased the formation

Figure 19 The lysozyme molecule. A photograph of a molecular model of lysozyme, taken in the Biochemistry Department at Sheffield University. The model shows the substrate bound in a cleft in the enzyme surface

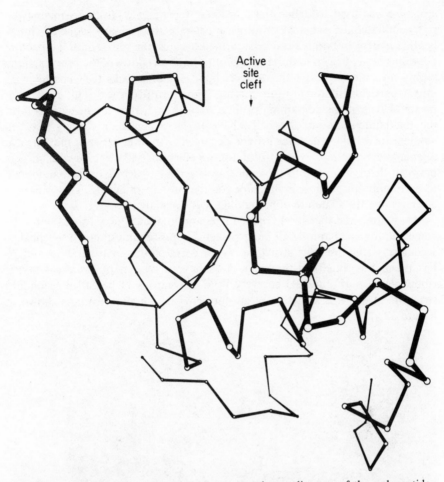

Figure 20 The lysozyme molecule. A computer-drawn diagram of the polypeptide chain showing the active site cleft. Reproduced by courtesy of Professor A. C. T. North, Astbury Department of Biophysics, The University of Leeds

of the inter-peptide H bond becomes less and less favoured due to the increasing size of the $T\Delta S$ term, and the random coil is the favoured form.

Hydrogen bonds can also be formed between the peptide $C=O$ groups and various side chain groups, or between the side chain groups themselves. Thus a wide variety of hydrogen bonds is possible in protein molecules and the types of bond actually formed in a given protein at a given temperature in a given aqueous environment will be the ones which, together with all the other possible interactions, allow the free energy of the whole system to reach a minimum value.

Much early work on protein structure was carried out on the fibrous proteins where α-helix and β-sheet structures are very important. The term *secondary structure* was used to define these ordered, repeating ways of organizing the polypeptide chain. Wherever proline appears in the primary sequence, inter-peptide hydrogen bonds cannot be formed because the nitrogen in the peptide bond has no hydrogen to donate. In addition the structure of the prolyl residue allows no rotation about the α-C—N bond and this locks the peptide chain into a particular configuration, forcing it to 'turn a corner'. This restriction means that proteins containing proline, even though they might tend to form the maximum amount of α-helix, have to have bends or corners between stretches of α-helix whenever proline occurs. The folding between the α-helical segments of a protein, so as to form a series of sausage-like helices with corners between them, gave rise to the concept of *tertiary structure*, which could lead to a roughly globular overall shape for the protein molecule. This was seen clearly when the structure of myoglobin was elucidated.

Myoglobin, and the closely related haemoglobin, have now turned out to be somewhat unusual among globular proteins in possessing so much α-helical or secondary structure. The structure of lysozyme, shown in Figures 19 and 20 has much less α-helix, although it does have a section of β-sheet, whereas ribonuclease A (Figure 21) has very little of what could be called secondary structure at all. The structure of another enzyme, chymotrypsin, is shown in Figure 22.

Figure 21 The ribonuclease molecule. A photograph of a molecular model of ribonuclease S built in the Biochemistry Department at Sheffield University and photographed by courtesy of Dr Pauline M. Harrison

Figure 22 The chymotrypsin molecule. A photograph of a molecular model of chymotrypsin built in the Biochemistry Department at Sheffield University and photographed by courtesy of Dr Pauline M. Harrison

As a result the terms 'secondary' and 'tertiary' structure have become less meaningful when applied to enzyme structures. One can only say that the primary structure folds up in such a way as to allow the maximum hydrophobic interaction that is consistent with the optimal formation of hydrogen bonds between the various side chains and the peptide backbone.

3. Electrostatic interactions

In addition to the hydrogen bond, whose formation involves electrostatic forces, the so-called salt bridge may contribute to the stabilization of protein structures. Here we are considering the possibility that a positively charged amino group $-NH_3^+$, either from the N-terminal residue or from a lysyl side chain, or a guanido group

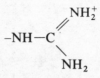

belonging to an arginyl residue or the imidazolium group,

of a histidyl residue, may be attracted electrostatically by a negatively charged carboxylate group, $-COO^-$, in the side chain of a glutamyl or aspartyl residue. All these charged groups are able to attract water molecules to form stable hydration shells around them if they protrude into the aqueous medium surrounding the protein. In order for such hydrated ions to form salt bridges it would be necessary to break the hydration shells:

$$\text{protein} - NH_3^+(H_2O)_n + \text{protein} - COO^-(H_2O)_m$$

$$\text{protein} - NH_3^+ \ldots {}^-OOC - \text{protein} + (n + m)H_2O$$

The enthalpy change ΔH for such a process is positive because more enthalpy is required to break the ion–water interaction than is gained by forming an ion-pair. This is partly because the electrostatic attraction between oppositely charged ions is only a small force in a medium of high dielectric constant like water. On the other hand, the entropy change will be also positive, due to the fact that the strongly orientated hydration shells are replaced by less ordered water. On balance, however, ΔF is positive and most charged groups are found on the surface of the protein, surrounded by hydration shells of water molecules. This is why such groups are said to be hydrophilic. There is one situation in which the salt bridge can be important. That is when charged groups, because of their location in the primary sequence, are forced into the non-polar interior of the protein core due to a multiple hydrophobic interaction of other nearby side chains. If there is no way in which the charged groups can be accommodated at the surface while allowing hydrophobic groups to lie inside the protein, then salt bridges will tend to form inside the non-polar core. In that environment the dielectric constant will be lower than in water and the attractive force between opposite charges correspondingly bigger. Increasing the temperature will stabilize such salt bridges, because of the favourable $T\Delta S$ term in the free energy equation, in just the same way that hydrophobic interactions are stabilized at increased temperatures. The effect of high salt concentrations upon salt bridges will be to weaken them, because high ionic strength in the aqueous environment will stablize the hydrated charges. This effect of high salt concentration is the opposite of that expected for hydrophobic interactions where a 'salting-out' effect encourages and stabilizes them.

Although salt bridges are not found to be very important in holding proteins together in the native structure, electrostatic interactions can become extremely important in denaturation. Proteins contain both positively and negatively charged groups in the pH range around neutrality. As the pH is lowered by adding acid to a protein solution, the mercaptide, $-S^-$, and carboxylate groups, $-COO^-$, will become progressively protonated, giving the uncharged sulphydryl, $-SH$, and carboxyl, $-COOH$, groups, while amino, imidazole and guanido groups will remain positively charged. The result is an increase in the net positive charge of the protein and an increasing tendency for these positive charges to repel each other. Alternatively, as pH is increased above neutrality, mercaptide and carboxylate ions remain negatively charged while amino and imidazole groups progressively lose their protons, with the result that the protein gains an increasing net negative charge. Once again repulsive forces can tear the native structure apart leading to a more extended 'random coil-like' structure. For every protein there is a pH at which the net charge is zero, owing to a balancing out of positive and negative charges. This value of pH is the *isoionic point* of the protein and it varies with the amino acid composition of the protein. At its isoionic point the repulsive forces between similarly charged side chain groups of the protein, will be balanced by and large by attractive forces between oppositely charged groups. It must be mentioned that if certain ions are present in solution they can bind to specific sites on the protein. The specificity of such ion-binding varies from protein to protein. For example, ribonuclease A can bind phosphate ions at specific sites on its surface, in which case the net negative charge of the protein is increased. This results in a lowering of the pH at which the net charge on the protein is zero to a value below the true isoionic point. This new pH value is called the *isoelectric point*. It is the pH at which the protein carries a net zero charge in a particular salt solution and will vary with the composition of the salt solution.

4. Electron delocalization

Another minor contribution to the forces stabilizing a particular three-dimensional structure can arise from resonance where two or more equivalent arrangements of the electrons are possible. Two situations exist in the case of proteins. The first involves the resonance between two forms of the peptide bond which, as we saw in Figure 16, leads to the planarity and rigidity of the

74

peptide bond:

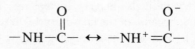

When inter-peptide hydrogen bonds are present, there is a greater possibility of electron delocalization:

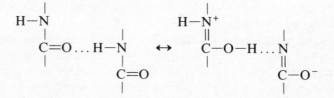

The longer the sequence of 'linked' peptide hydrogen bonds, the greater the resonance energy will be. This has given rise to the suggestion that, in extended systems of inter-peptide hydrogen bonding, an added stability will be conferred by the resonance energy. There are, however, rather strict geometric conditions to be met; and there is good reason to suppose that the resonance contribution from this source is only small in most proteins.

The phenomenon of 'aromatic stacking', on the other hand, is well established in proteins. When two aromatic ring systems, such as the phenyl rings of phenylalanine, are able to align themselves one over the other with the planes of the rings parallel, overlap of their pi orbitals is possible with a resulting delocalization of the electrons of both rings. Many instances of such stacking have been established in the protein structures studied so far and it appears to be an important, though minor, way of stabilizing them.

5. Van der Waals forces

This general term encompasses three classes of weak interactions. The first is called the *orientation effect* which is due to the attraction of the positively charged end of a dipole for the negatively charged end of another and vice versa. Of course, the positive ends of the two dipoles will also repel each other, as will the two negative ends. These dipole–dipole interactions may well average out to zero over a large number of polar molecules unless there is some other reason why they should all be properly orientated, as for example in the hydrogen-bonded water lattice.

The second class of interactions, the *induction effect*, is due to the induction of a dipole in a molecule owing to the proximity of another polar molecule. This fails to account for attractive forces between molecules which do not have dipoles, as does the orientation effect, although both may be important for polar molecules.

The major contribution to van der Waals forces in most circumstances, and the only one in non-polar molecules, is the third type of interaction, the *dispersion effect*, sometimes credited to its discoverer, London. London dispersion forces arise because of the instantaneous dipoles that exist momentarily in all molecules owing to the fluctuations in their electronic distribution. Such an instantaneous dipole can induce an oppositely oriented dipole in a neighbouring molecule, and so lay the basis for an attractive force between them. The force is weak and varies as the reciprocal of the sixth power of the distance between the molecules. It has been shown to be greater between like molecules than between unlike ones.

THE CENTRAL DOGMA

Between them, then, these five types of interaction together with the sequence of amino acids—the primary structure—are what determine the three-dimensional structure or conformation of a protein.[11] The 'central dogma' of molecular biology assumes that the conformation actually adopted by the protein in its native state is the one of lowest free energy, the most stable in the natural environment. This really is an assumption at the present time. Although it would undoubtedly be true at equilibrium, living systems are not at equilibrium and there may not always be time for a polypeptide chain to search out the most stable configuration. The assumption is certainly true for some protein molecules such as ribonuclease A, lysozyme and others, and the evidence for it will now be considered. On the other hand, it will remain a hypothesis until many more proteins have been studied in detail.

One strand of evidence supporting the central dogma comes from studies upon ribonuclease A. It will be recalled that this protein has a single polypeptide chain cross-linked to itself by four intra-chain disulphide bonds (Figure 14). When the disulphide bonds are reduced to SH groups by incubating the enzyme with mercaptoethanol, $HO \cdot CH_2 \cdot CH_2 \cdot SH$, or some other low molecular weight thiol, in the presence of 8 M urea the enzyme becomes inactive and physical studies show it to have a more open, random-coil-like structure. If, however, the mercaptoethanol and urea are removed by dialysis and the protein solution left exposed to air at pH 7–8, activity slowly returns. Ultimately the full original activity is regained and chemical analysis reveals that the four disulphide bonds have re-formed. This in itself is perhaps not too surprising, but further study shows all four disulphide bonds to be formed in exactly the same partnerships as before, even though theoretically any one SH group could have become linked to any of the other seven, giving a total of 105 possible results after re-oxidation, only one of which is actually found. Reoxidation in the presence of 8 M urea, on the other hand, gives no activity and the disulphide bonds formed are found to be 'incorrect'. The explanation seems to be as follows. After the first disulphide bond has been formed by the oxidation of any two SH groups that happened to be close enough to react, thiol–disulphide exchange occurs:

$$R_1-S-S-R_2 + R_3SH \rightleftarrows R_1-S-S-R_3 + R_2SH$$

$$\updownarrow$$

$$R_1SH + R_3-S-S-R_2$$

This allows the re-oxidizing protein to hunt through all 105 possible combinations until the 'correct' one is found. The implication of this experimental result is thus clear. The primary structure determines the thermodynamically most stable conformation, which is then locked in place by disulphide bond formation.

The experimental technique for determining which cysteinyl residues are linked in the native and in the reoxidized structures, is of fundamental importance in establishing the facts in this situation and deserves a brief mention. Before commencing sequence studies all disulphide bonds must be cleaved, so when the amino acid sequence has been determined one has no idea of the disposition of the disulphide bridges in the original native protein. It is therefore necessary to return to the native protein with its disulphides intact and cleave it into small peptides under conditions where the disulphide bridges will not become 'scrambled'. This means working at a pH below 6; and pepsin is the proteolytic enzyme usually chosen, because it normally works in the stomach at a pH of 2–3 and is unusual in being stable at low pH. Pepsin digestion of the native protein yields a mixture of small peptides, some of which contain disulphide bridges. These can be selected very elegantly from all the others by the diagonal technique of Hartley. The mixture of peptides is spotted onto a sheet of filter paper and electrophoresis is carried out at pH 6·5, separating the peptides according to their size and charge. The paper is dried and exposed to the vapour of performic acid so as to oxidize each disulphide-containing peptide to a pair of peptides each containing cysteic acid. Electrophoresis is now repeated under exactly the same conditions as before in a direction at right angles to the original. Most of the peptides will have the same electrophoretic mobility as before and will lie on a diagonal across the sheet. The pairs of peptides containing cysteic acid, on the other hand, are now smaller and more negatively charged than the original peptide with a disulphide in it. Consequently they lie off the diagonal, one above the other, and can be picked out with ease (Figure 23). Sequence studies on these pairs of peptides then establish which points of the sequence were linked by the disulphide bond.

THE QUATERNARY STRUCTURE OF PROTEINS

Although we have dealt with the ways in which a single polypeptide chain can fold up on itself and the forces which hold it in a given conformation, we have yet to look into the aggregation of polypeptide chains to form *oligomeric* proteins. Many enzymes now appear to be composed of subunits or protomers, each of which is usually a single polypeptide chain. There may be more than one type of protomer in a given oligomer. The type and mutual disposition of the subunits of an oligomeric protein is called its *quaternary structure*. The examples of three-dimensional protein structure so far seen in Figures 19–22 are all monomeric proteins, which show no tendency to aggregate under normal conditions. They have no quaternary structure but others like phosphoglycerate mutase (Figure 24) and aspartate transcarbamoylase (ATCase Figure

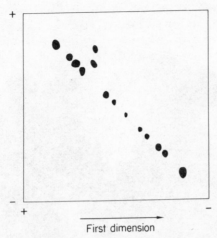

First dimension

Figure 23 A diagrammatic representation of the diagonal technique for
identifying those peptides in a mixture
that contain disulphide bridges. The
mixture of peptides is subjected to
paper electrophoresis in the horizontal
direction. The paper is then dried and
exposed to the vapour of performic
acid after which it is re-subjected to
electrophoresis under identical conditions, but in a vertical direction.
Peptides lying off the diagonal line
were originally linked by a disulphide
bridge

25) have. The tendency to form oligomers has been correlated convincingly
with the content of hydrophobic amino acids. Those proteins containing more
than about 30% of hydrophobic residues appear to be unable to tuck them all
away inside the core formed by the folding of a single chain. The result is that
the excess would be exposed to the aqueous medium were it not for the
possibility of more than one polypeptide coming together, hydrophobic cheek
by hydrophobic jowl, and thus excluding water. The main driving force behind
quaternary structure once again appears to be the hydrophobic interaction, as
it was in the case of tertiary structure. Hydrogen bonds, electrostatic interactions
and all the rest undoubtedly play a part in quaternary structure also, but
probably only a minor part.

In studying quaternary structure one has to find answers to the questions,
'How many subunits are there in a given protein molecule?' and 'Are the
subunits the same or different?'.

Some experimental methods capable of providing the answers have already
been described. Determination of the molecular weight of the native oligomeric
protein, either by ultracentrifugation or by gel-filtration, can be followed by

Figure 24 The tetrameric structure of phosphoglycerate mutase. The photograph shows a balsa-wood model of the molecule of yeast phosphoglycerate mutase. Each subunit is shaded differently. (By courtesy of Dr Herman Watson) (reproduced, with permission, from J. W. Campbell *et al.* in *Nature*, **240**, 137 (1972))

determination of the subunit molecular weight by the SDS–gel electrophoresis method. The latter may also reveal different types of subunit if their molecular weights are sufficiently different. As an example consider the enzyme β-galactosidase from the bacterium *E. coli*. Its molecular weight was found to be 540,000 by ultracentrifugation studies, whereas in SDS–gel electrophoresis only a single protein of molecular weight 135,000 was observed. These results clearly suggest that the native protein is a tetramer, consisting of four subunits of very similar molecular weight. One cannot be *certain* that the subunits are

Figure 25 The quaternary structure of aspartate transcarbamoylase (ATCase). This representation of the gross shape of the molecule shows no detail. The outlines of three catalytic subunits are shown at the front in the centre with three smaller regulatory subunits around the periphery. Behind these six subunits, another similar set of six can be seen, making twelve subunits in all

all identical from this evidence, only that they are of the same molecular weight.

Instead of using SDS–gel electrophoresis to determine subunit molecular weight, the analytical ultracentrifuge can be used if the measurement is carried out in the presence of substances such as SDS, urea ($NH_2 \cdot CO \cdot NH_2$) or guanidinium chloride

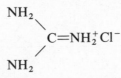

which generally cause oligomers to dissociate to their constituents. When this was done with β-galactosidase, using guanidinium chloride as the dissociating agent, the same result was found as with SDS–gel electrophoresis, a subunit molecular weight of 135,000.

On the other hand, some proteins are dissociated by specific reagents. In the case of aspartate transcarbamoylase (ATCase) the molecular weight of the native protein was found by ultracentrifugation to be $3 \cdot 1 \times 10^5$ in dilute phosphate buffer at pH 7·0. When the measurement was repeated in the presence of p-chloromercuribenzoate (Cl·Hg —⟨⟩— COO⁻), two protein components were found, both of smaller molecular weight, one of which was enzymically active ($M = 1.03 \times 10^5$) and the other inactive ($M \simeq 3 \times 10^4$). Later studies on the separated components, again in the ultracentrifuge but now with added 8 M guanidinium chloride, showed that the active component was actually a trimer of identical 'catalytic' or C-type chains, of molecular weight 3.4×10^4, whereas the inactive component was a dimer of identical 'regulatory' or R-type chains, each of molecular weight 1.7×10^4. The native oligomer actually consists of six C-type and six R-type chains as shown in Figure 25.

An accurate estimate of the number and type of amino terminal end groups present in the oligomer also yields information about the number of polypeptides in the protein molecule and whether or not they are the same. In cases where the N-terminal residues of the polypeptide chains are the same, and their molecular weights are indistinguishable, it is sometimes very difficult to be sure whether they really are the same or different. In such cases, the *finger-print* of the protein may help to provide an answer. The fingerprint is the pattern of peptides produced when the protein is digested with a specific proteolytic enzyme and then subjected to chromatography or electrophoresis in two perpendicular directions to separate the peptides (Figure 26). Trypsin is commonly used in preparing fingerprints as it only cleaves the peptide bonds following

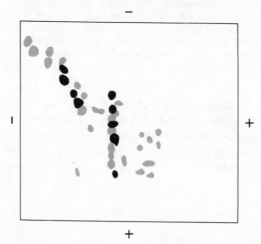

Figure 26 Peptide map (redrawn) of a tryptic digest of performic acid oxidized glyceraldehyde - phosphate - dehydrogenase (GPD) from pig muscle. The digest was subjected to paper electrophoresis first at pH 6·5 (horizontal direction) and then at pH 3·5 (vertical) (reproduced by permission of Academic Press Inc. (London) Ltd., from J. I. Harris and R. N. Perham in *Journal of Molecular Biology*, **13**, 880 (1965))

lysyl and arginyl residues. From a knowledge of the amino acid composition of the protein, the number of lysyl and arginyl residues in the measured native molecular weight M can be calculated. Assume that this number is N. Trypsin will therefore produce $(N + 1)$ different peptides if there is only one polypeptide chain per molecule. If there are two identical chains, each of molecular weight $M/2$, there will be only $N/2$ lysyl and arginyl bonds per chain and so the expected number of peptides will be $(N/2 + 1)$. Alternatively, two different chains will produce $(N + 2)$ peptides. Counting the peptides present in the fingerprint can sometimes distinguish the number and identity of chains in an oligomer, as was done for the enzyme glyceraldehyde phosphate dehydrogenase shown in Figure 26.

In the last analysis, the identity or otherwise of the chains in an oligomer may have to await the determination of the primary sequence. If only one type of polypeptide can be demonstrated by a variety of techniques and sequence studies give an unambiguous primary structure, one can safely assume that all the chains are identical.

X-RAY DIFFRACTION STUDIES ON THE THREE-DIMENSIONAL STRUCTURE OF PROTEINS

Even when the primary structure, molecular weight and the number and identity of polypeptide chains in the molecule of protein are known, the secondary, tertiary and quaternary structures of the protein are still far from elucidated. Hydrodynamic studies on the viscosity, sedimentation and diffusion of the protein can give some idea of its general three-dimensional shape, but no more than that it is roughly spherical, or else broadly ellipsoidal, or perhaps rod-like. A knowledge of the detailed three-dimensional structures already shown in Figures 19–25 had to await the application of X-ray crystallographic studies. In principle, such studies are able to describe the molecular structure of proteins in the greatest detail, allowing one to position all the atoms in the molecule. Certain conditions must be met for this to be possible. The first is that the protein must form crystals of about 1 mm length. That is not always easy to achieve. The crystal is placed in a diffractometer and bombarded with a narrow beam of X-rays which are diffracted according to the atomic spacings in the crystal to give a diffraction pattern which is photographed. Laborious analysis of the diffraction pattern and the intensities of the spots in it then allows the investigator to construct an electron density map of the protein molecule. That, however, can only be achieved if the so-called phase problem can be solved, which requires a repeat experiment using a second protein crystal identical with the first in every way, except that one or two additional *heavy atoms* are present in each protein molecule. The preparation of these *isomorphous heavy atom derivatives* is one of the major problems of X-ray crystallography. Needless to say, if one succeeds in binding a heavy ion, say the mercuric ion, into a specific site in a protein molecule, the chances are that the protein structure will alter sufficiently to invalidate the isomorphous condition.

Clearly, the evidence of Figures 19–25 shows that it is possible to overcome all the obstacles, but the work is time-consuming. The interested reader is referred to more specialist texts for the details and theory underlying the techniques of X-ray crystallography.[12]

When the three-dimensional structure of a protein is finally established, there is no more to be said on the subject of its structure. One must then turn to the problem of function. Of course it has to be assumed that the conformation of the protein observed in the crystal is conserved essentially unchanged in solution, and furthermore in solution in the living cell. These are big assumptions after what has been said about the sensitivity of structure to environment. Nevertheless the knowledge of how the protein molecule folds up in the crystal at least gives one a start in understanding how it might be folded up in a more biological environment. The crystals of enzymes studied so far have been shown to contain a large proportion of aqueous medium filling the spaces between enzyme molecules; and in one or two cases the enzyme molecules in the crystal have been shown to be enzymically active. All this gives one reason to hope that the picture drawn by X-ray crystallography provides a sound basis for understanding how enzymes work, but caution must be exercised in extrapolating from the crystal to the cell.

FACTORS AFFECTING THE CONFORMATION OF A PROTEIN

So far it has been possible to show that temperature, pH, ionic strength and the dielectric constant of the solution containing a protein may exert a profound effect upon its three-dimensional structure. The effect of the detergent SDS in disrupting structure has been mentioned, and other ionic detergents function in a similar way. Urea and guanidinium chloride at concentrations in the range 4–8 M also unfold protein structures, probably largely by interfering with the hydrophobic interaction, for which reason they have been called *chaotropic* reagents—they favour a more chaotic situation. Specific reagents are known for individual proteins, which will dissociate subunits (e.g. PCMB and ATCase), unfold structures and even bring about precipitation. An example of the latter is the use of mercuric ions in the specific precipitation of the enzymes papain and enolase in the course of their purification. In those cases the precipitated and inactivated enzymes can be solubilized and reactivated by removing the mercuric ions with a chelating agent such as EDTA.

Precipitation of proteins in this way must be brought about by a reversible interaction between the precipitant and the protein in a fashion which so disturbs the balance of non-covalent forces in the native, soluble protein that the new most stable structure is one where protein–protein interactions are favoured at the expense of protein–water interactions.

A class of general protein precipitants includes substances such as trichoracetic acid (TCA), picric acid (trinitrophenol), phosphotungstic acid, sulphosalicylic acid and perchloric acid. All these are potent agents for the precipitation of proteins, but in most cases the protein is irreversibly denatured in the process. Although they are of great value in deproteinizing biological fluids, they are of little value to enzymology on account of their destructive action. They probably function by binding to specific sites on the protein and causing changes in the local charge which result in large conformational changes leading to unfolding and precipitation.

The effect of very high concentrations (1–4 M) of certain salts, such as ammonium sulphate or potassium phosphate, in precipitating proteins is much more useful. The 'salting-out' of proteins in this way is usually accompanied by preservation of biological activity so that when the separated protein is re-dissolved and the excess salt removed by dialysis, biological activity returns. Salting-out probably occurs by virtue of the increase in ionic strength in the solution rather than by binding of the ions to specific sites in the protein molecule. It will be recalled that high ionic strength stabilizes hydrophobic interactions and destabilizes electrostatic interactions. Its effect on hydrogen bonds is less certain. In addition, the presence of solute ions at high concentration will compete with the hydrophilic groups of the protein for the available water, making the hydration shells of the hydrophilic groups less stable and encouraging protein–protein interactions leading to precipitation.

In addition to all the agents mentioned which cause dissociation, unfolding or precipitation, there is a whole gamut of more subtle modifiers of protein structure. These are, on the whole, site specific solutes in that they bind selectively and reversibly to certain proteins, perhaps only one, and not to others. When they bind, they do so by non-covalent interactions at a unique location on the surface of the protein molecule and produce changes in the conformation which, although not so drastic as to cause general unfolding or precipitation, nevertheless have far-reaching effects via the alteration in biological activity that they bring about in the protein. These are the substances of greatest interest to the enzymologist; to see how they function it is necessary to investigate the nature of specific binding sites in enzyme molecules.

Before turning to the structure of enzymes in the next chapter, as opposed to the general structure of proteins with which we have been concerned in this, it is worth pointing out that irreversible alteration to protein structures can come about in a number of 'chemical' ways. By 'chemical' one here means alteration to the primary structure. A wide range of chemical modifications to the side chains of a protein can be made in attempts to study its function. Some of these will be discussed later. There are, however, three ways in which proteins may become accidentally modified during isolation and handling; they can be avoided if due precautions are observed. The first is by atmospheric oxidation of the protein. This occurs at the side chains of the sulphur-containing amino acids, methionine and cysteine. In the former case the methionyl side chains, being hydrophobic, are usually buried deeply in the core of a globular protein. If they become oxidized to the sulphoxide, the structure may be modified. In the case of cysteine, the –SH group may protrude into water or be buried in the core, but, in either case, oxidation to the sulphenic acid (–SOH), or the di-sulphide can occur on exposure to air, leading to changes of charge or shape or both. If this is found to be troublesome, a simple remedy is available in the form of a protective reducing agent which can be added to the protein solution to mop up excess dissolved oxygen. Mercaptoethanol ($HO{\cdot}CH_2{\cdot}CH_2{\cdot}SH$), cysteine itself or, best of all, dithiothreitol (DTT; $HS{\cdot}CH_2{\cdot}(CHOH)_2{\cdot}CH_2{\cdot}SH$) can all serve as mild de-oxidants for the preservation of proteins in the reduced form. They should not be used on proteins that already contain disulphide bridges or these will be reduced with adverse effects. It is of interest that proteins containing cyst(e)inyl residues usually have *either* SH groups (cysteinyl residues) *or* SS groups (cystinyl residues) but not both.

The second avoidable type of modification to the primary structure is proteolytic digestion. Traces of proteolytic enzymes present in a protein solution will bring about cleavage of the polypeptide chain and loss of activity. Proteolytic contaminants may be inherent in the source from which a protein is isolated and if this is so, only rigorous purification will remove them. They may also arise from chance microbial contamination of protein solutions. Many microbes produce extracellular proteolytic enzymes, can survive cold room temperatures and thrive on the purified protein solutions that the enzymologist spends so much time in preparing! They can be removed by the expedient of filtration through a sterile membrane filter, and this is recommended whenever protein solutions are to be stored for more than a few hours.

Lastly, loss of amino or amido groups may come about, again by virtue of the presence of traces of contaminating enzymes. These may convert amino groups to keto groups, and carboxamido ($-CONH_2$) groups in the side chains of asparaginyl and glutaminyl residues to the corresponding carboxyl groups. Such modifications have been shown to be responsible for multiple electrophoretic forms of cytochrome c, and for the 'ageing' of aldolase but they are seldom easily demonstrated. They are sometimes suspected as the cause of mysterious spontaneous inactivation. Rigorous purification and careful exclusion of microbes may avoid such problems. On the other hand the hydrolytic de-amidation of asparaginyl residues may occur non-enzymically if the asparaginyl residue occurs in certain specific sequences. In that case there would appear to be little one can do to avoid de-amidation. The 'ageing' of aldolase appears to be a case of this type and the spontaneous hydrolysis of one of its asparaginyl residues has been suggested as a built-in mechanism to ensure the turnover of aldolase molecules in the living cell.

Chapter 4

Enzyme Structure and Function

All enzyme molecules are proteins, though the converse is not true. What special features make certain proteins act as enzymes? What underlies their specificity for certain substrates and how can their catalytic activity be controlled? The answers to all these questions lie in an understanding of the concept of the binding site. By virtue of its organized and involuted three-dimensional structure, a protein possesses regions on its surface where small solute molecules or ions can bind reversibly. Such solutes are called ligands, a term borrowed from organometallic chemistry. There may be many ligand binding sites on the surface of a protein, but each site usually possesses the power to bind only a limited range of ligands, by virtue of the character of the site. The term 'character' is here used to cover not only the three-dimensional shape of the site but also its charge characteristics and to what degree it is hydrophobic or hydrophilic. The character of a binding site is clearly a function of the amino acid side chains that are brought together there by the folding of the polypeptide chain.

Enzymes are distinguished from other proteins by having *active* sites. The substrate binds at the active site of its enzyme in much the same way as other ligands might do but, once bound there, a chemical reaction ensues because of the special nature of the enzyme active site.

LIGAND BINDING SITES

It has long been realized that in order to achieve catalysis the enzyme must become physically associated with its substrate, in an enzyme–substrate complex. The idea that the site of binding is a particular one with a complementary character to that of the substrate was embodied many years ago in the template hypothesis, or lock and key hypothesis, attributed to Emil Fischer. We now know, however, that specific binding of small molecules at particular sites is a property common to many proteins, not all of them enzymes. For example, plasma albumin binds fatty acids and other lipids and transports them through the blood, haemoglobin binds and transports oxygen in similar fashion, and repressor proteins bind co-repressors or inducers. In none of these cases, however, does chemical alteration to the ligand occur while it is bound, even though in some of them the binding is extremely tight and specific, so that one must assume a high degree of complementary character between the ligand and its site.

The peculiar property of an enzyme active site resides, at least in part, in the existence of catalytic groups within or adjacent to the binding site. Strategically placed side chains belonging to the enzyme are able to interact with the bound ligand so as to catalyse chemical change while it is bound.

A ligand binding site may be defined as that part of the protein where amino acid side chains provide a niche where the ligand can fit. The atoms of the side chains in the binding site come into contact with atoms of the ligand, that is to say they approach to within van der Waals radii. Electrostatic attractions, hydrogen bonds and hydrophobic interactions may be involved individually or collectively in providing the stabilizing forces that hold the ligand in the site. Once the ligand is bound, other protein side chain groups which are not involved in binding may be located close to the ligand. If these groups are of the right type and in the right place to catalyse the reaction of the bound ligand, then the site is an active site. The active site of an enzyme comprises both the binding groups and catalytic groups. There is only a limited selection of side chains available for providing binding and catalytic groups, namely the side chains of the 20 amino acids listed in Table 1 (Chapter 3).

Enzyme cofactors

Sometimes an enzyme is able to extend this range by binding cofactors which themselves possess additional reactive groups or properties. These in turn provide binding or catalytic functions when the substrate comes along. Tightly bound cofactors are sometimes called prosthetic groups. Cofactors fall into two broad groups, the metal ion cofactors and the organic cofactors. The former includes iron, copper, potassium, magnesium, zinc, cobalt, manganese and molybdenum ions. The organic cofactors encompass a wide range of molecular types, but all of them are related to the vitamins. Indeed it appears that the biological role of vitamins and trace metals is simply to provide, either directly or as precursors, the necessary cofactors for certain enzymes. Table 2 shows how the common vitamins are related to enzymic cofactors with an indication of the enzymes they serve. The study of many deficiency diseases has thus led via the vitamins and metal ion cofactors to a study of the enzymic processes that underlie normal metabolism and which fail to function in the deficient state.

Experimental detection and characterization of ligand binding sites

When a ligand L binds to a protein P a protein–ligand complex PL will be produced: $P + L \rightleftharpoons PL$. In order to demonstrate that binding has occurred a means of detecting PL is required. In cases where the spectral properties of either P or L, or both, are changed when the complex PL is formed, one may use a spectrometric method to detect PL. Various spectral properties have been used in this way. The *absorption spectrum* of P or L may show changes in either

Table 2

Enzymic cofactors derived from the vitamins

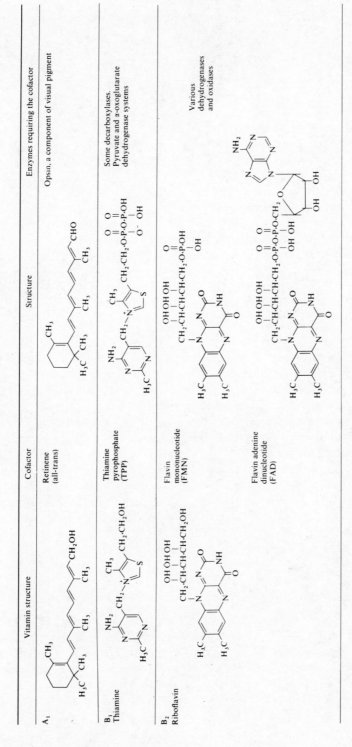

	Vitamin structure	Cofactor	Structure	Enzymes requiring the cofactor
A₁		Retinene (all-trans)		Opsin, a component of visual pigment
B₁ Thiamine		Thiamine pyrophosphate (TPP)		Some decarboxylases. Pyruvate and α-oxoglutarate dehydrogenase systems
B₂ Riboflavin		Flavin mononucleotide (FMN)		
		Flavin adenine dinucleotide (FAD)		Various dehydrogenases and oxidases

Table 2 (continued)

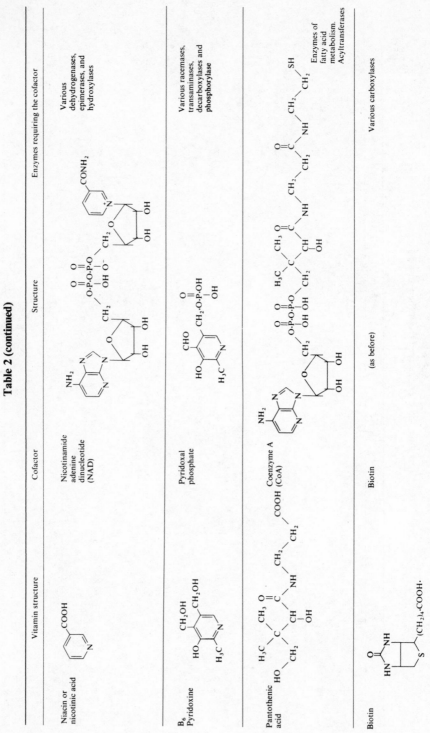

Vitamin structure	Cofactor	Structure	Enzymes requiring the cofactor
Niacin or nicotinic acid	Nicotinamide adenine dinucleotide (NAD)		Various dehydrogenases, epimerases, and hydroxylases
B₆ Pyridoxine	Pyridoxal phosphate		Various racemases, transaminases, decarboxylases and phosphorylase
Pantothenic acid	Coenzyme A (CoA)		Enzymes of fatty acid metabolism. Acyltransferases
Biotin	Biotin	(as before)	Various carboxylases

89

Table 2 (continued)

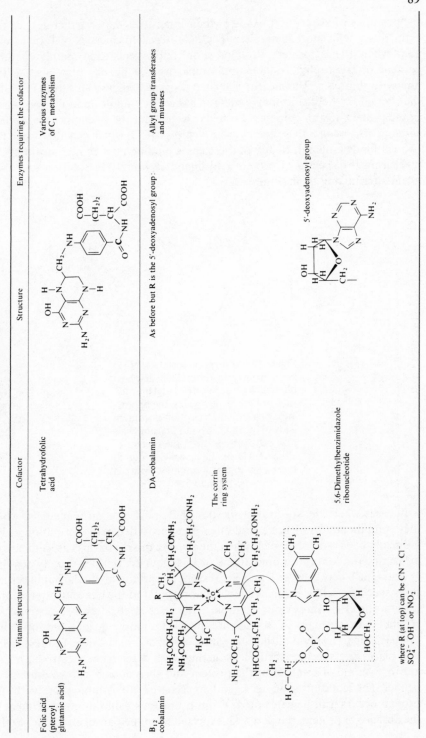

the positions of absorption bands or their intensities, that is either λ_{max} or the extinction coefficient ε, may alter. The *fluorescence* properties of P or L may alter when PL is formed. Proteins show fluorescence due mainly to their content of tryptophyl residues and some organic ligands also possess the ability to fluoresce. The *nuclear magnetic resonance* (n.m.r.) spectrum of either the protein or the ligand may undergo changes attendant upon the formation of the protein–ligand complex. Proteins also exhibit the phenomena of *optical rotatory dispersion* (ORD) and *circular dichroism* (CD) which can alter when the protein binds a ligand. In any of these cases binding may be demonstrated by measuring the spectra of protein and ligand *separately* in solution and then again after mixing them (Figure 27).

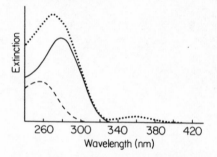

Figure 27 Absorption spectra of NAD (···), the enzyme glyceraldehyde-phosphate-dehydrogenase (—) and the mixture of both (···) showing the appearance of a broad absorption maximum with its peak at around 360 nm. This new peak, absent from the spectra of either NAD or the enzyme, is due to the formation of a complex between the two

In cases where the spectral properties of P and L remain unchanged when they combine to form PL it is still often possible to detect the formation of PL indirectly by *loss* of free P or L when the two are mixed. In this case a means of measuring the concentration of free P or L is required. It is not usually possible to distinguish between free P and PL, (except by spectral methods) but in certain cases specific electrodes are available that respond to the concentration of free L but not to PL. An obvious example is the glass electrode which responds to H^+, but there are many others that are specific for inorganic ions and some that can be made to respond to the concentrations of organic ligands. In these cases, one places the specific electrode, together with a reference electrode, into a known volume of solution of the protein and then adds a known amount of ligand. The concentration of ligand expected if no combination with the protein occurs, can be calculated. From a previous calibration experiment in the absence of protein, the potential expected from the pair of electrodes at this ligand concentration will be known. Any binding of ligand to protein will reduce the concentration of free ligand and hence the observed potential will be less than expected (Figure 28)

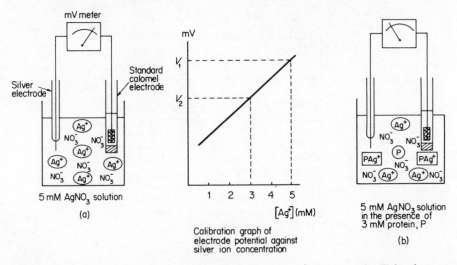

Figure 28 Demonstration of the binding of silver ions (Ag^+) to a protein (P) by the use of a silver electrode. A piece of silver wire acts as a reversible electrode towards free silver ions in solution. At a given temperature the potential developed between a silver electrode and a standard calomel electrode is a linear function of the free silver ion concentration as shown in the calibration graph. Thus 5 mM $AgNO_3$ solution gives a potential V_1 (graph and diagram (a)). The potential measured in the presence of 5 mM $AgNO_3$ and 3 mM protein (diagram (b)) is V_2 which, from the graph, indicates a free silver ion concentration of 3 mM. Hence the PAg^+ concentration must be 2 mM and the free protein concentration must be 1 mM. The dissociation constant, K_d, for the PAg^+ complex can be estimated:

$$PAg^+ \rightleftharpoons P + Ag^+; \qquad K_d = \frac{[P][Ag^+]}{[PAg]}$$

$$\therefore \quad K_d = \frac{(10^{-3})(3 \times 10^{-3})}{(2 \times 10^{-3})} = 1.5 \times 10^{-3} \text{ M}$$

Lastly, where neither a specific electrode nor a spectral change can be used, the detection of binding is still possible, provided that one has a method of measuring the total amount of ligand in a solution. Chemical or biological assays are usually available for the ligands of interest or, alternatively, radioactively labelled ligand can be used. In this case the technique of *equilibrium dialysis* is employed. The protein solution is placed on one side of a semipermeable membrane with the ligand solution on the other side and the two are left to reach equilibrium. The membrane is chosen so as to allow the ligand to diffuse through it into the protein solution but to prevent the protein from passing in the opposite direction (Figure 29). At equilibrium the concentration of *free* ligand must be the same on both sides of the membrane. If the ligand can bind to the protein there will be additional ligand in the form of PL on the

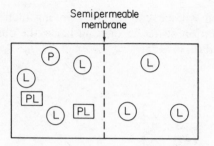

Figure 29 Equilibrium dialysis. At equilibrium, the total ligand concentration to the left of the membrane ($[L] + [PL]$) is found to be 5 mM, whereas on the right it is only 3 mM. The total protein concentration on the left is known to be 3 mM and at equilibrium this must be present as ($[P] + [PL]$). The dissociation constant K_d for the PL complex can be calculated as in Figure 28

protein side of the membrane. Measurement of the *total* ligand concentration on each side of the membrane will then reveal whether ligand binding has occurred because then there will be more on the protein side than on the other side. The simplest use of this technique employs a bag made of cellophane dialysis tubing into which the protein solution is placed. The bag is then closed and dropped into a solution of the ligand. After stirring for several hours the bag is removed and the total ligand concentration measured both inside the bag and in the solution remaining outside. In some cases the binding of ligand is so tight that the protein–ligand complex can be precipitated or separated from free P and L by ion exchange methods and the presence of L in the complex can be demonstrated.

Instead of using equilibrium dialysis one may achieve similar results by employing a variation upon the techniques of ultracentrifugation or gel filtration. In both cases use is made of the fact that the protein travels faster than the small ligand molecule. In the first case a sample of ligand-free protein is banded on top of a column of buffer containing the ligand. Ultracentrifugation moves the protein through the ligand solution, carrying with it bound ligand and leaving behind a deficit in the ligand concentration. In a similar way a sample of ligand-free protein can be applied to the top of a chromatographic column packed with a molecular sieve such as Sephadex suspended in a buffer containing the ligand. The same buffer–ligand solution is used to elute the column. Because the protein moves faster through the column than the small ligand molecule, a band of extra ligand moves with the protein while a zone of lower ligand concentration follows it. In each case measurements of the total ligand concentration at a number of points in the resulting pattern allow one to establish the characteristics of binding between protein and ligand.

The experimental use of any of these methods of detecting ligand binding is limited to some extent by two facts. In the first place proteins have large molecular weights, of the order of 100,000, and second is the fact that one cannot usually obtain solutions containing much more than about 100 mg of protein per millilitre. In other words the maximum obtainable concentration of protein can be expected to be of the order of 1 mM. This figure is clearly a generalization, but it illustrates the fact that in *molar* terms one cannot work with very concentrated protein solutions. This in turn leads to the problem that even if binding of ligand to protein does occur it may be difficult to detect unless it is *tight* binding. In the following example it is assumed that the dissociation constant K_d, of the PL complex is 10^{-1} M, i.e.

$$K_d = \frac{[P][L]}{[PL]} = 10^{-1} \text{ M} \tag{25}$$

The initial concentration of protein $[P]_0$ is assumed to be 10^{-3} M, that is, about as concentrated as possible, so that at all times

$$[P] + [PL] = 10^{-3} \text{ M} \tag{26}$$

From Equations 25 and 26 we can eliminate [P] and write

$$\frac{(10^{-3} - [PL])[L]}{[PL]} = 10^{-1}$$

$$\therefore \quad [L](10^{-3} - [PL]) = 10^{-1}[PL] \tag{27}$$

Let us calculate what ligand concentration will be needed to convert 10% of the protein into the complex PL, in other words for [PL] to be 10^{-4} M. The concentration of free ligand [L] must be given by (from Equation 27):

$$[L](10^{-3} - 10^{-4}) = 10^{-1}.10^{-4} = 10^{-5}$$

$$\therefore \quad [L] = 10^{-5}/(9 \times 10^{-4}) = 11 \cdot 1 \times 10^{-3} \text{ M}$$

The total ligand concentration is then given by $[L] + [PL] = 11 \cdot 1 \times 10^{-3} + 10^{-4} = 11 \cdot 2 \times 10^{-3}$. In other words, in this case the total ligand concentration must be more than ten times the total protein concentration in order to convert 10% of the protein into protein–ligand complex, and less than 1% of the total ligand is present in the bound form. A method that relied on a change in some spectral property of the protein might detect the conversion of 10% of P into PL, but methods that relied on loss of free ligand or the difference between total and free ligand concentrations would be looking for a difference of only 1%. On the other hand if $K_d = 10^{-3}$ M instead of 10^{-1} M, re-calculation for $[P]_0 = 10^{-3}$ shows that when $[PL] = 10^{-4}$ M as before, $[L] = 1.11 \times 10^{-4}$ and $[L] + [PL] = 2 \cdot 11 \times 10^{-4}$. In other words, a total ligand concentration equal to about 20% of the total protein concentration is now enough to cause

10% of the protein to be converted to the bound form and roughly half of the total ligand will be bound. Even lower values of K_d, representing even tighter binding of the ligand, give even more satisfactory results in terms of these experimental problems.

In the examples just quoted it was assumed that one molecule of protein was able to bind only one molecule of ligand, with a specified tightness of binding, or dissociation constant. This implies just a single specific binding site on the protein molecule. In many real situations, however, the protein molecule is able to bind more than one molecule of ligand. For example, the enzyme glyceraldehyde-phosphate-dehydrogenase (GPD) can bind four molecules of the coenzyme NAD, one for each of its four identical subunits. One must be able to measure not only the fact of binding but also how *many* molecules of ligand are bound to each protein molecule. In this particular case titration of the protein (GPD) solution with NAD can be carried out and the formation of the GPD–NAD complex can be monitored by following the absorption of light at 360 nm as shown in Figure 27. A graph of the extinction at 360 nm against amount of NAD added gives an end point as shown in Figure 30, when the

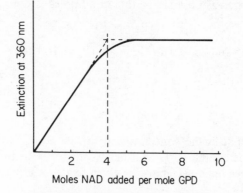

Figure 30 Titration of the enzyme glyceraldehyde - phosphate - dehydrogenase
(GPD) with NAD, followed spectrophotometrically at 360 nm

GPD is saturated with NAD, after which no further increase in E_{360} occurs when excess NAD is added. The number of moles of NAD added at the end point is divided by the number of moles of GPD present in the solution to give the number of binding sites per molecule of protein. The procedure works in this instance because GPD binds NAD very tightly, with a dissociation constant of about 10^{-6} M. On the other hand in cases where the binding is weak, with a dissociation constant greater than the available protein concentration (say about 10^{-3} M) no end point will be observed and the number of binding sites cannot be measured directly. In such cases an indirect method must be employed.

The Scatchard plot

The most widely used method is that of Scatchard where one measures the concentration of ligand bound to protein at a series of values of the concentration of free ligand in a solution containing a fixed total protein concentration of $[P]_0$. The concentration of protein-bound ligand, $[L_P]$, is most conveniently obtained as the difference between the concentrations of total ligand and free ligand. For each value of $[L]$ the ratio $[L_P]/[P]_0$ is calculated and is denoted by the symbol $\overline{N}$. It represents the average number of ligand molecules bound by the protein molecule. The fraction $\overline{N}/[L]$ is then plotted against $\overline{N}$ and a straight line is obtained as shown in Figure 31. Extrapolation of the line to the point

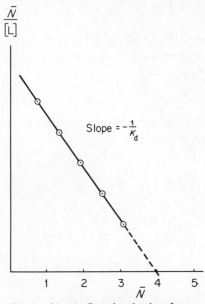

Figure 31 A Scatchard plot for a protein that can bind four molecules of a ligand (L) per molecule of protein, each with a dissociation constant of K_d

where it intersects with the $\overline{N}$ axis (i.e. where $\overline{N}/[L] = 0$) gives the number of binding sites per molecule of protein while the slope of the line gives the negative reciprocal of the dissociation constant, $-1/K_d$.

The theory of the Scatchard plot

Consider the simplest case where there is only one binding site per molecule:

$$P + L \rightleftharpoons PL; \qquad K_d = \frac{[P][L]}{[PL]}$$

$$\therefore \quad [PL]K_d = [P][L] \tag{28}$$

The total protein concentration $[P]_0$ is always equal to the sum of the concentration of P and PL, therefore

$$[P]_0 = [P] + [PL]$$

$$\therefore \quad [P] = [P]_0 - [PL]$$

Substituting for $[P]$ in Equation 28:

$$[PL]K_d = ([P]_0 - [PL])[L]$$

$$\therefore \quad [PL](K_d + [L]) = [P]_0[L]$$

$$\therefore \quad \frac{[PL]}{[P]_0} = \overline{N} = \frac{[L]}{K_d + [L]}$$

By a similar argument it can be shown that if there are n binding sites per molecule of protein, each with the same dissociation constant K_d, then

$$\overline{N} = \frac{n[L]}{K_d + [L]}$$

In the derivation of this equation the binding equilibria are as follows for the protein P containing n binding sites:

$$P + L \overset{K_d}{\rightleftharpoons} PL; \qquad K_d = \frac{n[P][L]}{[PL]}, \qquad \therefore \quad [PL] = \frac{n[P][L]}{K_d} \qquad (29)$$

$$PL + L \overset{K_d}{\rightleftharpoons} PL_2; \qquad K_d = \frac{(n-1)[PL][L]}{2[PL_2]}, \qquad \therefore \quad [PL_2] = \frac{(n-1)[PL][L]}{2K_d}$$

$$\therefore \quad [PL_2] = \frac{n(n-1)[P][L]^2}{2K_d^2} \qquad (30)$$

$$PL_2 + L \overset{K_d}{\rightleftharpoons} PL_3; \qquad K_d = \frac{(n-2)[PL_2][L]}{3[PL_3]}, \qquad \therefore \quad [PL_3] = \frac{(n-2)[PL_2][L]}{3K_d}$$

$$\therefore \quad [PL_3] = \frac{n(n-1)(n-2)[P][L]^3}{3.2K_d^3} \qquad (31)$$

$$\vdots$$

$$PL_{(n-1)} + L \overset{K_d}{\rightleftharpoons} PL_n; \qquad K_d = \frac{[PL_{(n-1)}][L]}{n[PL_n]},$$

$$\therefore \quad [PL_n] = \frac{n(n-1)(n-2)\cdots 3.2[P][L]^n}{n(n-1)(n-2)\cdots 3.2K_d^n} \qquad (32)$$

The appearance of n in the numerator of Equation 29 arises because the concentration of empty binding sites on P is $n[P]$. Similarly $(n-1)$ appears in the numerator of Equation 30 because the concentration of empty binding sites on PL is $(n - 1)[PL]$ and so on. The Figures 2 and 3 appear in the denominators of Equations 30 and 31 because the concentration of bound ligand in PL_2 and PL_3 is, respectively, $2[PL_2]$ and $3[PL_3]$.

The average number of ligand molecules bound per molecule of protein is now given by:

$$\overline{N} = \frac{\text{total concentration of bound ligand}}{\text{total concentration of protein, } [P]_0}$$

$$= \frac{[PL] + 2[PL_2] + 3[PL_3] + \cdots + n[PL_n]}{[P] + [PL] + [PL_2] + [PL_3] + \cdots + [PL_n]}$$

From Equations 29–32:

$$\overline{N} = \frac{\dfrac{n[P][L]}{K_d} + \dfrac{2n(n-1)[P][L]^2}{2K_d^2} + \dfrac{3n(n-1)(n-2)[P][L]^3}{3.2K_d^3} + \cdots + \dfrac{n[P][L]^n}{K_d^n}}{[P] + \dfrac{n[P][L]}{K_d} + \dfrac{n(n-1)[P][L]^2}{2K_d^2} + \dfrac{n(n-1)(n-2)[P][L]^3}{3.2K_d^3} + \cdots + \dfrac{[P][L]^n}{K_d^n}}$$

(33)

This seemingly complex equation reduces very neatly to a much simpler one by the use of the binomial theorem which states that:

$$(1 + x)^n = 1 + nx + \frac{n(n-1)x^2}{2} + \frac{n(n-1)(n-2)x^3}{3.2} + \cdots + x^n$$

We can recast Equation 33 as follows after cancelling $[P]$ out from both numberator and denominator:

$$\overline{N} = \frac{n\left(\dfrac{[L]}{K_d}\right) + \dfrac{2n(n-1)}{2}\left(\dfrac{[L]}{K_d}\right)^2 + \dfrac{3n(n-1)(n-2)}{3.2}\left(\dfrac{[L]}{K_d}\right)^3 + \cdots + n\left(\dfrac{[L]}{K_d}\right)^n}{1 + n\left(\dfrac{[L]}{K_d}\right) + \dfrac{n(n-1)}{2}\left(\dfrac{[L]}{K_d}\right)^2 + \dfrac{n(n-1)(n-2)}{3.2}\left(\dfrac{[L]}{K_d}\right)^3 + \cdots + \left(\dfrac{[L]}{K_d}\right)^n}$$

Let $[L]/K_d = x$.

$$\therefore \quad \overline{N} = \frac{nx\left[1 + (n-1)x + \dfrac{(n-1)(n-2)x^2}{2} + \cdots + x^{(n-1)}\right]}{1 + nx + \dfrac{n(n-1)x^2}{2} + \dfrac{n(n-1)(n-2)x^3}{3.2} + \cdots + x^n}$$

which is clearly the ratio of two binomial expansions:

$$\therefore \quad \overline{N} = \frac{nx(1 + x)^{(n-1)}}{(1 + x)^n} = \frac{nx}{(1 + x)}$$

(34)

Re-substituting for $x = [L]/K_d$, we obtain, as predicted:

$$\bar{N} = \frac{n\dfrac{[L]}{K_d}}{1 + \dfrac{[L]}{K_d}} = \frac{n[L]}{K_d + [L]} \tag{35}$$

To obtain the relationship between $\bar{N}/[L]$ and $\bar{N}$ we proceed as follows:
From Equation 35:

$$n - \bar{N} = n\left(1 - \frac{[L]}{K_d + [L]}\right) = n\left(\frac{K_d}{K_d + [L]}\right) \tag{36}$$

Dividing Equation 35 by 36:

$$\therefore \quad \frac{\bar{N}}{n - \bar{N}} = \frac{\dfrac{n[L]}{K_d + [L]}}{\dfrac{nK_d}{K_d + [L]}} = \frac{[L]}{K_d}$$

$$\therefore \quad \frac{\bar{N}}{[L]} = \frac{n - \bar{N}}{K_d} = \frac{n}{K_d} - \frac{\bar{N}}{K_d} \tag{37}$$

This shows that a plot of $\bar{N}/[L]$ versus $\bar{N}$ gives a straight line of slope $-1/K_d$.
When $\bar{N}/[L] = 0$ then

$$\frac{n}{K_d} = \frac{\bar{N}}{K_d} \quad \text{i.e.} \quad n = \bar{N}$$

so that the intercept when $\bar{N}/[L] = 0$ gives the value of n, the number of binding sites per molecule of protein.

In more complicated cases there may be more than one *type* of binding site on the protein.[13] For example a protein may possess four identical sites on its surface where the ligand binds tightly with a dissociation constant of K_1, and four other sites where the same ligand binds much less tightly with dissociation constant K_2. Provided K_1 and K_2 are sufficiently different the Scatchard plot will reveal this situation as shown in Figure 32. Clearly at low ligand concentration (when $\bar{N}$ is small, to the left of the diagram), the ligand binds preferentially at the tight sites and the Scatchard plot is a straight line of steep slope, but as the ligand concentration increases these sites will become saturated and the ligand will start to bind more and more to the weak sites, causing a deviation from the line. At very high ligand concentrations (to the right of the diagram), effectively all the tight sites will be occupied and the Scatchard plot will tend to a new straight line of shallower slope, representing the filling of the weak sites.

Linear Scatchard plots that extrapolate to an integral number of binding sites give a good indication that one is dealing with a specific set of identical binding sites. There is, however, a possibility that the sites may actually be

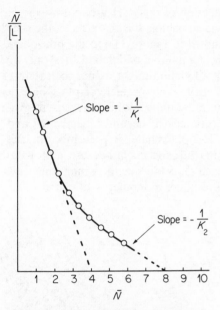

Figure 32 A Scatchard plot for a protein that binds four molecules of ligand (L) tightly with a dissociation constant K_1 and a further four less tightly with a larger dissociation constant K_2

different in terms of their protein structure, but able to bind the ligand with about the same dissociation constant.

Many instances are known where the binding of one ligand is prevented or weakened by the binding of another in solution. In the case of enzymes the catalytic activity towards the substrate may be reduced or removed by the presence of inhibitors that can be shown to bind to the enzyme. In these cases if there is some structural similarity between the substrate and the inhibitor there is a good case for supposing that they both bind at the same site and that occupation of the site by one therefore excludes the other. This situation is called competitive binding or competitive inhibition. The possibility still remains, of course, that they actually bind at different sites, or even nonspecifically, and the binding of one ligand in some way alters the overall structure of the protein so that it can no longer bind the other. This cannot be ruled out easily.

Chemical modification studies

Another way in which evidence may be obtained for the binding of a ligand at a specific site, is by carrying out a chemical modification of the protein. Imagine a protein which binds a ligand at a single specific site, and that this binding site contains the sulphydryl group, SH, of a cysteinyl residue. One would guess that covalent modification of this –SH group, for example by alkylation with iodoacetamide, would so change the character of the site that the ligand could no longer bind as shown in Figure 33. On the other hand, if the ligand could bind non-specifically at a number of different sites, only one of which contained the –SH group, then alkylation might reduce the extent of binding but should not prevent it. The enzyme glyceraldehyde-phosphate-dehydrogenase (GPD) again provides a good example. When the enzyme is modified with iodo-acetamide or iodoacetic acid it becomes inactivated and the binding of sub-strate, glyceraldehyde-3-phosphate, is prevented. In this case additional in-formation comes from the experiment because, although the enzyme possesses four SH groups in each of its four identical monomer units, it becomes inactive when only one SH group per monomer is alkylated.

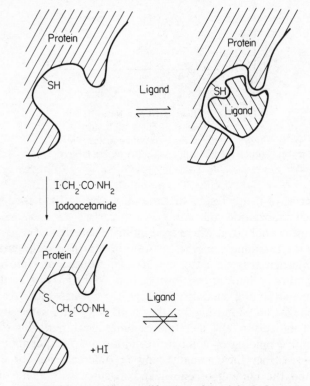

Figure 33 Covalent modification of a sulphydryl group in a ligand binding site prevents the binding of the ligand

If low concentrations of the alkylating agent are used, this SH group is the only one to react and the enzyme becomes inactivated. Sequence studies have revealed that the four cysteinyl residues lie in four distinct stretches of the polypeptide chain, but after inactivation with low concentrations of alkylating agent a unique cysteine, and only that one, becomes alkylated. This evidence strongly implicates a specific binding site containing the unique cysteinyl residue. Furthermore, when the substrate is added before the iodoacetic acid, inactivation and alkylation of the specific SH group do *not* occur. In other words *substrate protection* has occurred and is compelling evidence that the SH group actually lies in the binding site.

Matters are not always as straightforward as this. The reagent used to modify the protein may react with more than one residue in the polypeptide and indeed may in some cases react with more than one type of residue. Substrate protection may not be demonstrable if the binding of substrate is weak. Nevertheless chemical modification studies have often provided much useful information about binding sites and indeed about other aspects of protein structure. A list of some of the reagents used and the side chain groups with which they react, is given in Table 3. The list is by no means complete and for a thorough survey of the many reagents that have been developed the reader is referred to one of the excellent reviews of this topic.[14]

Affinity labelling

A specific binding site can be pinpointed in some cases by the technique of affinity labelling. The ligand itself is chemically modified in a way which does not greatly alter its overall shape but does make it chemically reactive. When the modified ligand is bound by the protein it reacts with a group in the binding site forming a stable covalent bond and labelling a particular amino acid side chain. Chemical degradation of the labelled protein can then reveal which amino acid has become modified and sequence studies can locate its position in the primary structure. Non-specific binding might be expected to label a number of different residues in the sequence whereas specific binding will lead to a single modification. A successful use of this technique is illustrated by the labelling of a single histidine residue at position 57 in the polypeptide chain of chymotrypsin when reacted with tosyl phenylalanine chloromethyl ketone (TPCK):

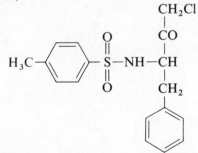

This substance closely resembles tosyl phenylalanine methyl ester, a good substrate for the enzyme, but has a reactive chlorine atom in its methyl group. The enzyme recognizes the molecule of TPCK as if it were a substrate by virtue of its affinity for the tosyl and phenyl groups but, once bound, the histidine side

Table 3

Chemical modification of amino acid side chains in proteins

Amino acid side chain	Reagents used	Modified side chain
Cysteinyl $-CH_2SH$	(1) Alkylating agents: Iodoacetic acid $I \cdot CH_2 \cdot COOH$	$-CH_2-S-CH_2-COOH$
	N-ethylmaleimide	
	(2) Heavy metal derivatives: p-chloromercuribenzoate (PCMB)	
	(3) Ellman's reagent: 5,5'-dithiobis[2-nitrobenzoic acid]	

Table 3 (continued)

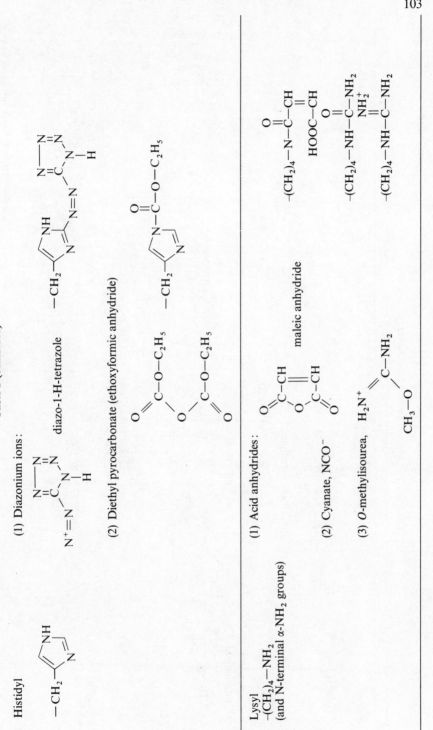

Histidyl
$-CH_2$—

(1) Diazonium ions:

diazo-1-H-tetrazole

(2) Diethyl pyrocarbonate (ethoxyformic anhydride)

Lysyl
$-(CH_2)_4-NH_2$
(and N-terminal α-NH_2 groups)

(1) Acid anhydrides:

maleic anhydride

(2) Cyanate, NCO^-

(3) O-methylisourea, H_2N^+

Table 3 (continued)

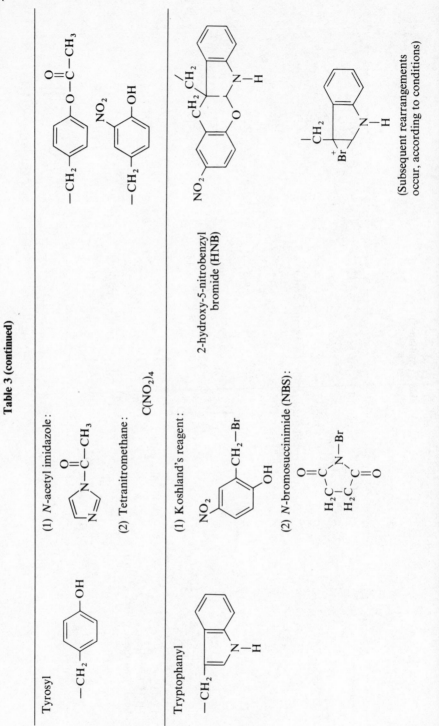

Tyrosyl

(1) N-acetyl imidazole:

(2) Tetranitromethane:

$C(NO_2)_4$

2-hydroxy-5-nitrobenzyl bromide (HNB)

Tryptophanyl

(1) Koshland's reagent:

(2) N-bromosuccinimide (NBS):

(Subsequent rearrangements occur, according to conditions)

Table 3 (continued)

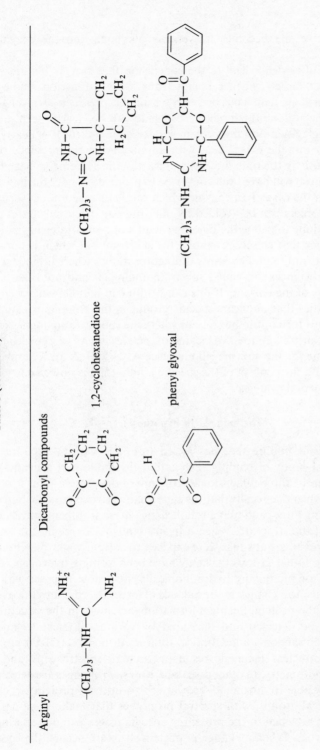

Arginyl	Dicarbonyl compounds

1,2-cyclohexanedione

phenyl glyoxal

chain in the active site becomes irreversibly alkylated and the enzyme is inactivated.

Many strands of evidence lead to the conclusion that *specific* binding sites exist but the best direct evidence comes from X-ray diffraction studies on crystals of the protein that have been grown in the presence of the ligand. Where this has been done the molecule of ligand can be clearly 'seen' fitting snugly into a single region on the protein surface or in a cleft. It is nearly impossible to carry out the experiment with an enzyme in the presence of its substrate(s) because by the time the X-ray diffraction measurements have been made the substrate(s) will have been converted to product(s) under the influence of the enzyme. In the case of an enzyme which catalyses reaction between *two* substrates the problem can be avoided by omitting one of the substrates. The best that can be done to define the binding site of a one-substrate enzyme is to choose a substance that closely resembles the substrate but which is not converted to product, and to find by X-ray diffraction studies where it binds to the enzyme. Such substances are called substrate analogues and are often competitive inhibitors of the enzyme. If this can be done in parallel with chemical modification studies that implicate specific groups in the binding or catalysis, and if those groups turn out to be close to where the substrate analogue is seen to bind, one can fairly claim to have located the active site. As will be seen later, this has been done for the enzyme ribonuclease A, as well as for a number of other enzymes, and the concept of the specific binding site is now firmly established by experimental evidence.

The induced-fit hypothesis

A highly significant finding that has arisen from the various experimental approaches to the study of binding sites is that the binding of a ligand may cause a change in the three-dimensional structure of the protein. This is not to be wondered at when one recalls that the stable native structure is the outcome of many conflicting forces and that a small change in environment may cause a rearrangement of the structure. When a ligand binds to a specific site on the surface of the protein, groups that were exposed to solvent water may become protected from it. Other groups that may have been forming hydrogen bonds with each other, or interacting hydrophobically, may now interact with the bound ligand, causing a change in the balance of forces that dictates the overall conformation of the protein. Such conformational changes in the structure of the protein can be detected and measured by various physical techniques including X-ray diffraction studies, optical rotatory dispersion (ORD), circular dichroism (CD), nuclear magnetic resonance (n.m.r.) spectroscopy and difference spectrophotometry. In other cases where no such changes are detectable it has proved possible to introduce *reporter groups* into the protein structure. These are chemical groups, with spectral properties that make them easy to detect, which are attached to the protein at specific places without altering its biological properties. When a change in protein structure occurs, the spectral properties of the reporter group are perturbed and can be detected and measured. Conformational changes can sometimes be detected by changes in the chemical reactivity of side chains in the protein. For example the SH groups of exposed cysteinyl residues react readily with alkylating agents, such as iodoacetic acid or N-ethylmaleimide, and with organomercurials like p-chloromercuribenzoate (PCMB). If the SH groups are 'masked' or hidden in the hydrophobic core of the enzyme they do not react unless the native structure is disrupted for example by urea or guanidinium chloride. In certain cases, notably the enzyme phosphoglucomutase, SH groups that react slowly in the native enzyme are found to become more reactive when the enzyme binds its substrate. Had the reverse been found it might have been argued that the binding of substrate had 'covered' the SH groups, but it is hard to explain the uncovering of SH groups by anything other than conformational change in the enzyme.

These and similar findings led Koshland to propose his 'induced-fit hypothesis' in 1958.[15] The hypothesis assumes that an exact complementary binding site may not exist as such until the ligand arrives at the locality. Some propensity to bind the ligand must of course pre-exist or the ligand would never find its way there, but the fit may not be a good one. Once the initial act of 'catching' the ligand has been achieved, however, the hypothesis supposes that the flexible protein proceeds to wrap itself around the ligand. Side chains that were not initially very close to the 'hook' that catches the ligand now approach and interact with other parts of the ligand, progressively tying it down more and more tightly and restricting its freedom, until the fit between ligand and protein is a good one and the binding site has been formed. A good analogy to this process is the shrugging-on of a coat by its owner, or the fitting of a glove to a hand. In neither case is the complementary character between coat or glove and body or hand very obvious when the coat lies across the back of a chair or the glove lies crumpled in a drawer. As soon as the owner begins to insinuate himself into the articles, they assume the correct fit. The glove analogy incidentally also illustrates very nicely the reasons for stereospecificity of ligand binding which is a feature of many enzyme–substrate interactions. A right-hand glove will not take your left hand and in a similar way a ligand binding site may accept only one of two stereoisomers, for example, L-alanine but not D-alanine.

Examples of induced fit have been provided by X-ray diffraction studies on the enzymes lysozyme, carboxypeptidase, elastase and lactate dehydrogenase. In each case the three-dimensional structure of the native enzyme is found to be slightly altered when a substrate analogue is bound at the active site. In the case of lysozyme, the active site region is shown in Figure 34. The whole enzyme molecule resembles a bird (Figures 19 and 20) with a broad cleft in the region between the two 'wings'. This cleft is the binding site for the substrate, a mucopolysaccharide made up of repeating NAG–NAM dimers where NAG is short for N-acetyl-D-glucosamine and NAM represents N-acetyl-D-muramic acid. Short stretches of a mucopolysaccharide chain, such as the tetramer NAG–

Figure 34 The active site region of lysozyme. A close-up photograph of the same molecular model shown in Figure 19, showing more detail of the active site region. In the centre, that portion of the substrate molecule is shown at which hydrolytic cleavage occurs. The close fit between protein and substrate is clearly seen

NAM–NAG–NAM, are not substrates but bind in the active site, as shown by X-ray studies, and act as inhibitors. When they do so the binding cleft in the lysozyme molecule is seen to be deeper and narrower than in their absence.

Active sites and allosteric sites

The active site of an enzyme is thus a ligand binding site where the ligand, in this case called the substrate, binds to the enzyme. The act of binding may be accompanied by structural changes in the enzyme which can bring up amino acid side chains that possess catalytic groups so that these groups then catalyse a chemical change in the bound substrate. For example, this is seen in the case of lysozyme where the narrowing and deepening of the binding cleft is accompanied by the movement of carboxyl groups belonging to an aspartyl residue at position 52 and a glutamyl residue at position 35. These groups approach close to the bound substrate analogue as shown in Figure 34 and are believed to catalyse the hydrolytic cleavage of a true substrate. Other examples of a similar sort are also known.

The role of catalytic groups in enzyme active sites probably contributes very largely to their being *active* sites rather than just binding sites. In many cases another important factor in this distinction can be the distortion of the structure of the substrate itself upon binding (see modern Rack hypothesis on p. 128). This is also illustrated by the case of lysozyme where the structure of one of the carbohydrate rings of the substrate analogue is distorted from its usual chair form into a less stable, and almost planar, configuration in the bound state.

The idea of induced-fit at an active site has a close parallel in the concept of the *allosteric site*, defined by Monod and his colleagues in 1963,[16] and indeed was a necessary historical precedent for the latter. An allosteric site is nothing more than a binding site on an enzyme where a ligand can bind and cause a structural change in the enzyme molecule. The allosteric site is distinct and separate from the active site of the enzyme but conformational changes produced by binding of a ligand at the allosteric site cause an alteration in the structure of the active site which may either enhance or reduce its ability to act as a catalyst. The ligand which binds at the allosteric site does not undergo any chemical change when it binds. It merely acts as an indirect activator or inhibitor of the enzyme, working via the conformational changes it induces when it binds. Such allosteric ligands are called *effectors* and their structure is often quite distinct from that of the substrate.

One of the best examples of an allosteric enzyme is provided by aspartate transcarbamoylase (ATCase) which catalyses the condensation of aspartate and carbamoyl phosphate to produce carbamoyl aspartate and inorganic phosphate:

$$NH_2-\overset{\overset{\displaystyle O}{\|}}{C}-O-PO_3^{2-} \ + \ \overset{+}{N}H_3-\overset{\overset{\displaystyle CH_2COO^-}{|}}{CH}\cdot COO^-$$

$$\overset{ATCase}{\rightleftharpoons} NH_2-\overset{\overset{\displaystyle O}{\|}}{C}-NH\cdot \overset{\overset{\displaystyle CH_2COO^-}{|}}{CH}\cdot COO^- \ + \ H_2PO_3^-$$

110

This is the first step in the synthesis of pyrimidines and one of the final products of the metabolic pathway that begins here is the molecule cytidine triphosphate (CTP):

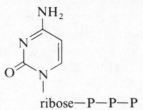

Cytidine triphosphate (CTP)

CTP is an allosteric inhibitor of the enzyme ATCase. Although CTP bears very little resemblance to either of the substrates, or the products of the enzymic reaction it is able to bind to the enzyme and inactivate it. As discussed in Chapter 3 (Figure 25) the ATCase molecule is composed of twelve subunits, six of one type called 'catalytic' subunits and six of another called 'regulatory' subunits. It has been shown that substrates bind to the catalytic subunit, which is indeed able to function as an enzyme on its own. CTP on the other hand cannot bind to these subunits and acts as an inhibitor of the whole assembly by binding to the regulatory subunits. It follows that it must exert its effect via a conformational change induced first of all in the regulatory subunit and then passed on to the catalytic subunits.

One should be clear that allosteric sites are not necessarily situated on separate protein subunits from active sites as in the extreme case of ATCase, but are always physically distinct and separate from them. In conclusion it can be said that the difference between an allosteric site and an active site is that the latter has the property of catalysing a chemical change in the ligand that binds there. It should be added that although binding of an effector at an allosteric site *must* produce a conformational change in the enzyme, binding of a substrate at an active site need not be accompanied by such a change.

Enzymes do not always possess an allosteric site. Some enzymes may have only the active site and no other sites where ligands can bind. Others may have sites where ligands other than the substrate can bind, but without producing any effect on the activity of the enzyme. On the other hand, some enzymes have more than one type of allosteric site so that their activity can be modified by a number of different effectors.

Specificity

The outline of the way in which an enzyme achieves specificity for its substrates and allosteric effectors must now be clear. Enzymic specificity for substrates is of three types. In the first place an enzyme possesses *reaction specificity*, that is to say it catalyses a particular type of reaction. For example it may catalyse the oxidation of primary alcohols to aldehydes (an alcohol dehydrogenase) or the transamination between amino acids and keto acids (a transaminase). Secondly the enzyme usually has a preference for one particular alcohol or one special pair out of all the possible (amino acid)–(keto acid) pairs. This kind of specificity is called the *substrate specificity* of the enzyme. In some instances the substrate specificity is very broad in that a number of chemically similar substrates can be handled. Examples of such broad specificity are to be found amongst the transaminases, the proteolytic enzymes or peptidases and in many other groups of enzymes. Conversely the substrate specificity is sometimes so narrow as to encompass only a single substrate. The best example of this is probably urease which catalyses the hydrolysis of urea to ammonia and carbonate. In spite of efforts to find substituted ureas that will serve as substrates for urease, not a single one has been found. Lastly enzymes may possess *stereospecificity*. That is to say, if the substrate can exist as stereoisomers only one of these will normally serve as a substrate for the enzyme. Almost without exception, peptidases only hydrolyse peptides made up of L-amino acids and spurn those containing D-amino acids, even though their substrate specificity as previously defined may be broad. Similarly the enzymes that are involved in carbohydrate metabolism usually act only upon the naturally occurring D-sugars and their derivatives.

It is easy to see that reaction specificity must arise from the particular nature of the catalytic groups within the active site, and that substrate specificity is a function of the contact residues and binding groups in the active site. Stereospecificity requires a little more thought. In order to distinguish between two enantiomorphic molecules (mirror images of each other) such as A and A′ shown in Figure 35, the enzyme must interact with the substrate by at least

(a)   (b)

Figure 35 The stereospecificity of binding sites. In (a) a ligand molecule A is shown interacting with a hypothetical binding site in which three groups x, y and z interact favourably (pecked lines) with the groups a, b and c on the ligand. In (b) the optical isomer A′ is shown as the mirror image of A. The same binding site is shown (not a mirror image) but it has been rotated through 180° so as to place the group x in a favourable position to interact with the group a on the molecule A′. The groups y and z in the site are now in the wrong positions to bind b and c respectively. A mirror image of the site would be needed to bind A′ at all three groups

three points of attachment. The enzyme active site shown in Figure 35 as having groups x, y and z within it, respectively able to bind the groups a, b and c of the molecule A, cannot bind a, b and c in the enantiomorphic A′, because the three groups x, y and z are wrongly arranged. If interaction between the active site and the substrate only took place at *two* points, say a and b on the substrate and x and y on the enzyme, then the enzyme would be able to bind either A or A′. The 'three-point attachment hypothesis' is also necessary to explain the incorporation patterns of radioactive groups into TCA cycle intermediates, even though they do not exhibit stereoisomerism, as was pointed out many years ago by Ogston. It thus appears that the snug fit required between enzyme and substrate in the active site calls for three or more points of contact between them in many if not all cases. What this probably means in practice is that the enzyme and its substrate contact each other not at one or two *points*, but over a fairly extensive *area*, perhaps made up of many individual points.

Before leaving the topic of specificity it is necessary to pose a question (to be answered later in the discussion of how enzymes achieve catalysis). It was seen earlier that enzymes catalyse both forward and reverse directions of a given chemical reaction so that the terms 'substrate' and 'product' only have meaning when we focus attention on a particular direction of reaction. The substrate in one direction is the product in the reverse direction and vice versa. Since the enzyme catalyses both directions it must be able to bind both 'substrate' and 'product'. As these two may be quite different in chemical structure, are we to think of the active site as having specificity for the 'substrate' or for the 'product'? Although the enzyme is flexible and the fit may be induced upon binding, and although the active site may be able to change its shape during catalysis, so as to accommodate the product too, it is expecting too much for there to be a perfect or near-perfect fit in both cases. The answer to this question will become clear later.

PROENZYMES, ISOENZYMES AND MUTATIONS

Certain enzymes are synthesized in the form of inactive precursors called proenzymes or zymogens. Most known examples of such proenzymes are synthesized within the cell for export to the extracellular environment and are the precursors of hydrolytic enzymes which would be dangerous if they were active inside the cell. The proteolytic enzymes trypsin and chymotrypsin, for example, which are secreted by the pancreas into the digestive system of mammals, are synthesized within the pancreatic cells as the inactive zymogens trypsinogen and chymotrypsinogen. In each case the zymogen consists of a single polypeptide chain cross-linked by disulphide bonds. As such they are quite stable until they reach the duodenum where they are activated by another proteolytic enzyme present there in small quantities. The activation process or zymogen–enzyme conversion involves a limited and highly specific proteolysis in which certain exposed peptide bonds in the zymogens are hydrolysed. Once a few molecules of the zymogens have become activated they can themselves activate others by virtue of their own new-found proteolytic activity. This auto-catalytic process finally ensures that all the zymogen molecules become activated. Recent studies on the structures of the two enzymes and their precursors have revealed striking similarities. The amino acid sequences are similar, their three-dimensional structures have a great deal in common and the bonds cleaved during activation are homologous in the two systems. Chymotrypsin, however, mainly hydrolyses peptide bonds on the C-terminal side of aromatic amino acid residues while trypsin is specific for the bonds formed by the basic amino acids lysine and arginine. The catalytic groups present in the active sites of trypsin and chymotrypsin are known to be a histidine imidazole group and a serine hydroxyl group in both cases. The active site regions of the molecules change their structure slightly during the activation process, but not so much that one can readily understand why the zymogens are inactive and the enzymes active. Although the process of activation from zymogen is not yet fully understood, the reasons for the differing specificity are clear enough and, as expected, arise from the slightly different characters of the two substrate binding sites. Clearly these two zymogen–enzyme systems are very closely related to each other and probably both evolved from a common 'primitive' proteinase. A third enzyme, elastase (along with its zymogen proelastase), is also found in the pancreas of the pig and it turns out to be a third member of this evolutionary family.

It appears that a large proportion of the original primary structure has been conserved in these three systems during their evolution. What is pictured as having happened is that the gene directing the synthesis of the 'primitive' ancestral zymogen became duplicated twice in the evolution of the mammals. Subsequently mutations in the three genes must have occurred independently leading to the eventual divergence of the gene products.

Many other cases of zymogen–enzyme conversion have been documented and it appears that this is one method of control of enzyme activity that has been widely adopted by organisms from the simplest to the most highly differentiated, in circumstances where the enzyme activity in question is required only outside the cell that makes it. It is, of course, an irreversible control mechanism. Once the zymogen has been activated it cannot be reformed from the resulting enzyme. Unlike allosteric control, which involves the *reversible* and *non-covalent* modification of enzyme activity, this form of control is exerted by *irreversible covalent* modification—the cleavage of peptide bonds. A third category of control mechanisms is also found in which *reversible covalent* modification of the enzyme occurs. In the case of muscle phosphorylase, for example, an inactive form, phosphorylase b, is converted to active phosphorylase a by the covalent attachment of phosphate groups to specific serine residues in the phosphorylase b. A specific enzyme system exists for this purpose; and another is present to remove the phosphate groups when phosphorylase activity is no longer required. The interlocking regulation of these systems is discussed later in Chapter 7.

114

Although the evolution of trypsin, chymotrypsin and elastase has led to enzymes with different specificities, another class of evolutionarily related enzymes exists in which the specificity is almost identical but the different enzymes possess different kinetic or regulatory properties. This is the class of isoenzymes. An example has already been given in Chapter 2 where the lactate dehydrogenase (LDH) isoenzymes were mentioned. In the human body five LDH isoenzymes are found in differing proportions in different tissues. The LDH molecule consists of four subunits which can be of two types, the H type and the M type. Heart muscle synthesizes only H type subunits so the predominant enzyme in heart has four identical H type subunits and can be denoted as H_4. In the same way white skeletal muscle enzyme is made up of identical M type subunits and is denoted by M_4. In other tissues, hybrids are found in which H and M subunits are both present in the tetramer. Three such hybrids, H_3M, H_2M_2 and HM_3 exist, giving rise to the five isoenzymes actually observed. The electrophoretic properties of the isoenzymes are different so that they can be separated using electrophoresis, as shown in Figure 36. In addition the kinetic behaviour of the isoenzymes is different, being closely tailored to the situation in which the enzyme must work.

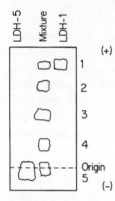

Figure 36 The electrophoretic separation of the isoenzymes of lactate dehydrogenase. The isoenzymes were separated by starch gel electrophoresis. On the left is shown the pure LDH-5 isoenzyme, made up of only muscle-type subunits (M_4). On the right is the LDH-1 isoenzyme comprising only heart-type subunits (H_4), while on the centre is a mixture of all five isoenzymes, LDH-2, -3, and -4 being the hybrids H_3M_1, H_2M_2 and HM_3 respectively (reproduced from C. I. Markert in *Science*, **140**, 1329–1330 (1963); Copyright 1963 by the American Association for the Advancement of Science)

The white skeletal muscle enzyme has a high Michaelis constant (K_m) for the substrate pyruvate, which means that when the muscle works and pyruvate concentration increases, the enzymic rate keeps on increasing and lactate can be produced to keep pace with it, thus removing pyruvate in the form of lactate which is of course exported from working muscle by the bloodstream. The heart muscle enzyme, on the other hand, has a low K_m for pyruvate so that when pyruvate concentration rises the enzyme is quickly saturated and the rate cannot increase to keep pace with it. In addition the enzyme becomes strongly *inhibited* by higher concentrations of pyruvate. This is an unexpected finding since one usually expects the rate of a chemical reaction to increase when the concentration of reactants increases. Nevertheless, it serves a useful purpose in the heart because under conditions of high pyruvate concentration it means that pyruvate cannot be converted to lactate and is forced into the TCA cycle where it is oxidized to provide energy. In addition the heart enzyme has a low K_m and a high V_m for lactate, so it is well placed to convert lactate to pyruvate.

These findings are entirely in keeping with the different physiological roles of heart muscle and white skeletal muscle. The latter works in powerful spurts and rapidly becomes anaerobic, exporting lactate to the other tissues and accumulating an oxygen debt. Heart muscle, on the contrary, works steadily and aerobically and is a net importer of lactate, converting it to pyruvate for oxidation. An interesting finding is that in birds, the nature of the LDH enzyme found in the breast muscles used for flight is a function of the flight pattern. Those like grouse and pheasant which only fly occasionally have an M type isoenzyme and the breast muscle quickly fatigues, whereas the hummingbird has an H type isoenzyme and is able to fly continuously without muscle fatigue.

Many other examples of isoenzymes are known and it must be supposed that, even where no biological reason for their existence is known, one will probably emerge in the same way that intense study of the LDH enzymes revealed the elegant adaptations of the same basic enzymic function to different physiological roles.

ACTIVATION, INACTIVATION AND INHIBITION OF ENZYMES

From the foregoing discussion it is clear that there is no single or unique solution in structural and chemical terms to the problem of enzymic catalysis of a particular reaction. On the one hand evolution has in many, or perhaps most, cases produced more than one polypeptide or aggregate of polypeptides capable of catalysing a given reaction. On the other hand, the enzymologist is able to alter the activity of a particular native enzyme by various treatments, some of which abolish, some of which reduce, and yet others of which may increase the native activity. This whole topic has been bedevilled by misunderstandings arising from terminology. Confusion may arise because there are so many ways in which an enzyme can be rendered more active or less active. In the former case the enzyme is universally agreed to have undergone 'activation' and the agent is an 'activator' whatever the molecular mechanism involved. On the other hand when an enzyme becomes less active it is sometimes said to have been 'inactivated', sometimes 'inhibited' and occasionally simply 'denatured'.

Changes in enzyme activity always result from alteration to the active site in one way or another. This can occur indirectly because an agent primarily causes alteration in the gross protein structure, which leads to a change in the active site by altering the relative positions of the groups within the site. Alternatively the agent may directly affect the structure of the active site either by physically obstructing it, or by chemical modification of the covalent or ionic state of groups within it. Lastly, of course, activity may be affected by combinations of these effects. Some order can be introduced into this complex state of affairs by considering three general types of agents that can increase or decrease enzyme activity.

1. Agents that alter enzyme activity by changing the covalent structure of the enzyme

⅃ class fall those agents that cause the cleavage of peptide bonds in ⅃y structure of the enzyme. We have already noted that limited cleavage can lead to activation, in the case of the zymogen to enzyme conversions, although in other cases it results in loss of activity. On the other hand, extensive cleavage of peptide bonds invariably leads to loss of activity because of a general disruption of the molecular architecture and this would be called inactivation or denaturation.

The reduction of disulphide bonds (SS) to sulphydryl groups (SH) is another example of covalent alteration that can sometimes lead to activation, as in the case of papain and streptococcal proteinase where an SH group is required in the active site, or in other cases may lead to inactivation as with ribonuclease A which loses activity when its four disulphide bonds are reduced. Conversely the oxidation of SH groups to disulphides brings about the appearance of activity in some cases and its disappearance in others.

There are many other ways to modify the covalent structure of an enzyme, leading to either loss or gain of activity. The agents involved may act either reversibly or irreversibly to increase or decrease the *amount* of active enzyme.

2. Agents that alter enzymic activity via changes in the enzymic environment

Within this class are included temperature, chaotropic agents, organic solvents and pH. They do not affect the covalent structure of the enzyme, but cause changes in its three-dimensional structure by altering the fine balance of forces that stabilizes the conformation. In the process, the ionic state of the protein may be altered owing to changes in the pK s of its ionizable side chains or, in the case of changes in pH, by direct alteration of the ionic equilibria. In most cases, provided the change is not too extreme, it can be reversed; but it is often necessary to carry out the reversal slowly and preferably in the presence of ligands that stabilize the native conformation, such as substrates or substrate analogues. For example, enzymes inactivated by the addition of urea to a final concentration of 6–8 M usually regain activity if the urea is dialysed away slowly, but often become irreversibly inactivated or 'denatured' if diluted rapidly. Changes in pH are usually reversible in their effect upon activity provided one stays within a limited range of pH. The range naturally depends upon the enzyme, but in most cases pH s above 9 or below 4 can be expected to lead to permanent loss of activity. Here again agents in this class act by altering the *amount* of active enzyme present.

3. Alteration of enzymic activity by the reversible, non-covalent binding of ligands

Ligands that produce an increase in activity when they bind to an enzyme are called activators, or sometimes cofactors, which have been discussed already. Ligands that lead to a loss of activity are always called inhibitors. Some ligands produce their effects on activity by binding at the active site whereas others do so by binding to an allosteric site. If a ligand is known to bind at an allosteric site it is called an *effector*. Positive effectors are allosteric activators whereas negative effectors are allosteric inhibitors.

Because they bind reversibly the effects of such ligands upon enzymic activity are concentration dependent in much the same way as the effects of substrates and products are. Some ligands have an 'all or none' effect in that they activate an otherwise inactive protein or alternatively totally inactivate an active enzyme. Such ligands act by affecting the *amount* of active enzyme. Others modulate or moderate the existing activity of an enzyme, increasing or decreasing it when they bind.

In passing, it is worth noting that the effect of pH can be considered in this class as well as in class 2 as described above. If the hydrogen ion is treated as a ligand, the effects of pH within the reversible range can be considered to arise from the effects of the hydrogen ion acting either directly at the active site or indirectly through allosteric sites.

To summarize, the terms 'activation' and 'activator' are used indiscriminately to describe processes and agents that lead to an increase in activity. The terms 'inhibition' and 'inhibitor' are usually limited to the processes and agents that lead to reversible loss of activity by the non-covalent binding of ligands, although, to complicate matters, 'irreversible inhibition' is also sometimes referred to. Finally the terms 'inactivation' and 'inactivator' or 'inactivating agent' are used to describe processes and agents leading to loss of activity by any other means.

CATALYSIS

The preceding discussion of enzyme structure has been in a sense only a preamble to the vital question of how enzymes actually achieve their catalytic effects. The preamble is a necessary one as it sets out the structural framework in which the function of catalysis occurs; and, as with all biological studies, structure and function are inextricably interwoven.

In order to understand how the enzyme molecule speeds up a chemical reaction it is necessary to look first in some detail at how the uncatalysed chemical reaction normally occurs, so as to illuminate the role of the catalyst. The rate of a reaction was shown in Chapter 2 to be related to the concentration of the reactants by an empirical rate law. For a simple unimolecular reaction, A $\rightarrow$ products, the rate is given by $v = k[A]$, while for a simple bimolecular reaction, A + B $\rightarrow$ products, the equation is $v = k[A][B]$, where k is a rate constant in each case. It must be quite clear that different unimolecular chemical reactions proceed at different rates because, and only because, they have different rate constants and similarly for bimolecular reactions. Furthermore, the rate of reaction usually increases with temperature, which means that the rate constant must increase with temperature. The relationship between the rate of reaction and the temperature gave the first clue as to what controls the rate of a reaction. Arrhenius found that the logarithm of the rate was proportional to $-1/T$ where T is the absolute temperature. This fact must be considered alongside the Boltzmann distribution law which states that in any system of particles the number, n, with a given energy E is proportional to $e^{-E/kT}$ where e is the base of natural logarithms and k is the Boltzmann constant. In other words the logarithm of n is proportional to $-1/T$. Clearly one explanation of why reaction rate increases as it does when the temperature is increased would be that only those molecules which have a certain minimum energy are able to react. As the temperature is raised the number of molecules possessing the minimum energy is increased and so the rate of reaction increases. Although this is a satisfactory way of accounting for unimolecular reaction rates, when bimolecular reactions are considered it is clear that for reaction to occur at all the reactants must first collide. The number of collisions per second between molecules of A and B can be calculated for any given temperature and is always found to be much greater than the number of molecules that react per second. In other words many collisions are unfruitful; the molecules merely bounce apart again without reacting. Once again, however, if the colliding molecules must possess between them a certain minimum energy before they can react, a ready explanation is at hand. At low temperatures few of the colliding pairs will possess the necessary energy but at higher temperatures more of them will. This concept of a minimum, or threshold, energy needed before chemical reaction can occur is now firmly established.

The transition state

Consider two molecules, one of A and one of B, both in their electronic, vibrational and rotational ground states but possessing translational energy due to their movement. If they are to react when they collide it is clear that a rearrangement of the electronic distribution in each molecule must take place, perhaps leading to a transfer of atoms from one molecule to the other. Certain conditions must be met before this can happen. For example when chloride ion reacts with methyl bromide to produce methyl chloride and bromide ion it is clear that the incoming chloride ion must first of all have sufficient translational energy to penetrate within the van der Waals radius of the carbon atom, otherwise it will just bounce off and retreat without altering the methyl bromide molecule. It must also approach from the right direction if it is to hit the carbon atom and not the bromine or one of the hydrogens. The best direction for it to approach from will be diametrically opposite the bromine atom so that as its electron cloud penetrates into that of the carbon atom, the bromide ion can leave unhindered from the other side as shown in Figure 37, while the umbrella of hydrogen atoms flips over through a planar intermediate form. This intermediate form is a half-way house, neither methyl bromide nor yet methyl chloride. It is called the *transition state* or *activated complex* of the reaction. A chloride ion approaching along the right direction but with insufficient translational energy may get close enough to deform the methyl bromide part of the way towards the transition state, but then it would be

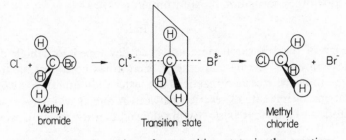

Figure 37 The formation of a transition state in the reaction between chloride ion and methyl bromide to produce methyl chloride and bromide ion

rejected again. Chloride ions approaching from any direction but the one shown, will need much *more* translational energy if they are to push the methyl bromide molecule into the transition state. It is possible to conceive of a chloride ion approaching with just enough energy to reach the transition state, but with none left over to complete the ejection of the bromide ion. In that situation the transition state could break down in either direction, to reform methyl bromide or to form methyl chloride, but this is such an unlikely eventuality that it can be ignored. In general the chloride ions will either have insufficient translational energy or an excess, so that if the transition state is formed at all it breaks down to form products.

Activation energy

In the example just considered the methyl bromide was thought of as stationary while the incoming chloride ion had to provide all the necessary translational energy to reach the transition state. In fact, of course, they both contribute translational energy in mutually reaching their transition state. This then is the origin of the threshold energy needed to achieve reaction. The smallest amount of translational energy with which the two reactants can approach and reach the transition state is called the *activation energy* of the reaction.

It turns out to be convenient to consider the formation of the transition state as if it were an ordinary chemical equilibrium:

$$A + B \rightleftharpoons AB^{\ddagger}$$

where $AB^{\ddagger}$ represents the transition state. This equilibrium can be thought of in the same way as any other, even though $AB^{\ddagger}$ is a highly unusual and very transient chemical species. In particular the equilibrium has an equilibrium constant defined in the usual way as:

$$K^{\ddagger} = \frac{[AB^{\ddagger}]}{[A][B]} \tag{38}$$

from which it can be deduced that there is an activation free energy, $\Delta F^{\ddagger}$, associated with the formation of $AB^\ddagger$ from A and B:

$$\Delta F^\ddagger = -RT \ln K^\ddagger \tag{39}$$

The theory of the transition state has been developed to a high degree of sophistication by H. Eyring and others[17] and it has been possible to learn a great deal about why certain reactions are faster than others. First of all it is clear that the overall rate of reaction between A and B will depend upon the concentration of $AB^\ddagger$ that exists. Eyring showed that the reaction rate is given by

$$v = \frac{RT}{Nh}[AB^\ddagger] \tag{40}$$

where R and T are the gas constant and absolute temperature, while N is Avogadro's number and $\mathbf{h}$ is the Planck constant. From the definition of $K^\ddagger$ given in Equation 38, $[AB^\ddagger] = K^\ddagger[A][B]$ and so Equation 40 can be re-written as:

$$v = \frac{RT}{Nh}K^\ddagger[A][B] = k[A][B]$$

where k is the familiar second order rate constant of the reaction: $A + B \rightarrow$ products. This gives a clear relationship between the rate constant k, and $K^\ddagger$:

$$k = \frac{RT}{Nh}K^\ddagger \tag{41}$$

which is of fundamental importance as we shall see. It reflects the starting assumption that fast reactions have large values of $K^\ddagger$ implying a relatively high concentration of the transition state at a given concentration of reactants. What is more important, however, is that Equation 41 allows us to *calculate* $K^\ddagger$ and hence $\Delta F^\ddagger$ using Equation 39. It follows that a fast reaction has a small activation free energy, $\Delta F^\ddagger$.

Returning from a consideration of transition state theory to the role of enzymes in catalysing chemical reactions, we may now ask what an enzyme can do to increase the apparent value of k the rate constant of a reaction. The answer is immediately clear from the above discussion. The enzyme must increase $K^\ddagger$ and hence reduce the activation free energy $\Delta F^\ddagger$.

Compare the uncatalysed reaction with what happens in the presence of an enzyme, E. For convenience we draw the following diagram (Figure 38) in which A, B and E are all present in solution together. On the top line the un-catalysed reaction proceeds as before via the transition complex $AB^\ddagger$ without the involvement of the enzyme. Alternatively, proceeding to the bottom line the enzyme forms the enzyme–substrate complex EAB, which then reacts with

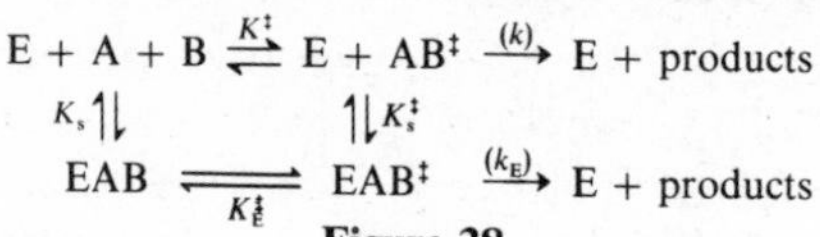

$$
\begin{array}{ccc}
\mathrm{E + A + B} \underset{}{\overset{K^\ddagger}{\rightleftharpoons}} \mathrm{E + AB^\ddagger} & \xrightarrow{(k)} & \mathrm{E + products} \\
\kappa, \updownarrow \qquad\qquad \updownarrow \kappa^\ddagger & & \\
\mathrm{EAB} \underset{K_E^\ddagger}{\rightleftharpoons} \mathrm{EAB^\ddagger} & \xrightarrow{(k_E)} & \mathrm{E + products}
\end{array}
$$

Figure 38

a new apparent rate constant k_E, proceeding via the transition state $EAB^{\ddagger}$. Thermodynamically speaking, since E, A and B can form $EAB^{\ddagger}$ via EAB, following a route down the left-hand side and along the bottom of the square, they must also be able to form $EAB^{\ddagger}$ via $E + AB^{\ddagger}$ following the other route. K_s is the equilibrium constant for the formation of the enzyme–substrate complex from $E + A + B$; $K^{\ddagger}$ has the meaning already described and $K_E^{\ddagger}$ represents the equilibrium constant for the formation of $EAB^{\ddagger}$ from EAB. The constant $K_s^{\ddagger}$ is a theoretical equilibrium constant for the binding of $AB^{\ddagger}$ to free enzyme. It is theoretical because, of course, $AB^{\ddagger}$ is a transient species, no sooner formed than broken down, but in principle it must be able to bind to E as already discussed. As in any such 'thermodynamic box' the product of the equilibrium constants K_s and $K_E^{\ddagger}$ must equal the product of $K^{\ddagger}$ and $K_s^{\ddagger}$:

$$K_s \cdot K_E^{\ddagger} = K^{\ddagger} \cdot K_s^{\ddagger} \qquad (42)$$

As we have already shown, $K_E^{\ddagger} \gg K^{\ddagger}$ because $k_E \gg k$ (see Equation 41), and so from Equation 42 it follows that $K_s^{\ddagger} \gg K_s$. In other words *the enzyme binds* $AB^{\ddagger}$, *the transition state, much more tightly than it binds the substrates* A *and* B. The importance of this conclusion cannot be overstated. It leads us to a deeper understanding of how catalysis is achieved by the enzyme and it also explains away the paradox that the enzyme appears to be specific for both its substrate and its product. One can hardly amplify these insights better than in the words of Linus Pauling:[18]

> I think that enzymes are molecules that are complementary in structure to the activated complexes of the reactions that they catalyse, [that is to the molecular configuration that is intermediate between the reacting substances...]. The attraction of the enzyme for the activated complex would thus lead to a decrease in its energy, and hence to a decrease in the energy of activation of the reaction and to an increase in the rate of reaction.

Transition state analogues

The answer to the question posed earlier, on specificity, 'Is the enzyme specific for substrate or product?' is clearly that it has some specificity for both, but binds most strongly to the transition state. Pauling made these observations in 1948 when there was no evidence to support his belief. But there is now experimental evidence from studies on a number of enzymes that the transition state is indeed bound orders of magnitude more tightly than substrate or product. For these enzymes, so-called *transition state analogues* have been synthesized. They are molecules that resemble in structure the putative transition state for the reaction catalysed by the enzyme, but which are stable. Such transition state analogues turn out to be very tightly bound to their enzymes and are in fact among the best inhibitors known for these enzymes.

Lysozyme is one enzyme for which a transition state analogue has been synthesized. As mentioned previously (p. 107) the substrate for lysozyme is a repeating polymer made of alternating N-acetyl-glucosamine (NAG) and N-acetyl-muramic acid (NAM) residues. There is good experimental evidence for believing that in the transition state one of the residues of the substrate is forced out of its usual chair conformation into an unstable half-chair conformation and that hydrolytic cleavage of the polymer chain occurs at this distorted residue. A derivative of the tetrameric substrate analogue molecule $(NAG)_4$ has been synthesized, in which the final residue is a lactone instead of the usual hexose ring:

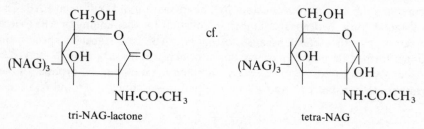

tri-NAG-lactone   cf.   tetra-NAG

The stable conformation adopted by such a lactone ring is the half-chair one and so it might reasonably be expected to act as a transition state analogue. In fact tri-NAG-lactone is found to bind about 3600 times more tightly to the enzyme than tetra-NAG.

Another good example is the enzyme triose-phosphate isomerase which catalyses the interconversion of glyceraldehyde-3-phosphate and dihydroxy-acetone phosphate, and for which an ene-diol anion transition state has been postulated:

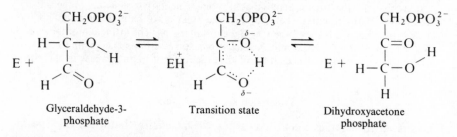

Glyceraldehyde-3-   Transition state   Dihydroxyacetone
phosphate   phosphate

The compound phosphoglycolohydroxamate has been found to be a potent inhibitor of triose-phosphate isomerase binding roughly 30 times more tightly than either substrate or product, in all probability because one of its stable structures resembles the transition state:

$$CH_2OPO_3^{2-}$$
$$\overset{\delta-}{C} \cdots O$$
$$\overset{\delta+}{N} \cdots \overset{H}{\underset{O}{}}$$
$$H$$

A stable tautomeric
form of phosphoglycolohydroxamate

The finding that transition state analogues are good inhibitors has value in the search for drugs and antimetabolites that will specifically inhibit certain enzymes. An intelligent guess is made as to the structure of the transition state for the enzyme-catalysed reaction, then a transition state analogue is synthesized and may be expected to be a very specific and potent inhibitor. But the experimental confirmation that the transition state is most tightly bound to its enzyme can lead us further than that. Quite apart from the possibility that the enzyme carries catalytic groups within its active site, it now appears that catalysis will result merely because of the tightness of fit and complementarity of binding between the active site and the transition state.

Exactly why this should be so in mechanistic terms is not immediately obvious, but an examination of the nature of the transition state and the processes leading to it throws light on the various ways in which enzymes can reduce the activation energy of a chemical reaction and bind its transition state, thereby leading to catalysis.

The transition state is a molecule in flux. It contains partial bonds, chemical bonds that are in the process of breaking or forming. The free energy required to reach this state from the normal electronic and vibrational ground state of the reactants, is the activation free energy, $\Delta F^{\ddagger}$. When the reactants are bound to the enzyme active site, $\Delta F^{\ddagger}$ is less than it is in free solution. Diagrammatically this comparison is shown in Figure 39 in a hypothetical activation diagram. The standard free energy F^{0} is plotted vertically against a theoretical 'reaction coordinate'. Progress along this reaction coordinate measures the movement of the reactants through the transition state to form products. The uncatalysed reaction, corresponding to the top line of the thermodynamic box in the previous figure (Figure 38), proceeds along the upper curve with a large activation free energy $\Delta F^{\ddagger}$. The catalysed reaction follows a lower route,

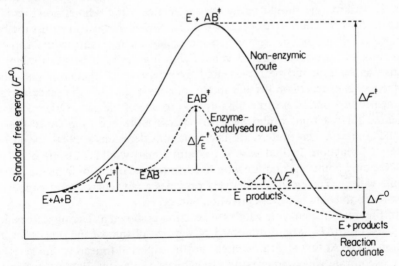

Figure 39 An activation diagram for the reaction, A + B → products, comparing the uncatalysed reaction with an enzyme-catalysed reaction. The difference between the activation energies, $\Delta F^{\ddagger}$ for the uncatalysed reaction and $\Delta F_{E}^{\ddagger}$ for the catalysed one, is illustrated

corresponding to the bottom line of Figure 38. Starting from the same position, the enzyme and reactants first form EAB the enzyme–substrate complex. (This reaction is shown as proceeding in one step for clarity, though it almost certainly takes place in at least two steps with either A or B binding first, followed by the other.) There will be a small activation energy ($\Delta F_1^\ddagger$) associated with this process, but usually very much less than the activation energies associated with the making and breaking of covalent bonds. Once the substrates are bound, the activation energy needed to bring about the formation of bound products is shown as $\Delta F_E^\ddagger$, clearly much smaller than $\Delta F^\ddagger$. Finally the products must dissociate from the enzyme to reach the same final situation as in the uncatalysed reaction. This process (again almost certainly a biphasic one at least) also has an activation energy or energies ($\Delta F_2^\ddagger$) associated with it, but as with substrate binding, such energies are usually small. Note that the overall standard free energy change, ΔF^0, is the same whichever route is followed, as required by thermodynamic restraints upon the reaction.

Figure 39 shows just one possible route for the enzyme-catalysed reaction. A whole range of alternative routes could be envisaged. The only requirements are that the entire route shall lie *under* the uncatalysed route, and that no single activation free energy shall be as great as $\Delta F^\ddagger$. Indeed it is entirely possible that a number of enzymes could have evolved, each catalysing the reaction and each following a slightly different route.

It should be pointed out that, in reality, Figure 39 is a highly artificial device. For the uncatalysed reaction it shows the lowest energy route that could be followed by molecules of A and B if they approached each other under optimal circumstances of orientation and with just sufficient energy to cross the reaction barrier. In fact one should really try to imagine a multidimensional contour map in which standard free energy would be plotted vertically against the bond lengths of all the bonds in the two molecules. Such a map would show the reactants and products lying in valleys with high ground between them. The transition state would be represented by a pass or col between the valleys. Some highly energetic molecules might, of course, go from the reactant valley to the product one by a route high up on the side of the pass. Others, less well oriented for reaction, would need to possess more energy because the pass would not be on their route at all. The same considerations apply to the enzyme-catalysed reaction. In that case the contour map would look quite different and the only points in common with the uncatalysed map would be the bottoms of the valleys. The enzyme effectively provides a new energy landscape with a new reaction route of lower activation energy.

In Figure 39 the enzyme-catalysed reaction is shown as having a free energy of activation ($\Delta F_E^\ddagger$) about one-third of the size of that of the uncatalysed reaction ($\Delta F^\ddagger$). At first sight this might not be expected to lead to the very large increase in rate which an enzyme can achieve. Equations 39 and 41, however, show that a very substantial rate increase can be achieved by quite a modest reduction in $\Delta F^\ddagger$.

Equation (39)

$$\Delta F^{\ddagger} = -RT \ln K^{\ddagger}$$

can be re-cast in the form:

$$K^{\ddagger} = e^{-\Delta F^{\ddagger}/RT} \tag{43}$$

from which Equation (41),

$$k = \frac{RT}{N\mathbf{h}} K^{\ddagger},$$

can be re-written as:

$$k = \frac{RT}{N\mathbf{h}} e^{-\Delta F^{\ddagger}/RT} \tag{44}$$

This last equation reveals the exponential relationship between k and $\Delta F^{\ddagger}$ that allows a small reduction in $\Delta F^{\ddagger}$ to produce a large increase in k.

Table 4 shows how some typical values for $\Delta F^{\ddagger}$ are related to the second order rate constant at 25 °C by Equation 44.

Table 4

$\Delta F^{\ddagger}$ (kcal/mol)	k (l/mol/s)
25	$3 \cdot 16 \times 10^{-6}$
20	$1 \cdot 38 \times 10^{-2}$
15	$63 \cdot 7$
10	2.91×10^{5}
5	$1 \cdot 33 \times 10^{9}$

In other words k, and hence the rate of reaction under any given set of conditions, increases by a factor of $4 \cdot 58 \times 10^3$ for every drop of 5 kcal/mol in $\Delta F^{\ddagger}$. An enzyme able to reduce $\Delta F^{\ddagger}$ from a value of 25 kcal/mol to 15 kcal/mol would actually increase the reaction rate by a factor of about *twenty million*.

Ways in which the enzyme can reduce the free energy of activation of a reaction

We are now in a position to appreciate what the enzyme must do in abstract terms, to catalyse a chemical reaction, and we must investigate what the physical mechanisms are whereby reduction in $\Delta F^{\ddagger}$ is achieved. They fall into a number of types.

The activation free energy, like any other free energy change, is made up of enthalpic and entropic components according to the familiar equation:

$$\Delta F^{\ddagger} = \Delta H^{\ddagger} - T\Delta S^{\ddagger} \tag{45}$$

where $\Delta H^{\ddagger}$ is the enthalpy of activation and $\Delta S^{\ddagger}$ is the entropy of activation.

Entropic factors: Propinquity and orientation

Consider first the entropic contribution, $\Delta S^{\ddagger}$. For any uncatalysed bimolecular reaction this is highly unfavourable because two molecules, *each* with three degrees of translational and three degrees of rotational freedom (i.e. a total of six degrees of translational and six of rotational freedom), must form but a single species with only three degrees of translational and three degrees of rotational freedom. The loss of translational and rotational freedom suffered by two reacting molecules when they enter the transition state entails a severe *decrease* in entropy and this acts as a deterrent to reaction. In other words $\Delta S^{\ddagger}$ is negative, and so the term $-T\Delta S^{\ddagger}$ in Equation (45) is positive and leads to a large value for $\Delta F^{\ddagger}$. Calculations have shown that there is a loss of about 35 entropy units (e.u.) when two reactants form a transition complex. At 37 °C this is equivalent to 9·45 kcal/mol and so contributes heavily to the observed values of $\Delta F^{\ddagger}$ of around 20 kcal/mol.

On the other hand when the enzyme-catalysed reaction is considered, the two reactants which have to enter the transition state are bound to the enzyme surface and have therefore already effectively lost their translational and rotational degrees of freedom. The value of $\Delta S^{\ddagger}$ is thus much smaller and consequently $\Delta F^{\ddagger}$ will be smaller. The unfavourable loss of entropy which attends the formation of a transition-complex has, in effect, been lost at the previous stage when the reactants bind to the enzyme surface, instead of having to be lost in the activation process. This consideration may well be one of the major factors underlying enzyme catalysis, at least for reactions involving more than one substrate. The effect has been termed the 'approximation effect' or the 'propinquity effect', because in essence it involves the tying down of two otherwise free reactants in close proximity on the enzyme surface. Looking at propinquity in another way, the effective local concentrations of the two reactants in the active site are very much higher than in free solution, which naturally increases their rate of reaction. Of course propinquity is much less important for enzymes that catalyse one-substrate reactions, but this type of enzyme is uncommon. This and other ways in which $\Delta S^{\ddagger}$ may be reduced in the enzyme catalysed reaction have been discussed at length by Lienhard[19] and by Jencks and his colleagues.[20]

A closely allied effect is the so-called 'orientation effect'. It was seen earlier how two colliding molecules must approach each other with the correct relative orientation in order to enter the transition state. According to some calculations the correct alignment of the electronic orbitals of the approaching molecules is highly critical. A small departure from the optimal alignment may mean that the energy required to enter the transition state is greatly increased. In free solution it is extremely unlikely that a collision between reactants will occur with the correct orientation, but on the enzyme surface where the reactants have been tied down in specific binding sites, the orientation of each with respect to the other is closely specified. It is argued that the evolution of enzymes is likely to have led to this relative orientation being very close to optimal. The enzyme is thought of as steering the two reactants into the correct orbital alignments in the active site and the expression *orbital steering* has been coined to describe this effect. Koshland and Hoare[21] have both argued the case for orbital steering very convincingly, although there is still controversy over the quantitative contribution that it can make to catalysis.

Enthalpic factors: Strain and general acid–base catalysis

Consider now the enthalpic contribution ($\Delta H^{\ddagger}$) towards the overall activation free energy. The transition state is like a molecule with some of its chemical bonds in an unusually extended or strained form. In order to distort the bonds in this way energy has to be provided. Anything that an enzyme can do towards straining or distorting the bound reactants in the active site may therefore bring them into a state approaching what they must achieve in the transition state. Clearly the strain must be applied at the right places and not just anywhere; but if, during the process of binding, the reactants have to undergo small changes of shape in order to fit snugly into the active site, and if these changes result in the extension of the bonds that are to undergo chemical reaction, those bonds are already part of the way towards achieving the transition state.

Take as an example a reactant molecule, A, containing three groups a, b and c attached to a central carbon atom (Figure 40). Imagine that the bond between a and the carbon atom is to be broken during reaction. Imagine further that there are functional groups x, y and z in the active site of the enzyme that have an inherent attraction for a, b and c respectively, and that the enthalpy of binding a to x, b to y, and c to z is in each case 5 kcal/mol when contact between each pair has been established. If the structure of the active site was exactly complementary to that of the reactant A, binding would occur with a net enthalpy change of -15 kcal/mol. If, however, the active site is not quite complementary, so that when b and c have become bound to y and z, a is not quite in the right position to interact optimally with x, then one possibility is

Figure 40 An illustration of how strain may be introduced into a substrate molecule upon binding to an enzyme active site. On the left the substrate molecule is positioned so that its groups b and c interact with y and z in the active site, but group a is too far away from x to interact favourably. On the right the bond between a and the central carbon atom has become stretched in order to allow a and x to interact

that the bond between a and the central carbon atom of the reactant molecule might become stretched so as to achieve a snug pairing of a with x. Let us imagine that this straining requires 3 kcal/mol. The molecule A will now bind to the active site with a net enthalpy change of only -12 kcal/mol. This is sufficiently large to ensure that A does bind at all three loci, but in a strained form in which the reactive bond has been weakened. The enthalpy of activation, $\Delta H^{\ddagger}$, will now be 3 kcal/mol less than it would be if reaction took place in free solution. Part of the intrinsic energy of binding to an active site that is not exactly complementary is being used to distort the substrate and bring it closer to the transition state before the actual activation process begins. Strain theories of this sort are often grouped under the umbrella title of the 'modern rack hypothesis'. An old theory of enzyme catalysis supposed that the enzyme literally tore its substrate apart by a Rack mechanism, but this concept proved too crude to account for the details of enzyme catalysis. The more sophisticated strain theories of the type described above thus became known as the modern rack.

Of course an alternative solution to the problem depicted in Figure 40, where there is incomplete complementarity between active site and substrate, is for the protein molecule to become distorted rather than the substrate. The group x could move closer to a, rather than the bond between a and its carbon atom being stretched or bent. This does of course occur in some cases and is covered by the induced fit hypothesis discussed earlier, but one must be clear that this contributes nothing to the lowering of $\Delta H^{\ddagger}$. The question of whether enzyme or substrate becomes distorted will be decided in each individual case by a balance of factors. Sometimes both may be distorted to reach a compromise, as with lysozyme. In any case the resulting enzyme–substrate complex will have the lowest accessible free energy.

Another way in which $\Delta H^{\ddagger}$ can be reduced is of course by the appropriate positioning of catalytic functional groups within the active site of the enzyme. Such groups may be able to donate or accept electrons or protons to or from the substrate in such a way as to weaken its reacting bonds before reaction actually begins. Many examples of this kind of effect have been studied and it is now clear that general acid or general base catalysis is a very important source of catalytic power in enzymes. Furthermore pairs of reactive groups belonging to the enzyme may be able to set up charge relay systems. In this case one group may for example donate a proton to one side of the reacting pair of molecules while another group accepts a proton from the opposite side during passage through the transition state. This mechanism reduces the activation energy by stabilizing the electric charges generated during reaction. An example of this source of catalytic power is found in the enzyme ribonuclease A which is discussed in detail at the end of this chapter.

In the course of donating electrons to the substrate, the reactive groups of the enzyme may actually become covalently bonded to the substrate thus forming a stable E–S compound. An example of this occurs in the mechanism of glyceraldehyde phosphate dehydrogenase (GPD):

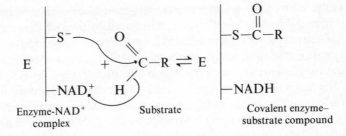

Enzyme-NAD⁺
complex   Substrate   Covalent enzyme–
substrate compound

The ionized sulphydryl group, $-S^-$ above, belonging to the enzyme attacks the aldehyde substrate, R·CHO, forming a stable thioester bond with the concomitant transfer of a hydride ion, H^-, to the bound coenzyme, NAD^+, forming NADH. Subsequently the thioester bond is broken by the second substrate, orthophosphate ion, regenerating the $-S^-$ group on the enzyme. Before this can happen it is probable that the NADH must be replaced by NAD^+ again:

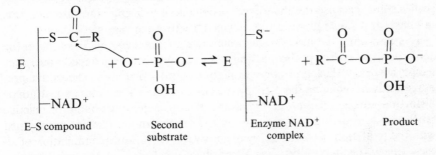

E–S compound   Second
substrate   Enzyme NAD⁺
complex   Product

This type of covalent catalysis, as it is sometimes called, is of importance in certain multisubstrate reactions where it is necessary to hold part of a substrate molecule on the enzyme while another part departs as one product, so that the retained part may then react with the second substrate to form the second product. It often occurs in 'ping pong' reactions which are discussed in Chapter 5.

Lastly, two processes which may contribute to enzyme catalysis, but which have not been well studied, will now be mentioned. The first of these may be loosely termed the environmental or dielectric effect. Some reactions are observed to occur best in organic solvents where the dielectric constant of the medium is low, whereas they do not occur so rapidly in water which has a high dielectric constant. Part of the reason for this may be that the attractive or repulsive forces between electric charges are much weaker in media of high dielectric constant than in those of low dielectric constant. Such reactions might be expected to occur more rapidly in a hydrophobic active site on an enzyme surface, because of shielding from the aqueous phase. In effect the enzyme may provide a kind of 'organic' or 'oily' micro-environment within the overall aqueous environment of the cell.

Finally, even when all the effects described above have been taken into account, and the value of $\Delta F^{\ddagger}$ has been reduced by binding the reactants to the enzyme active site, there still remains a finite free energy of activation, $\Delta F_E^{\ddagger}$. In other words, free energy still has to be provided from somewhere in order to nudge the reactants into and through the transition state. This can only come from the translational energy of solute molecules bombarding the enzyme–substrate complex because the substrates themselves have become tied down in the active site and have no translational energy left with which to enter the transition state. This has given rise to suggestions that the enzyme molecule may act as a kind of energy-transducer, channelling the energy derived from solvent bombardment through its own vibrational modes and into the bound substrates at the active site. Enzymes are flexible molecules and must be quivering and recoiling from solvent bombardment all the time in solution. Nevertheless suggestions of *specific* channelling of this vibrational energy into the reaction modes of the transition state, a kind of focusing of the random solvent bombardments at the active site, have not met with much enthusiasm. This is largely because there is no structural evidence to support such suggestions and because the theoretical problems of focusing vibrational energy, in phase, at a single place in the enzyme structure are very great.

To summarize, the mechanisms whereby enzymes achieve a lowering of the activation free energy of the reactions they catalyse can be grouped under four major headings. The first two, propinquity and orientation effects act predominantly in reducing the entropy of activation $\Delta S^{\ddagger}$. The other two, strain or distortion effects, on the one hand, and the strategic positioning of reactive functional groups in the active site on the other hand, are mainly concerned with the reduction of the enthalpy of activation $\Delta H^{\ddagger}$. Some combination of all these effects probably underlies all enzymic catalysis, but the contributions made by each effect will vary from one enzyme to another.

The distinction between active sites and other ligand binding sites, including allosteric sites, must now be clear. The former have the greatest affinity for the activated complex of the reaction they catalyse, whereas the latter bind a stable ligand tightly and have no tendency to induce it to change its chemical structure while bound.

The study of the mechanisms of enzymic catalysis clearly involves a large and ever-increasing number of experimental techniques. A prerequisite for complete understanding of any mechanism is a knowledge of the three-dimensional molecular structure of the enzyme and in particular of its active site. This is provided by sequence studies, chemical modification studies and X-ray crystallography with and without substrate analogues. Even when all this knowledge has been assembled, however, and one is in a position to make chemically intelligent guesses as to how catalysis *might* occur, one still needs to study the mechanism in the course of its occurrence. Since it is a very fast process, the next stage must involve the use of rapid reaction techniques employing spectroscopic methods that follow the changes both in enzyme structure and reactant structure as they occur. At the time of writing, this stage has been entered upon for a number of enzymes but it is still true to say that there is no single enzyme whose mechanism is fully understood.

Ribonuclease A is probably the most studied enzyme of all and a summary of our current knowledge of its mechanism is given below to illustrate some of the methods and problems that underlie the study of enzymes.

THE MECHANISM OF ACTION OF BOVINE PANCREATIC RIBONUCLEASE A (RNase A)

The reaction catalysed by RNase A, the hydrolysis of ribonucleic acid, proceeds in two steps (Figure 41). In the first, a phosphate ester bond between the oxygen belonging to the 5″-carbon atom of a ribose moiety, and the phosphorus atom is cleaved. The products of this step are two fragments of the original RNA molecule, one (the upper one in Figure 41) having a terminal 2′,3′-cyclic phosphodiester at its 3′-terminus, while the other has a free OH group on its 5′-CH$_2$ group (designated 5″ in Figure 41) to distinguish it from the other 5′-CH$_2$ group). The latter fragment is free to leave the enzyme surface, while the former undergoes Step 2, in which the cyclic phosphodiester is hydrolysed to a 3′-phosphomonoester, which then also leaves the enzyme.

The specificity of the enzyme is such that it shows a marked preference for cleaving the bond following a pyrimidine base (uracil or cytosine) in the RNA sequence as one goes from 5′- to 3′-termini. The base following the cleaved bond can be either a purine (adenine or guanine) or a pyrimidine (uracil or cytosine), with no marked preference for either.

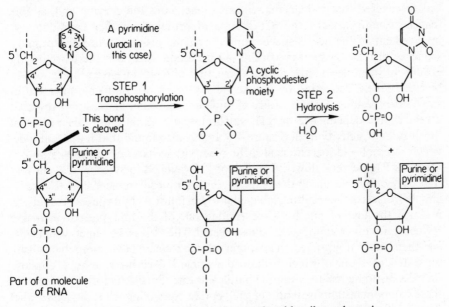

Figure 41   The two-step reaction catalysed by ribonuclease A

132

Various nucleotides, among them 3'-CMP (cytidylic acid), act as inhibitors of the enzyme. Sulphate, phosphate and cupric ions are also inhibitors. A substrate analogue, known as UpcA, whose structure is shown in Figure 42, acts as a good inhibitor of the enzyme as it mimics the dinucleotide portion of an RNA molecule but cannot be cleaved because the oxygen atom attached to the 5"-carbon atom is replaced by a methylene group (compare Figures 41 and 42).

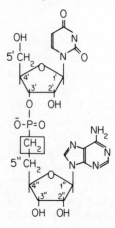

Figure 42   The structure of the ribonuclease A inhibitor UpcA

Early work on the inactivation of the enzyme by photooxidation showed that the loss of activity was paralleled by the loss of histidine residues, giving the first clue that histidine might be involved in the active site. Subsequently it was discovered that iodoacetic acid also inactivates the enzyme and, as this reagent usually attacks the SH groups of proteins, another candidate for involvement in the active site became a cysteine residue. When the primary sequence of the enzyme was worked out it became clear that there are no SH groups in the molecule (the primary structure of RNase is given in Figure 14), so clearly the inactivation by iodoacetic acid must be due to modification of some other residue. Further study of the iodoacetic acid inactivation revealed a remarkable dependence on pH which closely resembled the pH profile of the enzymatic activity itself (Figure 43). It was also found that iodoacetamide, which resembles iodoacetic acid in its reactivity towards SH groups, only inactivates RNase very slowly. The use of ^{14}C labelled iodoacetic acid showed that only one molecule of the reagent reacts with each molecule of RNase A, and hydrolysis of the resulting labelled protein, followed by amino acid analysis, revealed the loss of one histidine residue out of the four present in native RNase A, no other amino acid being affected. The ^{14}C-carboxymethyl–enzyme was then chromatographed on an ion-exchange resin and two components were separated, each one of which contained a single ^{14}C-carboxymethyl group and one less histidine residue than the native enzyme. Partial hydrolysis of each of these components and isolation of the peptide containing the radioactive label revealed that in one component the histidine residue at position 12 in the

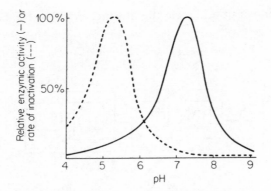

Figure 43 The dependence of enzymatic activity (solid line) and inactivation by iodoacetic acid (pecked line) on pH, for the enzyme ribonuclease A

sequence was labelled while, in the other, histidine 119 was labelled. Neither of the other two histidine residues in RNase A was ever found to carry a carboxymethyl label, nor was it possible to produce a modified RNase in which *both* histidines 12 and 119 were labelled. Clearly these two histidines must lie in some special structural situation such that both are susceptible to modification by iodoacetic acid (but not iodoacetamide) and in a mutually exclusive manner. Free histidine in solution, the other two histidine residues in RNase A, and histidine residues in most other proteins, react only extremely slowly with iodoacetic acid. All these facts, taken together with the pH profile of the inactivation and the observation that inactivation is prevented by substrates, analogues and inhibitors, all point to the conclusion that both histidines, 12 and 119, lie in the active site of the enzyme. Many other strands of evidence support this conclusion. For example the observation by Richards that brief exposure of RNase A to the proteolytic enzyme subtilisin at 0 °C results in the cleavage of a single peptide bond has led to much valuable work. The bond cleaved is the one between the alanine residue at position 20 and the serine at position 21 (see Figure 14), resulting in the modified enzyme RNase S which is still active. The amino terminal peptide consisting of residues 1 to 20, called S-peptide, remains tightly bound to the remainder of the molecule, called S-protein, but can be removed from it by gel filtration under certain conditions. Neither S-peptide nor S-protein is active on its own, but when mixed together in equimolar proportions they reform active RNase S. It is noteworthy that S-peptide contains His 12 whereas S- protein contains His 119.

The elegant studies of Crestfield and his colleagues on the formation and properties of active dimers of RNase also confirmed the conclusion that both His 12 and His 119 are in the active site. This work suggests that in the dimers the amino terminal portion of one RNase A molecule becomes bound to the C-terminal portion of its partner and vice versa. The two resulting composite active sites in the dimer (shown inside the dotted ovals below) can each react

134

with a single iodoacetic acid molecule, at either His 12 or His 119, as in monomeric RNase A :

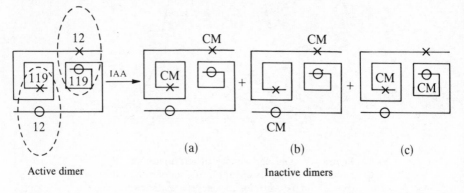

Active dimer   Inactive dimers

As shown above it is possible to form inactive dimers in which either His 12
in one site and His 119 in the other have reacted (a), or His 12 in both sites
(b), or His 119 in both sites (c). As a result when the inactive dimers are reconverted to monomers, (b) and (c) both form inactive monomers in which
either His 12 or His 119 is modified, but (a) gives rise to one *active* monomer
(from the lower one of the pair) and one inactive monomer in which now
both histidines are modified. Crestfield found that the various forms were
produced in the expected proportions.

The lysine residue at position 41 in the primary sequence was also found to
be involved in activity when it was found that reaction of this residue with
dansyl chloride or fluorodinitrobenzene led to loss of activity whereas the
other lysine residues could be modified without loss of activity.

The publication in 1967 of the three-dimensional structure of RNase A
(see Figure 21) confirmed these predictions when it was found that all three
residues are in fact clustered around a cleft in the otherwise roughly spherical
molecule.

A number of hypotheses about the mechanism of action of RNase have been
proposed over the years, but the earliest of these, due to Mathias and Rabin
and their colleagues, is probably nearest to the truth, although there is still no
conclusive proof. Mathias and Rabin proposed that the two histidine residues
at positions 12 and 119 in the sequence act in concert, one donating a proton to
the RNA substrate molecule and the other accepting a proton from it as shown
in Figure 44. The histidine shown as His_B starts with its imidazolyl ring in the
unprotonated form and initiates the reaction by activating the 2'-OH group of
the substrate. It does this by starting to abstract a proton, leaving the 2'-oxygen
with a partial negative charge enabling it to make a nucleophilic attack on the
phosphorus atom. At this stage the other histidine, His_A in Figure 44, which
starts in the protonated form, begins to donate a proton to the oxygen atom
of the 5''-CH_2OH group of the substrate. A transition state is reached in which
His_B has gained a partial positive charge and His_A is left with a partial negative

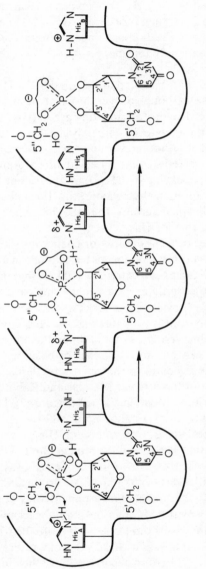

Enzyme–substrate complex   Transition state   Enzyme–product complex

Figure 44   The mechanism of ribonuclease action as proposed by Mathias and Rabin. The hypothetical mechanism is shown for Step 1 with RNA as the substrate. Note that the RNA substrate molecule in this diagram is upside-down as compared with Figure 41. Two imidazole groups belonging to histidine residues of the enzyme (His$_A$ and His$_B$) are shown in the active site. His$_A$ starts in the protonated form and loses its proton while His$_B$ starts in the unprotonated form and gains a proton. The phosphate group is shown with a single net negative charge distributed over the two unesterified oxygen atoms

charge; the phosphorus atom being in a pentacovalent form. Step 1 is completed when the $5''$-CH_2OH group is formed with the complete removal of the proton from His_A, while His_B gains complete possession of the proton from the $2'$-OH group, leaving the phosphorus atom as the cyclic phosphodiester. Mathias and Rabin assumed that the $5''$-CH_2OH then departed, its place being taken by a water molecule. Step 2 was then supposed to occur by a neat reversal of all the stages of Step 1, leading to the $3'$-phosphomonoester product and the protons back where they started, ready for another round of catalysis with a fresh substrate molecule.

Another hypothesis proposed that the initial attack on the $2'$-OH group was made, not by His_B, but by the carbonyl oxygen atom at position 2 of the pyrimidine ring of the substrate. The proton donating group (His_A in Mathias and Rabin's hypothesis) was thought to be either lysine 41, or the pair of histidine residues acting together. Another variant on the theme was a hypothesis that the same histidine (His_B) that accepted the proton from the $2'$-OH group gave it back to the leaving $5''$-CH_2O group. In that hypothesis lysine 41 was supposed to be involved in electrostatic attraction and orientation of the negative charge on the phosphate group. The other histidine residue was supposed to act in a supporting role for His_B.

X-ray crystallographic studies of RNase-inhibitor complexes show that the pyrimidine rings of the various substrate analogues tested all lie in a cleft, close to Phe 120, in a conformation that makes it extremely unlikely that the carbonyl oxygen on C_2 of the substrate pyrimidine ring could come close enough to activate the $2'$-OH group. His_{12}, however, lies very close to the $2'$-OH group and could readily deprotonate it, so it is probable that the His_B of Figure 44 is actually His_{12}. The other histidine residue, His_{119}, lies on the other side of the phosphate group, although it can occupy a number of positions depending upon which substrate analogue is bound in the active site. Perhaps the most convincing evidence in support of the original Mathias and Rabin hypothesis for Step 1 comes from studies with the analogue UpcA (Figure 42). The RNase–UpcA complex was studied by X-ray crystallography and the analogue was seen to be bound with His_{12} near the $2'$-OH group and His_{119} near the $5''$-O atom. It appears very unlikely that His_{12} could protonate the $5''$-O atom after deprotonating the $2'$-OH group because the former is too far away. Lysine 41 is fairly close to the free oxygen atoms of the phosphate group and could well be involved in electrostatic orientation of that group. It thus appears very likely that Step 1 at least occurs by the Mathias and Rabin mechanism. Step 2 is more controversial because there is room for a water molecule to attack the cyclic phosphodiester from the same side as the $2'$-O atom, rather than from the other side whence the $5''$-CH_2OH group departed. Step 2 therefore may not be a simple reversal of Step 1 if this possibility turns out to be correct—and there is some supporting evidence that it is correct.

This brief summary of some of the important pieces of evidence shows that there is still much to be learned about the detailed mechanism of even this, the most studied enzyme. It does illustrate some of the experimental approaches that are available to the enzymologist and highlights some of their limitations. It also clarifies to some extent the role of the enzyme in tying down its substrate in a 'correct' orientation in the active site while supplying an environment that encourages formation of the transition state.

Chapter 5

Enzyme Kinetics I: The Kinetics of Independent Active Sites

A question often asked by students of biology and biochemistry is, 'Why study enzyme kinetics?'. It is a valid question, especially in view of the seemingly esoteric and mathematical nature of much enzyme kinetics, but there is a valid answer to it. As biological catalysts, enzymes act by increasing the rate of the chemical reactions upon which living processes depend. If the case for enzymology as a valuable field of biological study is accepted, then the case for the study of enzyme kinetics follows naturally. An understanding of the structures and molecular mechanisms of enzymes satisfies the curiosity as to what they are and how they work, in much the same way as taking a watch apart and studying the cogs and levers that turn the hands does. But the value of a watch can only be understood by seeing it in action. The study of enzymes in action is the proper objective of enzyme kinetics.

In this and the succeeding chapters the essentials of enzyme kinetics will be covered, so as to present a picture of how the rate of an enzyme-catalysed reaction may vary in response to changes in the concentrations of substrates, products, activators and inhibitors. Temperature, pH, and ionic strength also affect the enzymic rate, as do other environmental factors such as whether the enzyme is present in a predominantly aqueous solution like the cytoplasm of the cell, or whether it exists in a predominantly non-aqueous medium such as a cell membrane.

An example of how such information is of value to the cell biologist comes from the study of glucose metabolism. The first step in glucose utilization inside the cell in most living organisms is its conversion to glucose-6-phosphate by reaction with ATP, which itself is converted to ADP:

$$\text{glucose} + \text{ATP} \rightleftharpoons \text{glucose-6-phosphate} + \text{ADP}$$

In the muscles of man and other animals this reaction is catalysed by the enzyme hexokinase. In the liver, however, two enzymes catalyse the reaction. One of them is hexokinase, as in the muscle, and the other is glucokinase. The distinction between them rests primarily on their substrate specificity. As its name implies, hexokinase will accept a number of hexoses as substrates, in addition to glucose. Glucokinase on the other hand only acts upon glucose. The question naturally arises as to why the liver goes to the trouble of synthesizing two enzymes where, at first sight, one would do. The reason for this apparent lack of economy (protein synthesis is energetically very expensive) became clear when the kinetics of the two enzymes were studied. Hexokinase has a very low K_m for glucose, about $37\,\mu M$. In other words the enzymic rate is half the maximal value at a very low glucose concentration. At glucose concentrations normally present in working muscle cells the enzyme is working as fast as it can so that small changes in glucose concentration do not affect the rate of glucose-6-phosphate formation. As a result any glucose that enters the working muscle cell is rapidly converted to glucose-6-phosphate and thereby locked in the cell, because glucose-6-phosphate is unable to cross the cell membrane and escape. Hexokinase, however, is inhibited by its product, glucose-6-phosphate. If the muscle cell is not actively using glucose-6-phosphate in glycolysis, and does not need any more glucose, no more will be phosphorylated. The kinetic properties of muscle hexokinase are clearly tailored to the needs of the muscle cell. Similarly the catabolic needs of the liver are served by liver hexokinase which, although very similar to muscle hexokinase is nevertheless a different isoenzyme. Liver also has an extra function to perform in that when excess glucose is present in the bloodstream it has to be converted to glycogen for storage. The first step in this process, as in glycolysis, is the formation of glucose-6-phosphate. If the liver were to rely on hexokinase it would be unable to carry out this second function and so it contains glucokinase, an enzyme with a much higher K_m (10 mM) and which is not so readily inhibited by glucose-6-phosphate. The glucokinase present in the liver can also catalyse a roughly threefold higher rate of phosphorylation when saturated with glucose. As the glucose concentration in the liver increases, the rate of the glucokinase-catalysed reaction goes on increasing to keep pace with it and is not retarded by an accumulation of glucose-6-phosphate.

Another example of how a study of the kinetics of an enzyme throws light on its biochemical role has already been mentioned in Chapter 4 where the different isoenzymes of lactate dehydrogenase (LDH) were seen to have different kinetic properties depending upon tissue of origin and the physiological function of that tissue.

In many enzymes, especially those involved in the regulation of metabolism, the kinetics are complicated by the existence of allosteric sites. The interaction between active sites and allosteric sites can lead to some remarkable kinetic results which are of great value to the cell. In a similar way oligomeric proteins can exhibit cooperative effects because of interactions between the active sites on different subunits. These interactions can also lead to complex kinetic behaviour. In this chapter we shall not consider such cooperative situations; they will be dealt with in Chapter 6. Here we shall be concerned with the simpler kinetics of independent active sites, that is to say of enzymes without allosteric sites, and with either a single active site per enzyme molecule or else with non-interacting active sites if there is more than one per molecule. Of course, much of what can be said about independent active sites can be carried over to the cooperative situations later.

ONE-SUBSTRATE ENZYME KINETICS

In Chapter 2 the simplest situation of enzyme kinetics was considered, where the enzyme has only one substrate and product is initially absent:

$$E + A \underset{k_{-1}}{\overset{k_1}{\rightleftharpoons}} EA \overset{k_2}{\rightarrow} E + P \tag{13}$$

It will be recalled that either of two simplifying assumptions can be made about this system, and both lead to a rate equation:

$$v_0 = \frac{V_m[A]_0}{K_m + [A]_0} \tag{12}$$

where $[A]_0$ is the initial total concentration of A, v_0 is the initial rate of reaction, and V_m and K_m are the maximum velocity and Michaelis constant of the enzyme under specified conditions. The two simplifying assumptions are either the rapid equilibrium assumption, in which case $K_m = k_{-1}/k_1$ and is equivalent to K_d, the dissociation constant of EA, or the steady state assumption in which case $K_m = (k_{-1} + k_2)/k_1$. In either case $V_m = k_2 e_0$ where e_0 is the total concentration of enzyme present. The steady state assumption is the more rigorous of the two and will be used in what follows unless otherwise stated. As already pointed out in Chapter 2 the steady state conditions are actually fulfilled in the living cell and so conclusions derived by steady state kinetics are applicable to cellular conditions. It is important to realize that K_m and V_m are constants only under a set of particular conditions. If, for example, the pH or temperature is changed, or if products are present, either or both of these 'constants' may alter. In a similar way the buffer solution in which the measurements are made must be specified when the constants are defined. In spite of these qualifications K_m and V_m are useful constants because, if they are measured under conditions approximating to those thought to apply in the living cell, they give one an idea of how fast the enzyme can work when saturated with substrate (V_m) and the concentration of substrate (K_m) needed for the rate to be half V_m. The substrate concentration needed to achieve the maximum rate is of course infinite, but when it is $5 K_m$ the rate is $0.83 V_m$, when it is $10 K_m$ the rate is $0.91 V_m$ and at $100 K_m$ the rate becomes $0.99 V_m$. These figures can be verified from Equation 12.

The effect of enzyme concentrations that approach those of substrate

In deriving Equation 12, either by the rapid equilibrium assumption or by the steady state approach, an additional assumption was made. The concentration of enzyme present, e_0, was taken to be so small that even if all of it were to combine with substrate A to form EA, the concentration, $[A]$, of free A left in solution would not be appreciably less than $[A]_0$ the initial total concentration of A. This is nearly always true in kinetic experiments carried out *in vitro* where very low enzyme concentrations are used. It allows one to use $[A]_0$ rather than $[A]$ in the rate equation, since $[A]_0$ will be known but $[A]$ will not. In the living cell, however, enzyme concentrations may sometimes approach those of substrates. The concentration of free A will be less than $[A]_0$ because a significant portion of it will be locked up in forming EA. Equation 12 is still valid provided that $[A]_0$ is replaced by $[A]$ as can be seen by inspecting the derivation of Equation 12 given in Chapter 2. V_m is still equal to $k_2 e_0$ and the rate of reaction will still be proportional to e_0 provided that $[A]$, rather than $[A]_0$, is kept constant. (This may be closer to the real situation inside a living cell owing to the operation of homeostatic mechanisms, which operate on *free* concentrations rather than total concentrations.)

The effect of product on the rate of enzyme-catalysed reaction

Although it is easy enough to perform *in vitro* experiments in the absence of product, in a real situation inside a living cell, of course, the product will always be present, even if it is being removed rapidly by other enzymes. In order to see what effect the presence of product, P, has upon the rate, the enzymic reaction must be written with an extra rate constant, k_{-2}, so as to allow product to react with the enzyme:

$$E + A \underset{k_{-1}}{\overset{k_1}{\rightleftharpoons}} EA \underset{k_{-2}}{\overset{k_2}{\rightleftharpoons}} E + P \qquad (46)$$

Equations such as Equation 13 and Equation 46 which describe how the enzyme is thought to combine with its substrate and product to form intermediate complexes, are called kinetic *models* or kinetic *mechanisms*. The steady state approach can be applied, assuming that we have an initial concentration $[A]_0$ of A, and $[P]_0$ of P, with the total concentration of enzyme e_0 as usual. In the steady state:

$$\left. \begin{aligned} \frac{d[E]}{dt} &= (k_{-1} + k_2)[EA] - (e_0 - [EA])(k_1[A]_0 + k_{-2}[P]_0) = 0 \\ \frac{d[EA]}{dt} &= (k_1[A]_0 + k_{-2}[P]_0)(e_0 - [EA]) - (k_{-1} + k_2)[EA] = 0 \end{aligned} \right\} \quad (47)$$

These two equations, for the rate of change in concentration of E and EA respectively, are identical. The overall rate is given by:

$$v_0 = \frac{d[P]}{dt} = k_2[EA] - k_{-2}(e_0 - [EA])[P]_0$$

$$= (k_2 + k_{-2}[P]_0)[EA] - k_{-2}e_0[P]_0 \qquad (48)$$

From Equation (47):

$$[EA] = \frac{(k_1[A]_0 + k_{-2}[P]_0)e_0}{(k_{-1} + k_2 + k_1[A]_0 + k_{-2}[P]_0)}$$

substituting for [EA] in Equation 48:

$$v_0 = \frac{d[P]}{dt} = \frac{(k_2 + k_{-2}[P]_0)(k_1[A]_0 + k_2[P]_0)e_0}{(k_{-1} + k_2 + k_1[A]_0 + k_{-2}[P]_0)} - k_{-2}e_0[P]_0$$

$$= \frac{(k_1k_2[A]_0 - k_{-1}k_{-2}[P]_0)e_0}{(k_{-1} + k_2 + k_1[A]_0 + k_{-2}[P]_0)} \tag{49}$$

This last step is somewhat curtailed but the result can be obtained by putting both terms on the right-hand side over a common denominator, then multiplying out terms in the numerator, whereupon all but two cancel out. The result given in Equation (49) is quite different from the familiar Michaelis equation (Equation 12) and cannot be simplified to resemble it. It can be simplified a little with a useful side result as will be shown shortly. Before doing so, however, it is worth noting that both Equations 49 and 12 have a similarity of form. If one writes them both out in terms of the rate constants this similarity can be seen more clearly. Equation 12 then becomes, for the steady state assumption where $K_m = (k_{-1} + k_2)/k_1$ and $V_m = k_2 e_0$:

$$v_0 = \frac{k_2 e_0[A]_0}{(k_{-1} + k_2)/k_1 + [A]_0} = \frac{k_1 k_2 e_0[A]_0}{k_{-1} + k_2 + k_1[A]_0}$$

$$\therefore \quad \frac{v_0}{e_0} = \frac{k_1 k_2[A]_0}{(k_{-1} + k_2) + k_1[A]_0} \tag{12a}$$

Equation 49 can be re-written in a similar form as:

$$\frac{v_0}{e_0} = \frac{k_1 k_2[A]_0 - k_{-1}k_{-2}[P]_0}{(k_{-1} + k_2) + k_1[A]_0 + k_{-2}[P]_0} \tag{49a}$$

In these two cases, by writing the left-hand side as the initial rate divided by the total enzyme concentration, one can see that the *numerator* of the right-hand side is derived from the *net flow* of substrate from left to right in the enzymic reaction. Since there is no back reaction in the kinetic model used to derive Equation 12a, its numerator just reflects the two forward rate constants k_1 and k_2 (see Equation 13). In deriving Equation 49, the kinetic model allows back flow and this is reflected in the appearance now of the negative term involving k_{-1} and k_{-2} (see Equation 46).

In a similar way the *denominators* of the right-hand side in these two rate equations (12a and 49a) reflect the *distribution* of the total enzyme e_0 among the different forms of enzyme. Thus in both cases $(k_{-1} + k_2)$ in the denominator reflects the contribution of [E] to the total; $k_1[A]_0$ reflects the contribution of [EA] in Equation 12a where EA can only be formed by the step whose rate constant is k_1; in Equation 49a the contribution of [EA] is represented by $k_1[A]_0 + k_{-2}[P]_0$ because [EA] can now also be formed by the step whose rate constant is k_{-2}.

Returning now to the simplification of Equation 49, consider what would happen if the concentration of P is equal to zero. Equation 49 reduces to:

$$v_0 = \frac{k_1 k_2 [A]_0 e_0}{k_{-1} + k_2 + k_1 [A]_0} = \frac{k_2 e_0 [A]_0}{(k_{-1} + k_2)/k_1 + [A]_0} = \frac{V_m^A [A]_0}{K_m^A + [A]_0}$$

where V_m^A is the maximum rate in the forward direction in the absence of P and K_m^A is the Michaelis constant for A. Both these constants can be determined as described before by measuring v_0 at a range of values of $[A]_0$ in the absence of P, and plotting $1/v_0$ against $1/[A]_0$. Similarly when the concentration of A is zero:

$$v_0 = -\frac{k_{-1} k_{-2} [P]_0 e_0}{k_{-1} + k_2 + k_{-2} [P]_0} = -\frac{k_{-1} e_0 [P]_0}{(k_{-1} + k_2)/k_{-2} + [P]_0}$$

$$= -\frac{V_m^P [P]_0}{K_m^P + [P]_0}$$

The negative sign merely refers to the fact that the reaction is going in the reverse direction from P to A. In this case K_m^P is the Michaelis constant for the product P, while V_m^P is the maximum rate of reaction in the absence of A. Both these constants can be determined experimentally as before by measuring the rate of the reverse reaction at a range of values of $[P]_0$ in the absence of A and plotting $1/v_0$ against $1/[P]_0$. Using all these relationships,

$$K_m^A = \frac{k_{-1} + k_2}{k_1}, \qquad K_m^P = \frac{k_{-1} + k_2}{k_{-2}}, \qquad V_m^A = k_2 e_0 \quad \text{and} \quad V_m^P = k_{-1} e_0$$

the original rate equation (49) can be re-written:

$$v_0 = \frac{V_m^A K_m^P [A]_0 - V_m^P K_m^A [P]_0}{K_m^A K_m^P + K_m^P [A]_0 + K_m^A [P]_0} \tag{50}$$

From this it can be seen that even if $[P]_0$ is kept constant while $[A]_0$ is varied, the plot of v_0 against $[A]_0$ will not be a hyperbola because there is an extra term in the numerator:

$$v_0 = \frac{V_m^A K_m^P [A]_0 - \text{constant}}{K_m^P [A]_0 + \text{constant}} = \frac{V_m^A [A]_0 - \text{constant}}{[A]_0 + \text{constant}}$$

cf. $v_0 = \dfrac{V_m^A [A]_0}{[A]_0 + \text{constant}}$ for a hyperbola.

Furthermore a plot of $1/v_0$ against $1/[A]_0$ at constant $[P]_0$ will not give a straight line, as it would if P were absent. So long as the product is present the kinetic behaviour of the enzyme becomes more complex owing to the phenomenon of *product inhibition*. This involves more than the fact that the net rate of production of P from A is reduced because of the back reaction. There is also the fact that P combines with some of the free enzyme and so makes less available for combining with A. (This type of product inhibition is more readily appreciated in multisubstrate enzyme kinetics where the presence of a single product only may tie up enzyme without leading to any back reaction.) These two phenomena are clearly reflected in the rate equation (Equation 49). The negative term $-k_{-1}k_{-2}[P]_0$ in the numerator clearly represents the back reaction, while the term $k_{-2}[P]_0$ in the denominator reflects the binding of E by P. If they were both absent we should have the familiar hyperbolic Michaelis equation. Both terms reduce the value of v_0 below what it would be in the absence of P.

This finding illustrates why one always avoids the presence of product, and therefore always measures *initial* rates of reaction, when performing experiments to determine K_m and V_m values. Once they are known for both A and P, on the other hand, the rate in the presence of both can be predicted from Equation 50. Before doing so, however, there is a simple test that can be applied to check that the values of the kinetic constants, K_m^A, K_m^P, V_m^A and V_m^P which have been determined are actually correct.

The Haldane Relationships

Consider what would happen if the initial concentrations of A and P were equilibrium concentrations, so that the overall reaction $A \rightleftharpoons P$ was at equilibrium. The rate of reaction would clearly be zero. In order for v_0 in Equation 49 to equal zero, the numerator must be zero because all terms in the denominator must be positive and less than infinite. Therefore:

$$k_1 k_2 [A]_{eq} = k_{-1} k_{-2} [P]_{eq}$$

$$\therefore \quad \frac{k_1 k_2}{k_{-1} k_{-2}} = \frac{[P]_{eq}}{[A]_{eq}} = K_{eq} \tag{51}$$

This means that once three of the rate constants have been fixed, the fourth is automatically decided. Incidentally this result proves that so long as the reaction $A \rightleftharpoons P$ has a finite equilibrium constant, that is $K_{eq} < \infty$, we are quite unjustified in writing an enzymic reaction in which $k_{-2} = 0$, such as Equation 13, given at the beginning of this section.

The same approach applied to Equation 50 shows that:

$$V_m^A K_m^P [A]_{eq} = V_m^P K_m^A [P]_{eq}$$

$$\therefore \quad \frac{V_m^A K_m^P}{V_m^P K_m^A} = \frac{[P]_{eq}}{[A]_{eq}} = K_{eq} \tag{52}$$

Consequently, by measuring K_{eq} for the overall reaction, an independent check on the values of the kinetic constants K_m^A, K_m^P, V_m^A and V_m^P can be applied, using Equation 52.

The relationships between the equilibrium constant and rate constants (Equation 51) or kinetic constants (Equation 52) are called *Haldane relationships* after their discoverer.[29]

The effect of isomerization of central complexes

One is now in a position to describe the kinetic behaviour of the simple one-substrate enzyme whatever the concentrations of its substrate and product, provided, of course, that it actually does function according to the model depicted by Equation 46:

$$E + A \underset{k_{-1}}{\overset{k_1}{\rightleftharpoons}} EA \underset{k_{-2}}{\overset{k_2}{\rightleftharpoons}} E + P \tag{46}$$

Close inspection of this equation reveals a disturbing lack of symmetry in that although A can combine with E to form EA, P is assumed to combine with E to also form EA rather than forming EP. This is a rather limited model for the kinetic mechanism of the enzyme. Perhaps one should have considered a more general model such as the following:

$$E + A \underset{k_{-1}}{\overset{k_1}{\rightleftharpoons}} EA \underset{k_{-2}}{\overset{k_2}{\rightleftharpoons}} EP \underset{k_{-3}}{\overset{k_3}{\rightleftharpoons}} E + P \tag{53}$$

or an even more complicated one:

$$E + A \underset{k_{-1}}{\overset{k_1}{\rightleftharpoons}} EA \underset{k_{-2}}{\overset{k_2}{\rightleftharpoons}} EX \underset{k_{-3}}{\overset{k_3}{\rightleftharpoons}} EP \underset{k_{-4}}{\overset{k_4}{\rightleftharpoons}} E + P \tag{54}$$

Fortunately when the steady state method is applied to such models, the rate equation always comes out to be identical with Equation 50, where K_m^A and K_m^P, V_m^A and V_m^P have the same kinetic meaning as before.

Although the rapid equilibrium assumption cannot be applied to Equation 46 it can apply to Equations 53 and 54 if the binding and release of A and P are much faster than the interconversions of the central complexes EA, EX and EP. In that special case, however, Equation 50 still applies. On the other hand, the exact meaning of each of the kinetic constants in terms of the *rate* constants of which they are constructed, will differ for the different models. For example V_m^A for the model of Equation 46 is equal to $k_2 e_0$, whereas for the model of Equation 53 it is $k_2 k_3 e_0 / (k_2 + k_3 + k_{-2})$ when treated by the steady state approximation but reduces again to $k_2 e_0$ in the special case of rapid equilibrium. A mechanism may have any number of central complexes (that is to say, enzyme-containing intermediates that isomerize to form each other) without affecting the 'structure' of the rate equation in terms of kinetic constants (rather than rate constants). There may be only one such as EA in Equation 46 or a string of them such as EA, EX, EP in Equation 54 and one could never tell from the dependence of v_0 on $[A]_0$ and $[P]_0$. It is worth pointing out that in many enzyme-catalysed reactions intermediate forms such as EX almost certainly have a real existence as stable entities distinct from EA the enzyme–substrate complex and EP the enzyme–product complex.

The effect of enzyme isomerization

The kinetic models considered so far (Equations 46, 53 and 54) assume that there is only one form of enzyme, E, that can combine with substrate and product. As enzymes are flexible and may undergo changes in conformation during the catalytic reaction it is reasonable to consider the possibility that two isomeric forms of the enzyme exist, one, let us say E, that binds the substrate A and another, say F, that binds product P. These two forms clearly must be in equilibrium, so that the kinetic model would be as follows:

$$E + A \underset{k_{-1}}{\overset{k_1}{\rightleftharpoons}} EA \underset{k_{-2}}{\overset{k_2}{\rightleftharpoons}} FP \underset{k_{-3}}{\overset{k_3}{\rightleftharpoons}} F + P \underset{k_{-4}}{\overset{k_4}{\rightleftharpoons}} E + P \tag{55}$$

The rate equation for this model can be shown by the methods used previously to be:

$$\frac{v_0}{e_0} = \frac{V_m^A K_m^P [A]_0 - V_m^P K_m^A [P]_0}{K_m^A K_m^P + K_m^P [A]_0 + K_m^A [P]_0 + K^{AP} [A]_0 [P]_0} \tag{56}$$

where V_m^A, V_m^P, K_m^A, K_m^P have the same kinetic significance as before although they each have a different composition in terms of the rate constants. There is, however, a new term in the denominator, in which the product of the concentrations $[A]_0$ and $[P]_0$ is multiplied by a new kinetic constant denoted as K^{AP}. This has no simple kinetic significance comparable to that of V_m or K_m, and is a complex function of the rate constants. When either A or P is absent, the rate equation simplifies in the same way as before, to a hyperbolic function, because the term $K^{AP} [A]_0 [P]_0$ disappears. This means that one can determine V_m^A, V_m^P, K_m^A and K_m^P in the usual way from Lineweaver–Burk plots of kinetic data obtained in the absence of A or P, but now there is an extra constant K^{AP} to be determined before an accurate prediction can be made of the enzymatic rate in the presence of A and P together.

If one were unaware that the enzyme mechanism actually involved an isomerization, one might determine the maximum velocities and Michaelis constants for substrate and product and attempt to substitute them in Equation 50 in order to obtain an estimate of enzymatic rates in the presence of both substrate and product. This would lead to error because, of course, Equation 56 is the correct one in this case. Caution must be exercised after V_m s and K_m s have been determined. A check can easily be made to see whether Equation 50 actually does apply and, if not, then Equation 56 may be used to estimate K^{AP}, though this is not as straightforward as estimation of the other kinetic constants.

The effects of inhibitors and activators

There is no difference in principle between activator and inhibitor so one may begin by considering a ligand X which, on binding to the enzyme, causes a change in its activity, either increasing it if X is an activator or reducing it in the case of an inhibitor. X may bind either to the free enzyme E, or to the enzyme substrate complex EA, or to both. Once bound it may affect either the ability of the enzyme to bind the substrate, or the catalytic power of the enzyme, or both. The effect of X will be considered in the absence of product and the general model of Equation 57 will be used:

$$E + A \underset{k_{-1}}{\overset{k_1}{\rightleftharpoons}} EA \overset{k_2}{\rightarrow} E + P$$

$$EX + A \underset{k_{-5}}{\overset{k_5}{\rightleftharpoons}} EXA \overset{k_6}{\rightarrow} EX + P \tag{57}$$

As we are considering only the kinetics of 'isolated' active sites in this chapter, it may seem unnecessary to consider the possibility of both A and X being bound to the enzyme at the same time, on the grounds that if A occupies the active site then X cannot, and vice versa. Nevertheless the ternary complex EXA is included in Equation 57 because one cannot rule out the possibility that there is room for both. Indeed, both *must* be present together if X is a non-allosteric activator. The Haldane relationship for this model requires that $K_1K_4 = K_3K_5$ or alternatively that $k_1k_4k_{-5}k_{-3} = k_{-1}k_3k_5k_{-4}$. The steady state rate equation for this general model (Equation 57) is complex and mathematically unwieldy. The Michaelis plot will not be hyperbolic and Lineweaver–Burk plots will not be linear except in special cases. In many real cases on the other hand, linear Lineweaver–Burk plots *are* found experimentally.

The initial rate v_0 is measured in a series of experiments in which product is absent and substrate concentration is increased steadily while the concentration of X is held constant. The series is then repeated in the absence of X. The data are plotted, in the usual reciprocal form of the Lineweaver–Burk plot. When linear plots are obtained, they fall into one of three categories. The first category comprises those cases where the plots intersect on the vertical axis as shown in Figure 45. Changing the fixed concentration of X in these cases alters the slope of the plot but does not alter the value of V_m, the rate achieved at infinite substrate concentration (i.e. when $1/[A]_0 = 0$). If X is an inhibitor, increasing its concentration increases the slope; the reverse is true if X is an activator. Effects of this type are termed competitive, on the basis of the experimental observation that the effect of the ligand X can be nullified by a sufficiently high concentration of substrate. The second class of effects is the *non-competitive* category (Figure 47) in which the Lineweaver–Burk plots intersect on the horizontal axis. Once more changing the concentration of X alters the slope of the plot but now it is the Michaelis constant K_m which remains unaffected. Again if X is an inhibitor the slope increases with increasing concentration of X and *vice versa* for an activator. The third category comprises those cases where both V_m and K_m are altered, and is a *mixed* category (Figure 49(a)). One special type of effect is the *uncompetitive* case in which X does not affect the slope of the Lineweaver–Burk plot although it does alter the points of intersection on the axes (Figure 49(b)).

The terms competitive and non-competitive are used to describe the experimentally observed kinetic effect of the ligand X and do not imply anything about the molecular mechanism by which X alters the enzymic activity. A number of different molecular mechanisms can give rise to each type of kinetic effect.

Competitive effects

1. Purely competitive effects

(a) *The ligand X binds only to the free enzyme* A very simple mechanism leading to purely competitive inhibition is the one where X binds only to the enzyme E and prevents binding of the substrate. The kinetic model is the following simplification of Equation 57:

$$E + A \underset{k_{-1}}{\overset{k_1}{\rightleftharpoons}} EA \overset{k_2}{\rightarrow} E + P$$
$$+$$
$$X$$
$$k_3 \big\downarrow\big\uparrow k_{-3}$$
$$EX \tag{58}$$

The concentration of free X is assumed to be equal to its total concentration, [X]. The steady state equations are:

$$\frac{d[E]}{dt} = 0 = [EA](k_2 + k_{-1}) + k_{-3}[EX] - [E](k_1[A]_0 + k_3[X]) \tag{59}$$

and

$$\frac{d[EA]}{dt} = 0 = k_1[E][A]_0 - [EA](k_2 + k_{-1}) \tag{60}$$

From these, the following ratios may be deduced:

$$\frac{[E]}{[EA]} = \frac{k_2 + k_{-1}}{k_1[A]_0};$$

$$\frac{[EX]}{[EA]} = \frac{[E]}{[EA]}\left(\frac{k_1[A]_0 + k_3[X]}{k_{-3}}\right) - \frac{(k_2 + k_{-1})}{k_{-3}} = \left(\frac{k_2 + k_{-1}}{k_1[A]_0}\right)\frac{k_3[X]}{k_{-3}}$$

The rate is conveniently expressed as a specific rate, or rate per unit enzyme concentration.

$$\frac{v_0}{e_0} = \frac{k_2[EA]}{[E] + [EA] + [EX]} = \frac{k_2}{[E]/[EA] + 1 + [EX]/[EA]}$$

Therefore, by substituting from above for the ratios [E]/[EA] and [EX]/[EA], we obtain:

$$\frac{v_0}{e_0} = \frac{k_2}{1 + \dfrac{(k_2 + k_{-1})}{k_1[A]_0} + \dfrac{(k_2 + k_{-1})}{k_1[A]_0}\dfrac{k_3[X]}{k_{-3}}}$$

$$\therefore \quad v_0 = \frac{k_2 e_0}{1 + \left(\dfrac{k_2 + k_{-1}}{k_1[A]_0}\right)\left(1 + \dfrac{k_3[X]}{k_{-3}}\right)}$$

$$= \frac{V_m}{1 + \dfrac{K_m}{[A]_0}\left(1 + \dfrac{[X]}{K_3}\right)} \tag{61}$$

where

$$V_m = k_2 e_0, \qquad K_m = \left(\frac{k_2 + k_{-1}}{k_1}\right), \qquad K_3 = \frac{k_{-3}}{k_3}$$

Taking reciprocals:

$$\frac{1}{v_0} = \frac{1}{V_m} + \frac{K_m}{V_m}\left(1 + \frac{[X]}{K_3}\right)\frac{1}{[A]_0} \qquad (62)$$

The plot of $1/v_0$ versus $1/[A]_0$ at a constant value of $[X]$ is therefore a straight line of slope $K_m(1 + [X]/K_3)/V_m$. The intercept on the $1/v_0$ axis (i.e. when $1/[A]_0 = 0$) is $1/V_m$ whatever finite value is given to $[X]$. This corresponds exactly to the situation shown in Figure 45. The intercept on the $1/[A]_0$ axis

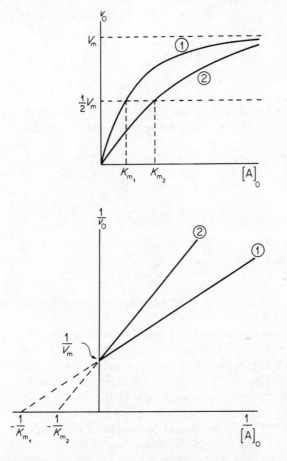

Figure 45 The effect of a *competitive inhibitor* on the kinetics of an enzyme. The upper graph shows a Michaelis plot, the lower one a Lineweaver–Burk plot. In both cases (1) and (2) represent the plot obtained in the absence and presence of the inhibitor X respectively

(i.e. when $1/v_0 = 0$) is found as follows from Equation 62:

$$0 = \frac{1}{V_m} + \frac{K_m}{V_m}\left(1 + \frac{[X]}{K_3}\right)\frac{1}{[A]_0}$$

$$\therefore \quad \frac{1}{[A]_0} = -\frac{1}{V_m}\Big/\frac{K_m}{V_m}\left(1 + \frac{[X]}{K_3}\right) = -\frac{1}{K_m\left(1 + \frac{[X]}{K_3}\right)}$$

When $[X] = 0$ this is, as expected, $-1/K_m$. In the presence of X the apparent Michaelis constant, shown as K_{m_2} in Figure 45, is $K_m(1 + [X]/K_3)$ where K_3, the *inhibitor constant* (sometimes denoted K_i), is nothing more than the dissociation constant of EX.

Equation 62 also shows that when $[A]_0$ is constant, a plot of $1/v_0$ against $[X]$ is also a straight line:

$$\frac{1}{v_0} = \left(\frac{1}{V_m} + \frac{K_m}{V_m[A]_0}\right) + \frac{K_m[X]}{V_m K_3[A]_0}$$

If the value of $1/v_0$ is plotted against $[X]$ for two different values of $[A]_0$, given by $[A]_1$ and $[A]_2$, the two resulting straight lines will intersect when:

$$\frac{1}{V_m} + \frac{K_m}{V_m[A]_1} + \frac{K_m[X]}{V_m K_3[A]_1} = \frac{1}{V_m} + \frac{K_m}{V_m[A]_2} + \frac{K_m[X]}{V_m K_3[A]_2}$$

$$\therefore \quad \frac{1}{[A]_1} + \frac{[X]}{K_3[A]_1} = \frac{1}{[A]_2} + \frac{[X]}{K_3[A]_2}$$

$$\therefore \quad \frac{1}{[A]_1}\left(1 + \frac{[X]}{K_3}\right) = \frac{1}{[A]_2}\left(1 + \frac{[X]}{K_3}\right)$$

Since $1/[A]_1 \neq 1/[A]_2$, this result can only be true if $(1 + [X]/K_3) = 0$, that is when $[X] = -K_3$.

This provides a graphical method of estimating the inhibitor constant K_3 and is due to Dixon.[31] The plot of $1/v_0$ against $[X]$ at a constant value of $[A]_0$ is known as a Dixon plot. Two such plots at different constant values of $[A]_0$ intersect at a value of $[X] = -K_3$ as illustrated in Figure 46.

It is worth mentioning in passing, that the simple model considered here assumes only that X and A both bind in a mutually exclusive manner to the free enzyme E. Clearly this would be the case if they both bound at the *same* site, but it is sometimes overlooked that they could bind at *different* sites if the binding of one caused a conformational change in enzyme structure that prevented binding of the other.

An exactly complementary situation to the one considered above will lead to 'purely competitive activation'. Consider that the enzyme can only bind A

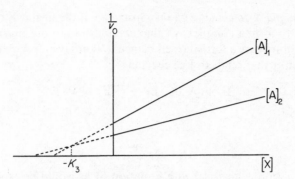

Figure 46 A Dixon plot for the determination of the inhibitor constant K_3 for a purely competitive inhibitor X. $[A]_1$ and $[A]_2$ represent the constant concentrations of the substrate A in two separate series of experiments in which $1/v_0$ is plotted against increasing values of $[X]$, the concentration of inhibitor

if X is also bound, so that the following simplification of Equation 57 applies:

$$E$$
$$+$$
$$X$$
$$k_3 \Big\downarrow\Big\uparrow k_{-3}$$
$$EX + A \underset{k_{-5}}{\overset{k_5}{\rightleftharpoons}} EXA \overset{k_6}{\rightarrow} EX + P$$

Application of the steady state assumption as before leads now to the rate equation:

$$v_0 = \frac{V_m}{1 + \dfrac{K_m}{[A]_0}\left(1 + \dfrac{K_3}{[X]}\right)} \tag{63}$$

where

$$V_m = k_6 e_0, \qquad K_m = \left(\frac{k_6 + k_{-5}}{k_5}\right); \qquad K_3 = \frac{k_{-3}}{k_3}$$

The equation and its reciprocal are entirely equivalent to Equations 61 and 62 except that the bracket $(1 + [X]/K_3)$ is replaced by $(1 + K_3/[X])$. Thus as $[X]$ is increased, the rate now increases instead of decreasing. When $[X] = 0$, the rate is zero as expected. Provided X is present, the rate in the presence of infinite substrate concentration $(1/[A]_0 = 0)$ is given by V_m as before, but here a note of caution must be introduced. The concentration $[X]$ is the concentration of free X. At infinite substrate concentration free X will only be present provided there is a molecular excess of X over E, so there must be sufficient of X to saturate the enzyme if truly competitive kinetics are to be observed. A plot of $1/v_0$ against $1/[X]$ may be employed to determine K_3 graphically. The intersection point of two plots at different concentrations of A will intersect at $1/[X] = -1/K_3$.

(b) *The ligand* X *combines with the substrate* If the ligand X combines with substrate A to form a complex AX that cannot bind to the enzyme, the model for the reaction bears a formal resemblance to that given in Equation 58 above for an inhibitor that binds to free enzyme:

$$\mathrm{E} + \mathrm{A} \underset{k_{-1}}{\overset{k_1}{\rightleftharpoons}} \mathrm{EA} \overset{k_2}{\rightarrow} \mathrm{E} + \mathrm{P}$$
$$+$$
$$\mathrm{X}$$
$$k_3 \downarrow \uparrow k_{-3}$$
$$\mathrm{E} + \mathrm{AX}$$

The rate equation is identical with Equation 61 as would be expected, and the ligand X functions as a purely competitive inhibitor. Conversely it is possible that the ligand X must combine with the substrate before the latter can bind to the enzyme:

$$\mathrm{E} + \mathrm{A}$$
$$+$$
$$\mathrm{X}$$
$$k_3 \downarrow \uparrow k_{-3}$$
$$\mathrm{E} + \mathrm{AX} \underset{k_{-5}}{\overset{k_5}{\rightleftharpoons}} \mathrm{EAX} \overset{k_6}{\rightarrow} \mathrm{E} + \mathrm{P} + \mathrm{X}$$

Here the model formally resembles that given above for an activator that combines with the free enzyme and the rate equation is identical with equation 63. X is clearly an activator.

2. Partially competitive effects

The two simple situations considered in Section 1(a) above are extreme cases in the sense that the enzyme is unable to bind substrate at all if, in the first case, the inhibitor is bound or, in the second case, the activator is not also bound. The possibility should be considered that the ligand X *affects* the binding of substrate by the enzyme, but not absolutely, as in Sections 1(a) and 1(b) above. Now the kinetic model is not as simple. In fact the only simplification of Equation 57 is that k_2 and k_6 are now identical since the binding of X does *not* affect the catalysis, only the binding of substrate:

$$\mathrm{E} + \mathrm{A} \underset{k_{-1}}{\overset{k_1}{\rightleftharpoons}} \mathrm{EA} \overset{k_2}{\rightarrow} \mathrm{E} + \mathrm{P}$$
$$+ \qquad\qquad +$$
$$\mathrm{X} \qquad\qquad \mathrm{X}$$
$$k_3 \downarrow \uparrow k_{-3} \qquad k_4 \downarrow \uparrow k_{-4} \quad k_2$$
$$\mathrm{EX} + \mathrm{A} \underset{k_{-5}}{\overset{k_5}{\rightleftharpoons}} \mathrm{EXA} \overset{k_2}{\rightarrow} \mathrm{EX} + \mathrm{P}$$

Steady state treatment of this model once again leads to a complex rate equation that does not give rise to linear Lineweaver–Burk plots except in special cases. One such special case is when the rapid equilibrium assumption applies, that is to say when the breakdown of EA and EXA to E and EX and product is so slow that it does not effectively perturb an equilibrium set up between E, EA, EX, EXA, A and X. In that case the individual rate constants can be replaced by dissociation constants:

$$K_1 = \frac{[E][A]_0}{[EA]}$$

$$
\begin{array}{ccccc}
E + A & \underset{}{\overset{K_1}{\rightleftharpoons}} & EA & \overset{k_2}{\rightarrow} & E + P \\
+ & & + & & \\
X & & X & & \\
K_3 \updownarrow & & \updownarrow K_4 & & \\
EX + A & \overset{K_5}{\rightleftharpoons} & EXA & \overset{k_2}{\rightarrow} & EX + P
\end{array}
$$

$$K_3 = \frac{[E][X]}{[EX]}$$

where

$$K_4 = \frac{[EA][X]}{[EXA]}$$

$$K_5 = \frac{[EX][A]_0}{[EXA]}$$

The rate equation can be written as:

$$\frac{v_0}{e_0} = \frac{k_2([EA] + [EXA])}{[E] + [EA] + [EX] + [EXA]} = \frac{k_2\left(1 + \dfrac{[EXA]}{[EA]}\right)}{\dfrac{[E]}{[EA]} + 1 + \dfrac{[EX]}{[EA]} + \dfrac{[EXA]}{[EA]}} \tag{64}$$

The ratios [EXA]/[EA] etc. are found from the definitions of K_1–K_5 above, to be as follows:

$$\frac{[EXA]}{[EA]} = \frac{[X]}{K_4}; \qquad \frac{[E]}{[EA]} = \frac{K_1}{[A]_0}; \qquad \frac{[EX]}{[EA]} = \frac{[EX]}{[E]}\frac{[E]}{[EA]} = \frac{[X]}{K_3}\frac{K_1}{[A]_0};$$

from which Equation 64 becomes:

$$\frac{v_0}{e_0} = \frac{k_2\left(1 + \dfrac{[X]}{K_4}\right)}{1 + \dfrac{K_1}{[A]_0} + \dfrac{[X]}{K_4} + \dfrac{K_1[X]}{K_3[A]_0}} = \frac{k_2\left(1 + \dfrac{[X]}{K_4}\right)}{\left(1 + \dfrac{[X]}{K_4}\right) + \dfrac{K_1}{[A]_0}\left(1 + \dfrac{[X]}{K_3}\right)}$$

$$\therefore \quad \frac{v_0}{e_0} = \frac{k_2}{1 + \dfrac{K_1}{[A]_0}\left\{\left(1 + \dfrac{[X]}{K_3}\right)\middle/\left(1 + \dfrac{[X]}{K_4}\right)\right\}}$$

$$\therefore \quad v_0 = \frac{V_m}{1 + \dfrac{K_1}{[A]_0}\left\{\left(1 + \dfrac{[X]}{K_3}\right)\middle/\left(1 + \dfrac{[X]}{K_4}\right)\right\}} \qquad \text{where } V_m = k_2 e_0 \tag{65}$$

Taking reciprocals:

$$\frac{1}{v_0} = \frac{1}{V_m} + \frac{K_1}{V_m}\left\{\frac{\left(1 + \dfrac{[X]}{K_3}\right)}{\left(1 + \dfrac{[X]}{K_4}\right)}\right\}\frac{1}{[A]_0} \tag{66}$$

From Equations 65 and 66 it can be seen that when $[X] = 0$, the equation reduces to the simple Michaelis equation with $K_m = K_1$. When $[X]$ is finite the slope of the Lineweaver–Burk plot is altered and will be increased if $K_3 < K_4$ or decreased if $K_3 > K_4$. These two conditions thus define whether X is an inhibitor ($K_3 < K_4$) or an activator ($K_3 > K_4$). This seems to be unexpected until one realizes that because of the thermodynamic restraints upon the system, the Haldane relationship $K_1 K_4 = K_3 K_5$ exists. If $K_3 < K_4$ it follows that $K_5 > K_1$. The presence of X in this case pulls the enzyme into a form (EX) that has a lower affinity for substrate because the dissociation constant of EXA (K_5) is greater than that of EA (K_1). Conversely if $K_3 > K_4$, then $K_5 < K_1$ and the free enzyme has a lower affinity for substrate than EX, so that X is an activator. As expected, whatever the value of $[X]$, and whether X is an activator or an inhibitor, the maximum rate (i.e. when $[A]_0 = \infty$) is V_m. The effect of X is therefore competitive in that it does not affect V_m and only affects K_m. This case, however, is described as the *partially competitive* case, because when $[X]$ is increased indefinitely the slope of the Lineweaver–Burk plot increases or decreases to a limiting value. The slope is given by Equation 66 as

$$\frac{K_1}{V_m} \left\{ \frac{\left(1 + \dfrac{[X]}{K_3}\right)}{\left(1 + \dfrac{[X]}{K_4}\right)} \right\}$$

When $[X]$ approaches infinity this expression approaches

$$\frac{K_1}{V_m} \left\{ \frac{\left(\dfrac{[X]}{K_3}\right)}{\dfrac{[X]}{K_4}} \right\}$$

which is equal to $(K_1/V_m)(K_4/K_3)$. Because of the Haldane relationship this is equal to K_5/V_m. This contrasts with the fully competitive cases described in Section 1 above where the slope of the Lineweaver–Burk plot approaches infinity for the inhibitor, or K_m/V_m for the activator, when $[X]$ is increased indefinitely.

The intercept, I, of the Lineweaver–Burk plot on the $1/[A]_0$ axis is given by:

$$I = -\frac{\left(1 + \dfrac{[X]}{K_4}\right)}{\left(1 + \dfrac{[X]}{K_3}\right)} \cdot \frac{1}{K_1} = -\frac{K_3(K_4 + [X])}{K_1 K_4(K_3 + [X])} \tag{67}$$

When $[X] = 0$ this reduces to $I = -1/K_1$ and allows the accurate determination of K_1. When $[X] = \infty$, $I = -K_3/K_1 K_4$ which is equal to $-1/K_5$ (Haldane relationship). Thus by extrapolating the value of the intercept I to infinite concentration of X an *estimate* of K_5 can be obtained.

Unfortunately the simple approach of the Dixon plot which allows one to determine the inhibitor or activator constant K_3 in the purely competitive case, does not work in the partially competitive case. For one thing there are now *two* constants K_3 and K_4, and for another the plot of $1/v_0$ against [X] is not linear, as inspection of Equation 66 shows.

Non-competitive effects

In the non-competitive case, the effect of the ligand X is to alter V_m without affecting K_m. This type of behaviour might be expected if the binding of X to the enzyme did not affect its ability to bind substrate, but altered the catalytic efficiency of the active site. Even this simplification of the general model (Equation 57) does not lead to linear Lineweaver–Burk plots unless, in addition, the rapid equilibrium assumption is made, so that Equation 57 becomes:

$$
\begin{array}{ccccc}
\mathrm{E} + \mathrm{A} & \overset{K_1}{\rightleftharpoons} & \mathrm{EA} & \overset{k_2}{\rightarrow} & \mathrm{E} + \mathrm{P} \\
+ & & + & & \\
\mathrm{X} & & \mathrm{X} & & \\
K_3 \updownarrow & & \updownarrow K_3 & & \\
\mathrm{EX} + \mathrm{A} & \overset{K_1}{\rightleftharpoons} & \mathrm{EXA} & \overset{k_6}{\rightarrow} & \mathrm{EX} + \mathrm{P}
\end{array}
$$

The Haldane relationship requires that if the binding of X does not affect the binding of A, neither does the binding of A affect the binding of X. In other words only the two dissociation constants K_1 and K_3 are needed. Proceeding as before we find that the rate equation therefore is:

$$
\begin{aligned}
v_0 &= \frac{k_2 e_0}{\left(1 + \dfrac{K_1}{[\mathrm{A}]_0}\right)\left\{\left(1 + \dfrac{[\mathrm{X}]}{K_3}\right) \middle/ \left(1 + \dfrac{k_6[\mathrm{X}]}{k_2 K_3}\right)\right\}} \\
&= \frac{k_6 e_0}{\left(1 + \dfrac{K_1}{[\mathrm{A}]_0}\right)\left\{\left(1 + \dfrac{[\mathrm{X}]}{K_3}\right) \middle/ \left(\dfrac{k_2}{k_6} + \dfrac{[\mathrm{X}]}{K_3}\right)\right\}}
\end{aligned}
\tag{68}
$$

The two alternative arrangements of Equation 68 allow one to express the rate in terms of $V_m = k_2 e_0$, the maximum rate in the absence of X, or alternatively in terms of $V'_m = k_6 e_0$ the maximum rate in the presence of an excess of X.

Two extreme cases of Equation 68 arise when either k_2 or $k_6 = 0$. If $k_6 = 0$, X must be an inhibitor, since the enzyme is inactive when it is bound, and Equation 68 reduces to:

$$
v_0 = \frac{V_m}{\left(1 + \dfrac{K_1}{[\mathrm{A}]_0}\right)\left(1 + \dfrac{[\mathrm{X}]}{K_3}\right)}
\tag{69}
$$

This equation is similar to Equation 61 for a purely competitive inhibitor except that the bracket $(1 + [\mathrm{X}]/K_3)$ affects V_m instead of K_m (Figure 47).

In the other extreme case $k_2 = 0$ and X is an activator because the enzyme

156

is only active when it is bound. Equation 68 reduces in this case to:

$$v_0 = \frac{V'_m}{\left(1 + \dfrac{K_1}{[A]_0}\right)\left(1 + \dfrac{K_3}{[X]}\right)} \tag{70}$$

This equation resembles that of the purely competitive activator (Equation 63) except that again the bracket $(1 + K_3/[X])$ affects V_m instead of K_m. Both these special cases (k_2 or $k_6 = 0$) are termed 'purely non-competitive'. As with the purely competitive cases the slope of the Lineweaver–Burk plot varies between K_1/V_m and ∞, depending upon the concentration of X.

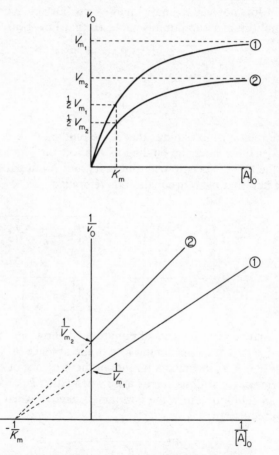

Figure 47 The effect of a *non-competitive inhibitor* on the kinetics of an enzyme. The upper graph shows a Michaelis plot and the lower one a Lineweaver–Burk plot. In both cases (1) and (2) represent the plot obtained in the absence and presence of the inhibitor X respectively

Inversion of Equations 69 and 70 leads to:

$$\frac{1}{v_0} = \frac{\left(1 + \dfrac{[X]}{K_3}\right)}{V_m}\left(1 + \frac{K_1}{[A]_0}\right) \tag{71}$$

or

$$\frac{1}{v_0} = \frac{\left(1 + \dfrac{K_3}{[X]}\right)}{V_m}\left(1 + \frac{K_1}{[A]_0}\right) \tag{72}$$

so that the slope of the Lineweaver–Burk plot is given by $K_1(1 + [X]/K_3)/V_m$ or $K_1(1 + K_3/[X])/V_m$, for an inhibitor or activator respectively. The intercept on the $1/[A]_0$ axis is found by setting $1/v_0 = 0$. In either case (Equation 71 or 72) this leads to $1/[A]_0 = -1/K_1$ and allows one to determine K_1. The intercept on the $1/v_0$ axis is found by setting $1/[A]_0 = 0$ and for the inhibitor (Equation 71) is given by $1/v_0 = (1 + [X]/K_3)/V_m$. When $[X] = 0$ this gives $1/V_m$ (Figure 47). The value of K_3 can be determined from this vertical intercept at finite values of $[X]$ once V_m is known. Alternatively, the Dixon plot may be used to determine K_3 graphically, since when $[A]_0$ is constant a plot of $1/v_0$ against $[X]$ will be linear for an inhibitor, and a plot of $1/v_0$ against $1/[X]$ will be linear for an activator. In the case of an inhibitor, Equation 71 shows that the point of intersection of two such plots at two values of $[A]_0 = [A]_1$ and $[A]_2$, is given by:

$$\frac{\left(1 + \dfrac{[X]}{K_3}\right)}{V_m}\left(1 + \frac{K_1}{[A]_1}\right) = \frac{\left(1 + \dfrac{[X]}{K_3}\right)}{V_m}\left(1 + \frac{K_1}{[A]_2}\right)$$

$$\therefore \quad \left(1 + \frac{[X]}{K_3}\right) = 0 \quad \text{and} \quad [X] = -K_3$$

In contrast to the case of a purely competitive inhibitor (Figure 46) in this case the two lines intersect on the horizontal axis because, when $[X] = -K_3$, Equation 71 shows that $1/v_0 = 0$ whatever the value of $[A]_0$, as shown in Figure 48.

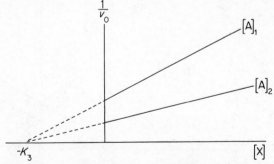

Figure 48 A Dixon plot for the determination of the inhibitor constant K_3 for a purely non-competitive inhibitor X. $[A]_1$ and $[A]_2$ are two constant concentrations of the substrate A in two series of experiments in which $1/v_0$ is plotted against increasing values of $[X]$, the concentration of inhibitor

158

In intermediate situations where neither k_2 nor k_6 is equal to zero, X will be an activator if $k_2/k_6 < 1$ and an inhibitor if $k_2/k_6 > 1$. These cases are 'partially non-competitive' and the slope of the Lineweaver–Burk plot never reaches infinity at any concentration of X. A plot of $1/v_0$ against either [X] or $1/[X]$ at constant $[A]_0$ will be non-linear and, as in the partially competitive cases, difficulty will be experienced in determining the constant K_3.

Mixed effects

A number of possibilities may give rise to mixed competitive and non-competitive effects of the type shown in Figure 49(a), where both V_m and K_m are affected by the ligand X. It is worth noting, however, that a mixture of fully competitive cases with non-competitive effects is not possible because in the fully competitive situations X combines with only the free enzyme, whereas in all non-competitive situations X must combine with both the free enzyme and the enzyme–substrate complex. Mixed effects can only arise from a combination of the partially competitive case (Section 2 above) with one of the non-competitive cases.

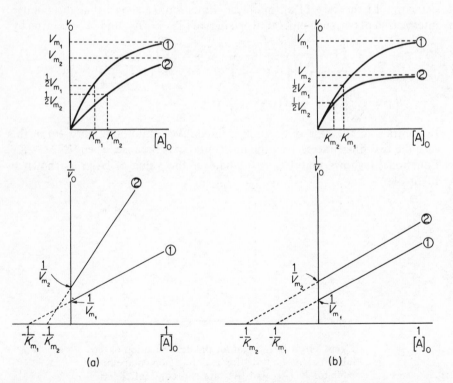

Figure 49 Mixed inhibition of an enzyme by an inhibitor X. The upper graphs show the Michaelis plots and the lower ones show Lineweaver–Burk plots for general *mixed* inhibition (a) and the special case of *uncompetitive* inhibition (b)

The uncompetitive kinetics shown in Figure 49(b) arise only rarely, but can be produced if X binds only to the enzyme–substrate complex or alternatively if X binds in a non-competitive manner to both forms of an enzyme for which the catalytic constants k_2 and k_6 are very much larger than k_{-1}:

$$
\begin{array}{ccccc}
\text{E} + \text{A} & \underset{k_{-1}}{\overset{k_1}{\rightleftharpoons}} & \text{EA} & \overset{k_2}{\rightarrow} & \text{E} + \text{P} \\
+ & & + & & \\
\text{X} & & \text{X} & & \\
k_3 \downarrow \uparrow k_{-3} & & k_3 \downarrow \uparrow k_{-3} & & \\
\text{EX} + \text{A} & \underset{k_{-1}}{\overset{k_1}{\rightleftharpoons}} & \text{EXA} & \overset{k_6}{\rightarrow} & \text{EX} + \text{P}
\end{array}
$$

In this case the steady state rate equation for the reaction in the absence of X is given by

$$
v_0 = \frac{V_m}{1 + \dfrac{K_m}{[\text{A}]_0}} \quad \text{where } K_m = \frac{k_2 + k_{-1}}{k_1} \quad \text{and} \quad V_m = k_2 e_0
$$

If $k_2 \gg k_{-1}$, $K_m \approx k_2/k_1$. The effect of X on k_2 is therefore bound to affect K_m in the same way as V_m, although it has no effect on K_1. In fact the slope of the Lineweaver–Burk plot, which is given by K_m/V_m will remain unaffected since it approximates to e_0/k_1.

The effect of pH on enzymic reaction rates

When the initial velocity v_0 is measured at a series of different pH s, all other conditions such as substrate concentration, temperature and salt concentration being held constant, and an activity/pH curve is plotted, the usual result is the classical bell-shaped curve shown in Figure 50(a). Sometimes either the ascending or descending limb of the curve may be missing in which case the result resembles Figure 50(b) or (c). These curves bear a strong similarity to titration curves and there is a temptation to assume that as the pH is increased ionizable groups on the enzyme surface are becoming de-protonated, leading to changes in the active site and thus affecting the catalytic properties of the enzyme. This assumption may be correct but there are other possible explanations. When,

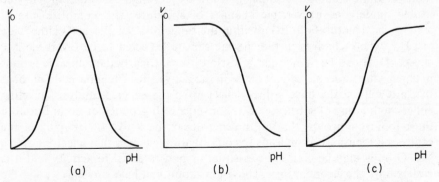

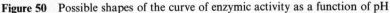

Figure 50 Possible shapes of the curve of enzymic activity as a function of pH

however, the points of inflection in the activity/pH profile are taken to be the pK_a s of groups essential to enzyme activity, the interpretation is almost certainly incorrect. The so-called 'optimum pH', the pH at which v_0 is at a maximum, also requires careful definition and cautious interpretation.

Hydrogen ions may be involved in the enzymic reaction in three different ways. They may be required or produced during the course of the overall reaction as with many dehydrogenases where the reaction catalysed may be represented as follows:

$$AH_2 + NAD^+ \rightleftharpoons A + NADH + H^+$$

In this case the hydrogen ion could be treated in theory as a second substrate or product in just the same way as any other reactant. In addition, in practice, the substrate or product may itself ionize with the uptake or release of protons as the pH is changed. If the ionized substrate is a better or worse substrate than the unionized form, this will clearly affect the observed catalytic rate, but can usually be taken account of by measurements made in the absence of enzyme.

The third way in which protons may affect an enzymic reaction is by their combination with or release from the enzyme itself. Normally there will be many ionizable groups at or near the surface of the enzyme molecule and increasing the pH will lead to the de-protonation of these groups. The pK_a of an ionizable group is the pH at which its de-protonation is half complete. In the range from one pH unit below to one pH unit above the pK_a, the group is in the process of losing its proton. Outside that range, to all intents and purposes, the group is either entirely protonated or entirely deprotonated, so that changes in pH have no effect upon its state of ionization. Although the pK_a of a group such as the carboxyl group, $-COOH$, is normally found to lie within a fairly narrow range of values when the group is part of a small freely soluble molecule, the pK_a may be greatly affected by its location if the group is situated in a protein molecule. Such environmental effects make it difficult to predict the pK_a s of ionizable groups in proteins. Furthermore, if the conformation of a protein changes for any reason, the pK_a s of the groups in the protein may alter.

The effects of changing pH upon enzyme structure, and hence activity, are therefore quite complex. At either very low or very high pH values the enzyme usually undergoes irreversible changes of structure with complete loss of activity, or denaturation. Even within the range of a few pH units either side of pH 7, where changes in the enzyme can be expected to be reversible, the effects of pH can be manifested in various ways. Clearly the ionizable groups in an enzyme molecule cannot all be in the active site. The ionic state of those that are will usually have a direct effect upon the enzyme catalysis, affecting either the binding of substrate or its conversion to product, or both. But even those ionizable groups that are nowhere near the active site may affect the catalysis if their ionization leads to conformation changes that affect the active site. Groups may be classed as essential or non-essential to activity and it is clearly only the essential ones whose ionization will have any effect upon the measured velocity of reaction.

In addition to pH-dependent ionization of the free enzyme, one must bear in mind the possibility of ionization of one of the enzyme-containing intermediates, such as the enzyme–substrate complex, or even ionization of the enzyme-bound transition state. The pK_a s of the ionizing groups may not be the same in the enzyme-containing intermediate as in the free enzyme because of the influence of the substrate.

Most theoretical treatments of the effect of pH on enzymic reaction velocity assume that there are at least two essential ionizable groups and that their ionic state may affect any or all of the rate constants involved in the kinetic mechanism. One group is assumed to be protonated and the other unprotonated in the most efficient form of the active site. In practice the proton is treated as a ligand that has multiple effects. The rate equation can be shown to reduce to the form:

$$v_0 = \frac{\tilde{V}_m[A]_0}{\tilde{K}_m + [A]_0}$$

where both $\tilde{V}_m$ and $\tilde{K}_m$ are pH dependent 'constants'. Information about the pK_a s of the groups in the active site can be obtained by measuring $\tilde{V}_m$ and $\tilde{K}_m$ at different pH s and plotting $\log \tilde{V}_m$, $\log \tilde{K}_m$, and $\log(\tilde{V}_m/\tilde{K}_m)$ against pH. The result is a set of graphs each made up of a series of straight lines of slope either zero or $+1$ or -1, with curved portions connecting them, as shown in Figure 51. Adjacent straight line portions intersect, as shown in the figure, at the pK_a values of groups involved in the activity.

For a more detailed review of the theory underlying pH effects the reader is referred to Chapter 5 of the book by Laidler and Bunting.[27] In passing it is worth noting that the so-called 'optimum' pH of an enzyme needs very careful definition if it is to mean anything. The pH at which the initial rate reaches a maximum will depend upon whether the substrate concentration is sufficient to saturate the enzyme, as well as on the temperature and buffer composition of the reaction mixture.

The effects of temperature on enzymic reaction rates

As with pH, so the reaction rate of an enzyme-catalysed reaction varies in a complex way with temperature. In general a plot of initial rate as a function of temperature first increases and then decreases, passing through a maximum which is sometimes referred to as the 'optimum temperature'. The reason for this behaviour is not far to seek. All chemical reactions tend to increase in rate with increase in temperature. A measure of the temperature sensitivity of a reaction is given by its Q_{10}, which is defined as the ratio of the reaction rate at a given temperature to the rate at a temperature $10\,°C$ lower. Most reactions have Q_{10} s of about 2, although Q_{10} for a given reaction can itself vary with temperature and with the conditions under which reaction occurs. What this means is that reaction rates roughly double for every increase in temperature of $10\,°C$. Enzymically catalysed reactions are no exception to this general rule-of-thumb, which explains the observed increase in initial rate as temperature is

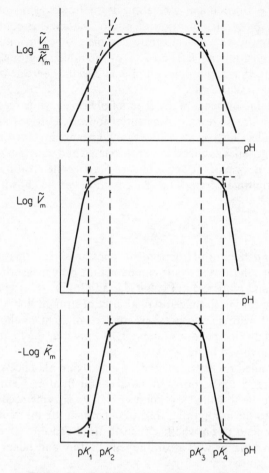

Figure 51 Typical plots of the logarithms of kinetic constants ($\tilde{V}_m$, $\tilde{K}_m$ and $\tilde{V}_m/\tilde{K}_m$) as a function of pH. The points of intersection of the linear portions give the pK_A values of groups involved in the enzymic mechanism (pK_1, pK_2, pK_3, pK_4)

increased from about $0\,°C$ to around $30\text{–}40\,°C$. After this, however, further increase in temperature is usually found to lead, eventually, to a fall in v_0. This is because enzymes become denatured at high temperatures and the Q_{10} for the denaturation process can be much larger than 2 at temperatures above normal ambient values. Two processes are always competing when one measures initial rates in an enzymic reaction: the catalytic reaction itself, and the denaturation of the enzyme. If reaction is initiated by adding the enzyme to the buffered solution of substrate, and if the enzyme is kept at a low temperature before adding it to the substrate, these two processes commence from the moment of mixing. At low reaction temperatures the denaturation is usually negligible, (although there *are* enzymes which are 'cold-labile') and one observes only the catalytic reaction. As the reaction temperature increases, the competing effect of denaturation becomes significant until it removes a measurable portion of the enzyme activity within the time necessary to establish the 'initial' velocity. Of course, if v_0 were truly the *initial* velocity and could be measured instantaneously with the addition of enzyme, the denaturation effect could be avoided. This is quite impracticable because a finite mixing time is required, a few milliseconds are always required to establish the steady state, and a finite recording time is needed in order to draw a tangent and determine the initial rate. For all these reasons, the effect of denaturation at higher temperatures is highly dependent upon the technique used to measure v_0. As a consequence the so-called 'optimum' temperature of reaction will also depend heavily upon the method used to assay the enzyme, as well as upon other conditions such as pH and substrate concentration.

In the living cell the same kind of considerations apply, except of course that the enzyme is working continuously (initial rates are of only passing interest to the cell!) and denaturation is occurring all the time. The 'optimum' temperature for the cell to operate at then becomes much more a question of how to get the fastest catalytic rate without having to expend too much free energy in re-synthesizing enzymes that have become denatured. Any resemblance between the 'optimum' temperatures observed *in vitro* and *in vivo* will be largely fortuitous.

Useful information may still be derived from studies on the temperature dependence of enzymic reaction rates, provided one works in the low-temperature region where denaturation is insignificant. In that case the dependence of v_0 upon T the absolute temperature, can be used to derive information about the process of activation occurring during passage through the enzyme-bound transition state, provided that the individual steps of the catalytic reaction can be properly distinguished. Each rate constant in the overall enzymic reaction will be temperature dependent (see Chapter 4) and the net effect of temperature upon v_0 may be quite complex.

The effects of ionic strength and dielectric constant on enzymic reaction rates

The ionic strength, μ, and dielectric constant, ε_r, of a solution may affect the conformational state of a protein dissolved in the solution, as discussed in previous chapters. These effects are complex and usually unpredictable in terms of the effect they will have upon the activity of an enzyme. Even where no changes in conformation are produced by changes in μ and ε_r, there may be effects upon the enzymic activity due to alterations in the pK_as of essential groups, or to changes in ionization or reactivity of the bound substrate. For all these reasons it is important that the ionic strength and dielectric constant should be maintained at constant values when studying enzyme kinetics. It should be remembered that charged substrates may make an important contribution to the ionic strength if buffers of low ionic strength are used. When attempting to study the role of an enzyme in the living cell it should be borne in mind that the values of these two parameters may be different in the cellular environment from their values in a dilute aqueous buffer such as is usually used for *in vitro* studies on enzymes.

KINETIC MODELS

It will be clear from what has gone before that the study of steady state enzyme kinetics falls into two parts. On the one hand there is the experimental measurement of initial reaction rates at different concentrations of substrates, products and inhibitors or activators, and the manipulation of this experimental data by various methods of plotting. On the other hand one imagines the various ways in which the enzyme may combine with the ligands that affect its rate, and one constructs kinetic models based upon these ideas. Mathematical analysis of the models then leads to predictions about how the rate of reaction will depend upon ligand concentrations. Finally the two parts come together when the observed experimental results are compared with the behaviour predicted by the various kinetic models. If the observations do not agree with the predictions, the kinetic model on which the predictions were based is clearly inadequate. If, however, agreement is found between experiment and hypothesis one must be quite clear that in this, as in all branches of science, the hypothesis is not thereby proved to be correct. That would only be the case if *every possible* kinetic model had been devised and all except one had failed to agree with experiment. Unfortunately it often turns out that several kinetic models yield the same rate equations. In other cases the differences between the predicted behaviour for two different models may be so small that the accuracy of the experimental data does not allow one to choose between them on the basis of the observed steady state kinetics. In that case alternative experimental approaches such as the study of pre-steady state kinetics must be adopted in order to try to distinguish which model is correct. Nevertheless steady state kinetic studies serve two useful purposes. They establish the actual response of the enzymic rate to changes in ligand concentrations and changes in pH, temperature and ionic strength. In addition they may allow one to *eliminate* certain otherwise plausible kinetic models and thus help to elucidate the framework in which the molecular mechanism of catalysis occurs at the enzyme active site.

The structure of rate equations

Whatever kinetic model is being considered, the equation relating the initial rate of reaction to the concentrations of reactants, products, inhibitors and activators, that is to say the rate equation, always has the same general form.

Because the rate v_0 is always directly proportional to the total enzyme concentration, e_0, it is highly convenient to express the rate as a *specific rate*, that is the rate per unit enzyme concentration, or v_0/e_0. When this is done the rate equation is always found to be a ratio:

$$\frac{v_0}{e_0} = \frac{\text{numerator}}{\text{denominator}}$$

As was shown earlier for the simple cases of Equations 12 and 49, the numerator is a mathematical expression relating to the *flows* of substrate and product, whereas the denominator represents the *distribution* of the total enzyme amongst different liganded forms. In what follows it may be helpful to bear this generalization in mind.

TWO-SUBSTRATE ENZYMES

Up till now all the discussion of enzyme kinetics has assumed that there is only one substrate and one product in an enzymic reaction. Although this simplifies the situation and allows a clearer understanding of the fundamental behaviour of enzymes, nevertheless in real life most enzymes catalyse reactions between two (or more) substrates to produce two (or more) products. In many cases the reaction may be treated as if only one substrate were involved if the other is always present at a saturating concentration. Hydrolytic reactions fall into this category because the second substrate is the solvent, water. In others there is no escaping the fact that two substrates are involved. Most of these reactions can be considered as group-transfer reactions in which something is transferred from one substrate to the other:

$$GX + Y \rightleftharpoons X + GY \tag{73}$$

The thing transferred is denoted by G; it may be a chemical group such as an amino, methyl or hydroxyl group, or it may in some cases be an electron or a pair of electrons, or an ion, for example a proton or hydride ion. For convenience all two-substrate reactions will be treated as if they can be represented by Equation 73, but the substrate GX will be denoted as A, and Y as B. Similarly the product X will be denoted P, and GY as Q. Thus substrate A and product Q are the ones carrying the group that is transferred.

Random or branched mechanisms

As soon as one attempts to devise kinetic models for two-substrate enzymes the problem arises as to which substrate binds first to the enzyme. The possibility that both bind simultaneously can be dismissed as exceedingly unlikely on purely statistical grounds because this would involve a three-body collision in the reaction mixture. Clearly until there is evidence to the contrary, it must be assumed that either substrate can bind first, which leads to the branched or random general mechanism for two-substrate reactions given below. The terms 'branched' and 'random' are synonymous, implying, in the first case, that the mechanism has alternative routes from substrate to product and, in the second case, that the order of binding of substrates is not fixed.

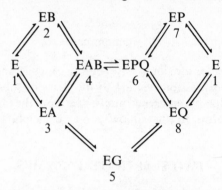

$$(74)$$

If B binds first, A must follow to form the ternary complex EAB which then undergoes conversion to another ternary complex EPQ with the release of products in either order. Alternatively, if A binds first the ternary complex EAB may be formed by the arrival of B on the enzyme surface, but another possibility is that the group G may be retained by the enzyme while the product P departs, followed by the binding of B to form Q on the enzyme. This possibility is represented by the lowest pathway in Equation 74, in which no ternary complex is formed but a 'modified enzyme' EG occurs.

The possibility of alternative pathways of reaction is clearly shown by Equation 74 and is an ever-present consideration in dealing with two-substrate enzymes. Nevertheless, in particular enzymic reactions the actual mechanism may be much simpler. If, for example, the substrate B *must* bind first and the product P *must* leave last, the model can have no alternative pathways and reduces to a much simpler one:

$$E \rightleftharpoons EB \rightleftharpoons EAB \rightleftharpoons EPQ \rightleftharpoons EP \rightleftharpoons E$$

The general mechanism (Equation 74) gives rise to a complex rate equation involving 22 individual rate constants, one for each step, and in which the concentrations of the substrates appear raised to powers higher than one. Even if products are absent, and one of the substrates (A) is kept at a constant concentration, the dependence of the initial rate on the concentration of B is given by:

$$\frac{v_0}{e_0} = \frac{c_1[B]_0^3 + c_2[B]_0^2 + c_3[B]_0}{c_4[B]_0^3 + c_5[B]_0^2 + c_6[B]_0 + c_7} \tag{75}$$

where the symbols c_1–c_7 are constants made up of rate constants and the concentration of A, and where $[B]_0$ is the concentration of B.

The mathematical derivation of Equation 75 using the steady state assumption involves writing down the steady state equations for the eight enzyme-containing intermediates E, EA, EB etc., together with the conservation equation, $e_0 = [E] + [EA] + [EB] + $ etc., and solving the resulting simultaneous equations in order to eliminate the unknown concentrations [E], [EA],

[EB], etc., just as for the simpler models considered previously for one-substrate enzymes. The labour may be reduced by using determinants, but is still considerable and will not be attempted here.

Fortunately it is possible to avoid most of this labour by employing a routine devised by King and Altman[32] and developed by Wong and Hanes.[33] This approach allows one to take any kinetic model, no matter how many substrates or products are involved, and to write down the rate equation by inspection, using a set of empirical rules based on the algebraic and matrix treatment of the steady state simultaneous equations.

The kappa notation

Each single step in a complex kinetic model has an associated rate constant, and the rate at which the step proceeds is given by the product of the rate constant and the concentrations of the reactants involved in it. For example in Equation 74 the top left-hand step from E to EB may be represented as:

$$\underset{1}{E} + B \xrightarrow{k_1} \underset{2}{EB}; \qquad v = k_1[E][B]$$

For convenience the step itself may be represented by a symbolic kappa (κ) which is equal to the product of the rate constant and the concentration of a substrate, if one is involved in the step. For this example, as the step is the one between the enzyme-containing intermediates labelled 1 and 2, the kappa would be $\kappa_{12} = k_1[B]$. The order of the subscripts indicates the direction of the step. For the reverse step,

$$\underset{1}{E} + B \underset{k_{-1}}{\leftharpoonup} \underset{2}{EB},$$

the kappa would be $\kappa_{21} = k_{-1}$. In this case the kappa is simply equal to the rate constant because no substrate is involved as a reactant in the step. The reaction rate of any step is thus the kappa for that step multiplied by the concentration of the enzyme-containing intermediate undergoing reaction. This device is convenient because it separates the unknown concentrations of enzyme-containing intermediates, from the known substrate concentrations.

Sequences of steps that form part of a kinetic model can be represented schematically by sets of arrows, such as the following, from Equation 74:

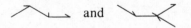

The same sequences can be represented mathematically by products of kappas; for the two schematic examples given above these kappa-products would be $\kappa_{12}\kappa_{24}\kappa_{46}$ and $\kappa_{24}\kappa_{46}\kappa_{86}\kappa_{67}$.

Two distinct classes of sequences or kappa-products need to be considered. Those sequences that lead from free enzyme via substrate binding, conversion to, and release of product and back to free enzyme are called *cyclic* because they represent possible cycles of enzyme catalysis. Examples of cyclic kappa-products in Equation 74 are $\kappa_{12}\kappa_{24}\kappa_{46}\kappa_{67}\kappa_{71}$ and $\kappa_{12}\kappa_{24}\kappa_{46}\kappa_{68}\kappa_{81}$. All other kappa-products are called *non-cyclic*.

The structural rules of Wong and Hanes[33]

The empirical rules devised by King and Altman and formalized by Wong and Hanes are as follows:

Rule 1. The specific rate of reaction (i.e. v_0/e_0) is given by the ratio of two sums of kappa products. In a mechanism involving n enzyme-containing intermediates, all terms in the numerator are kappa-products of n kappas, and all terms in the denominator are kappa-products of $(n - 1)$ kappas.
Rule 2. All kappa-products in the numerator are cyclic, but no cycle made up of directly opposing kappas is present. All kappa-products in the denominator are non-cyclic.
Rule 3. No kappa-product in either numerator or denominator contains more than one kappa originating on any one enzyme-containing intermediate.
Rule 4. (a) All terms in the denominator are positive but there are an equal number of positive and negative terms in the numerator.
(b) Each positive numerator term contains the concentration of each substrate, as well as a forward rate constant from each obligatory intermediate in the mechanism.
(c) Each negative term in the numerator contains the concentration of each product as well as a reverse rate constant originating on each obligatory intermediate.

These rules and their application to actual kinetic models are best illustrated with reference to a simple branched mechanism. For this purpose consider the model proposed by Ingraham and Makower[34] and given in Equation 76 for the two substrate reaction $A + B \rightarrow$ products. This is a simplification of the general mechanism (Equation 74):

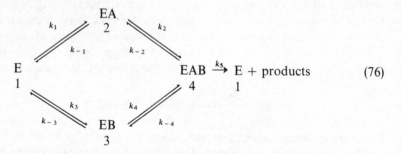

$$(76)$$

The model is as simple as it can be for a mechanism with alternative pathways for the addition of substrates to the enzyme; and it contains no pathway from products back to substrates. In the latter respect it is thermodynamically suspect, but provided one only considers the initial rate in the absence of products it is a viable model. There are four enzyme-containing intermediates in the model so *Rule 1* tells one that the rate equation will be of the form:

$$\frac{v_0}{e_0} = \frac{\kappa_{cd}\kappa_{fg}\kappa_{hj}\kappa_{kc} + \kappa_{cu}\kappa_{wx}\kappa_{xy}\kappa_{zc} + \cdots}{\kappa_{jk}\kappa_{lm}\kappa_{ho} + \kappa_{rs}\kappa_{tu}\kappa_{wx} + \cdots} \tag{77}$$

where c, d, f, etc. are as yet undefined. One merely expects a series of terms in the numerator, each of which is a kappa product containing four kappas, while in the denominator the terms each contain three kappas.

Rule 2 enlarges on the numerator terms. They are all to be *cyclic* kappa-products. This has been taken care of in Equation 77 as will be seen, because each term has its first kappa originating on 'c' and its last kappa terminating on 'c'. One has only to substitute '1' for 'c'. In no case does a kappa product contain directly opposing kappas such as κ_{cd} and κ_{dc}. *Rule 2* also enlarges on the denominator terms a little in that they are all to be non-cyclic as indeed they are in Equation 77. *Rule 3* goes further in describing the terms and forbids 'path-splitting'. By this it is meant that if each kappa product is drawn out schematically in the form of a sequence of steps, then divergent sequences are forbidden. Thus the kappa product represented by

$$\kappa_{12} \nearrow^{\kappa_{24}} \kappa_{41}$$
$$\searrow_{\kappa_{43}}$$

would be forbidden because both κ_{43} and κ_{41} originate on the same enzyme-containing intermediate (4). Note that *convergent* sequences are allowed, such as

$$\kappa_{12} \searrow^{\kappa_{24}} \kappa_{41}$$
$$\nearrow_{\kappa_{34}}$$

The three sections of *Rule 4* finally pin down the terms decisively. *Rule 4(a)* leads us to expect both negative and positive terms in the numerator, but reference to *Rule 4(c)* shows that the negative terms must contain product concentrations. In the model being considered, products must be absent and so the negative terms will all disappear. *Rule 4(b)* needs some amplification as it refers to *obligatory* intermediates. These are intermediates through which reaction must pass on the way from substrates to products. In Equation 76, E and EAB are clearly obligatory since all paths contain them, whereas EA and EB are not obligatory; one can get from E on the left to E on the right, without passing through EA if one follows the lower route and without passing through EB, by following the upper pathway.

With the exceptions discussed already, all possible kappa-products will appear in the rate equation. For the model shown in Equation 76, the rate equation can thus be drawn out schematically as in Equation 78. For convenience the denominator terms are grouped into those where the (non-cyclic) sequence ends on E, those ending on EA, those ending on EB and those ending

170

on EAB.

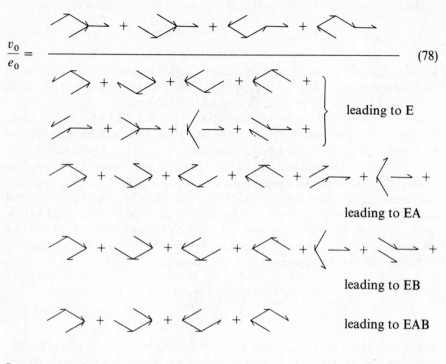

$$\frac{v_0}{e_0} = \text{(78)}$$

leading to E

leading to EA

leading to EB

leading to EAB

Sequences such as

in the first line of the denominator, appear strange until one recalls that free enzyme E occurs at both ends of the model. This particular sequence is a convergent one, starting on EA and leading to E at the left, and also starting at EB and leading to E at the right. In terms of kappas, Equation 78 therefore becomes:

$$\frac{v_0}{e_0} = \frac{\kappa_{12}\kappa_{24}\kappa_{34}\kappa_{41} + \kappa_{13}\kappa_{34}\kappa_{24}\kappa_{41} + \text{etc.}}{\kappa_{34}\kappa_{42}\kappa_{21} + \kappa_{24}\kappa_{43}\kappa_{31} + \text{etc.}} \tag{79}$$

$$= \frac{k_1 k_2 k_4 k_5 [\mathrm{A}]_0^2 [\mathrm{B}]_0 + k_3 k_4 k_2 k_5 [\mathrm{B}]_0^2 [\mathrm{A}]_0 + \text{etc.}}{k_4 k_{-2} k_{-1} [\mathrm{A}]_0 + k_2 k_{-4} k_{-3} [\mathrm{B}]_0 + \text{etc.}} \tag{80}$$

When all the terms are collected together Equation 80 becomes:

$$\frac{v_0}{e_0} = \frac{c_1 [\mathrm{A}]_0 [\mathrm{B}]_0 + c_2 [\mathrm{A}]_0^2 [\mathrm{B}]_0 + c_3 [\mathrm{A}]_0 [\mathrm{B}]_0^2}{c_7 + c_8 + c_9 [\mathrm{B}]_0 + c_{10} [\mathrm{A}]_0 [\mathrm{B}]_0 + c_{11} [\mathrm{A}]_0^2} \tag{81}$$
$$+ c_{12} [\mathrm{B}]_0^2 + c_{13} [\mathrm{A}]_0^2 [\mathrm{B}]_0 + c_{14} [\mathrm{A}]_0 [\mathrm{B}]_0^2$$

where the constants c_1–c_3 and c_7–c_{14} are as follows:

$$c_1 = k_1 k_{-3} k_2 k_5 + k_{-1} k_3 k_4 k_5$$

$$c_2 = k_1 k_2 k_4 k_5$$

$$c_3 = k_2 k_3 k_4 k_5$$

$$c_7 = k_{-1} k_{-2} k_{-3} + k_{-1} k_{-3} k_{-4} + k_{-1} k_{-3} k_5$$

$$c_8 = k_1 k_{-2} k_{-3} + k_1 k_{-3} k_{-4} + k_1 k_{-3} k_5 + k_{-1} k_{-2} k_4 + k_{-1} k_4 k_5$$

$$c_9 = k_{-1} k_2 k_3 + k_{-1} k_3 k_{-4} + k_{-1} k_3 k_5 + k_{-3} k_2 k_{-4} + k_2 k_{-3} k_5$$

$$c_{10} = k_1 k_{-3} k_2 + k_{-1} k_3 k_4 + k_1 k_2 k_{-4} + k_3 k_{-2} k_4 + k_2 k_4 k_5$$

$$c_{11} = k_1 k_{-2} k_4 + k_1 k_4 k_5$$

$$c_{12} = k_2 k_3 k_{-4} + k_2 k_3 k_5$$

$$c_{13} = k_1 k_2 k_4$$

$$c_{14} = k_2 k_3 k_4$$

If the concentration of substrate B is held constant, then Equation 81 becomes equal to:

$$\frac{v_0}{e_0} = \frac{i[A]_0^2 + j[A]_0}{l[A]_0^2 + m[A]_0 + k} \qquad \text{where} \quad i = c_2[B]_0 \tag{82}$$

$$j = c_1[B]_0 + c_3[B]_0^2$$

$$k = c_7 + c_8 + c_9[B]_0 + c_{12}[B]_0^2$$

$$l = c_{11} + c_{13}[B]_0$$

$$m = c_{10}[B]_0 + c_{14}[B]_0^2$$

This rate equation, like that of the general two-substrate model (Equations 74 and 75) is quite unlike anything seen so far for one substrate enzymes in that it contains $[A]_0^2$ and cannot be transformed to a simple linear relationship between $1/v_0$ and $1/[A]_0$, or any of the other linear relationships mentioned in Chapter 2. A plot of v_0 against $[A]_0$ for an enzyme obeying this rate equation will not be hyperbolic and may not even be hyperboloid (resembling a hyperbola). When the constants i, j, k, l, and m assume certain relative sizes, the rate curve may be sigmoid, or may pass through a maximum, or both, as shown in Figure 52.[35]

As a result, a two-substrate enzyme will only produce the hyperbolic type of Michaelis plot when certain limitations apply to its kinetic mechanism. In practice these limitations turn out to be very wide and hyperbolic rate curves are indeed found for many enzymes. On the other hand, non-hyperbolic rate curves are also found especially for those enzymes concerned with metabolic regulation. Of course there may be other fundamental causes of this behaviour as well as the possibility of alternative pathways discussed here. Another possible cause involving oligomeric enzymes will be discussed in Chapter 6.

172

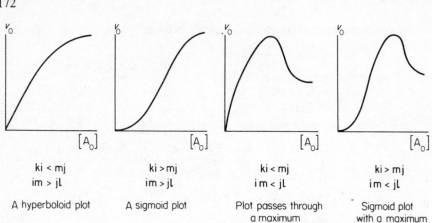

ki < mj	ki > mj	ki < mj	ki > mj
im > jl	im > jl	im < jl	im < jl
A hyperboloid plot	A sigmoid plot	Plot passes through a maximum	Sigmoid plot with a maximum

Figure 52 The various shapes of a plot of v_0 against $[A]_0$ for an enzyme obeying the rate equation

$$\frac{v_0}{e_0} = \frac{\mathbf{i}[A]_0^2 + \mathbf{j}[A]_0}{\mathbf{l}[A]_0^2 + \mathbf{m}[A]_0 + \mathbf{k}}$$

where v_0 is the initial rate at a substrate concentration $[A]_0$; $\mathbf{i}$, $\mathbf{j}$, $\mathbf{k}$, $\mathbf{l}$ and $\mathbf{m}$ are kinetic constants.[35] One kinetic model that will obey this rate equation is the Ingraham–Makower model[34] shown in Equation 76 in the text

One general situation where a two-substrate enzyme *must* give rise to a hyperbolic rate equation is when the concentration of one substrate is infinite, while the concentration of the other is finite. For example inspection of Equation 81 shows that when $[B]_0$ is infinite, only those terms containing $[B]_0^2$ will have any significance; all others will be negligible by comparison. The rate equation then simplifies to:

$$\frac{v_0}{e_0} = \frac{c_3[A]_0[B]_0^2}{c_{12}[B]_0^2 + c_{14}[A]_0[B]_0^2} = \frac{c_3[A]_0}{c_{12} + c_{14}[A]_0}$$

$$\therefore \quad \frac{v_0}{e_0} = \frac{\dfrac{c_3}{c_{14}}[A]_0}{\dfrac{c_{12}}{c_{14}} + [A]_0}$$

$$\therefore \quad v_0 = \frac{V_m[A]_0}{K_m + [A]_0} \qquad \text{where} \quad V_m = \frac{c_3}{c_{14}}e_0 = k_5 e_0$$

$$K_m = \frac{c_{12}}{c_{14}} = \frac{k_{-4} + k_5}{k_4}$$

This equation only strictly applies if $[B]_0$ is infinite, but it is a situation that is approached when the concentration of $[B]_0$ is 'saturating' and explains why hydrolytic enzymes, for example, can be treated as 'one-substrate' enzymes for kinetic purposes as mentioned earlier.

The structural rules of Wong and Hanes can be applied to any kinetic model, with any number of substrates, products and even activators or inhibitors. In fact one of their earliest applications was in deriving the steady state rate equation for the general case of a one-substrate enzyme with an inhibitor, Equation 57, mentioned earlier.

No account has been taken of the possibility of enzyme isomerization in the general mechanism or its simplification. This would be a complicating factor but it can be accommodated by including extra steps in the mechanism.

The degree of a rate equation

Whenever a substrate can combine with more than one enzyme-containing intermediate, alternative pathways for reaction exist and the steady state rate equation will contain the concentration of that substrate raised to a power greater than one. The rate equation is then said to be of *degree* greater than one with respect to that substrate. For example in the case of the Ingraham–Makower model (Equation 76) the rate equation is of second degree with respect to both A and B as shown by Equation 81, where $[A]_0^2$ and $[B]_0^2$ appear (as well as $[A]_0$ and $[B]_0$). The degree is the *highest* power to which the substrate concentration is raised in the rate equation. It can be shown very easily that the *degree* of a rate equation is actually the number of enzyme-containing intermediates that the substrate in question can combine with. In the case of Equation 76 inspection of the model shows that A can combine with either E or EB while B can combine with either E or EA. In the case of the general mechanism (Equation 74) inspection shows that B can combine with E, EA or EG, and Equation 75 shows that the rate equation for this model is indeed third degree with respect to $[B]_0$.

The reason for this follows directly from the structural rules. The degree with respect to a given substrate, say A, cannot be greater than the power to which that substrate's concentration is raised in any one kappa-product, because kappa-products are added or subtracted but not multiplied together. The power to which $[A]_0$ is raised in any one kappa-product must be equal to the number of kappas in that product which contain $[A]_0$. But *Rule 3* says that no kappa-product can contain *more than one* kappa originating on any one intermediate. So if the power of $[A]_0$ is greater than one it means that A combines with more than one intermediate. The number of intermediates that A *can* combine with is the greatest power to which $[A]_0$ can be raised in any one kappa-product and since all possible kappa-products are present in the rate equation, this will be the overall degree of the rate equation with respect to A.

This conclusion means that if the degree of the rate equation can be established for a given enzyme, with respect to all its substrates (and products), then certain deductions about its kinetic mechanism may be made. In practice, it is difficult to distinguish experimentally between second and third degree behaviour but it is possible to distinguish between first and higher degree behaviour by the simple expedient of a Lineweaver–Burk plot or some similar linear transformation of the experimental data. First degree behaviour leads to linear Lineweaver–Burk plots, whereas higher degree behaviour gives curved Lineweaver–Burk plots. All the branched simplifications of the general mechanism (Equation 74) can be distinguished from each other and from it by degree tests as described by Wong and Hanes.[33] If a two-substrate enzyme functions by any branched mechanism, it should be possible to decide which is the correct model, although estimation of the kinetic constants of its rate equation is not usually possible with any accuracy at the present time.

One exception to the last statement is for the special case of a branched mechanism in which the lower route of Equation 74, through EG, is missing, and in which all steps are rapid except the interconversions of the two ternary complexes:

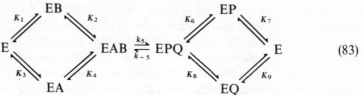

$$(83)$$

This model can be treated by the rapid equilibrium assumption and the rate equation turns out to be first degree with respect to all substrates and products (see below). In that respect it can be distinguished from all other branched mechanisms, but it cannot be distinguished by degree tests from any of the several unbranched or linear simplifications, all of which are of course also first degree with respect to all substrates and products. A different approach is needed in deciding between the various models that give rise to first degree rate equations.

Ordered or unbranched mechanisms

In contrast to the random mechanisms considered above, all the unbranched simplifications of the general mechanism (Equation 74) are *ordered* mechanisms, in the sense that the order of addition of substrates and release of products is fixed. There are two types of ordered mechanisms. In the first type both the substrates must bind to the enzyme before any product is released. These are called *sequential* mechanisms. In the second type, the *non-sequential* mechanisms, a product is released before the second substrate has been bound. This type of mechanism is exemplified by the lowest route in the general mechanism (Equation 74) and involves the substituted enzyme EG where G denotes the group that is transferred from substrate A to substrate B, via the enzyme:

$$E \underset{}{\overset{A}{\rightleftharpoons}} EA \underset{P}{\rightleftharpoons} EG \underset{}{\overset{B}{\rightleftharpoons}} EQ \underset{Q}{\rightleftharpoons} E$$

Cleland[36] has introduced a nomenclature and a pictorial method of representing ordered mechanisms which will be helpful in further discussions of the various types. The nomenclature is quite general, embracing enzymes with one, two and three or more substrates or products. The terms 'Uni', 'Bi', 'Ter' etc. are used to identify the number of substrates involved in the forward reaction and the same terms are used to identify the number of products. The reactions we have been considering in this chapter would be called Uni Uni for the simple one-substrate, one-product case, or Bi Bi where two substrates and two products are concerned, but the terminology can be extended to Uni Bi, or Ter Bi, and other cases. The sequential mechanism is called Ordered Bi Bi in the case where a ternary complex EAB is formed, and is depicted as follows:

$$
\begin{array}{ccccc}
A & B & P & Q & \\
\downarrow & \downarrow & \uparrow & \uparrow & \\
\hline
E & EA & (EAB & EPQ) & EQ & E
\end{array}
\qquad Ordered\ Bi\ Bi' \qquad (84)
$$

The horizontal line is taken to indicate the route followed by the enzyme molecule, first binding A to form EA, then B to form EAB which converts to EPQ, then loses P and finally Q to regenerate free enzyme. As in the case of the Uni Uni reaction considered earlier, it makes no difference to the form of the rate equation if EPQ is absent and EAB is the only ternary complex, breaking down directly to release P and Q. On the other hand there may be a number of ternary complexes such as EAB, EXY, EPQ in the central region of the diagram and the form of rate equation will be identical again. Isomerization of so-called central complexes makes no difference to the rate equation. Isomerization of the free enzyme on the other hand, does lead to a different rate equation as in the Uni Uni case considered earlier, and this would be called the Iso Ordered Bi Bi case and depicted as follows, where two forms of the enzyme exist, E, F, one of which binds A and the other Q:

$$
\begin{array}{cccccc}
A & B & & P & Q & \\
\downarrow & \downarrow & & \uparrow & \uparrow & \\
\hline
E & EA & (EAB & FPQ) & FQ & F & E
\end{array}
\qquad Iso\ Ordered\ Bi\ Bi \qquad (85)
$$

Another sequential case which leads to a different rate equation is the one where no central complex exists. This situation was first suggested by Theorell and Chance to account for the kinetic behaviour of alcohol dehydrogenase and the mechanism bears their names, often abbreviated to the T–C mechanism:

$$
\begin{array}{ccccc}
A & B & P & Q & \\
\downarrow & \searrow\nearrow & & \uparrow & \\
\hline
E & EA & & EQ & E
\end{array}
\qquad Theorell–Chance\ (T–C) \qquad (86)
$$

It is extremely unlikely that this mechanism is a 'real' one because it implies that as soon as the second substrate, B, binds to the enzyme the first product P departs instantly. Nevertheless if the concentration of the ternary complex is kinetically negligible, this model is a close approximation to reality and may apply in certain cases. Once again if two isomeric forms of enzyme take part in the reaction a different rate equation applies and we have the Iso Theorell–Chance mechanism:

$$
\begin{array}{ccccc}
\text{A} & \text{B} & \text{P} & & \text{Q} \\
\downarrow & \searrow\nearrow & & \uparrow & \\
\hline
\text{E} \quad \text{EA} & & \text{FQ} & \text{F} & \text{E}
\end{array}
\qquad \textit{Iso Theorell–Chance} \qquad (87)
$$

The ordered Bi Bi and Theorell–Chance mechanisms, together with their Iso variants are the only important ordered sequential models that need to be considered for two-substrate, two-product enzymes. The ordered non-sequential cases are all called Ping Pong mechanisms in Cleland's terminology because products and substrates alternate in their binding to and release from the enzyme:

$$
\begin{array}{ccccc}
\text{A} & \text{P} & \text{B} & \text{Q} & \\
\downarrow & \uparrow & \downarrow & \uparrow & \\
\hline
\text{E} \quad \text{EA} & \text{EG} & \text{EQ} & \text{E}
\end{array}
\qquad \textit{Ping Pong Bi Bi} \qquad (88)
$$

In this particular case (Ping Pong Bi Bi) isomerization of E or of EG makes no difference to the form of the rate equation.

The order of binding and release of substrates and products in all these ordered models will be taken to be as follows. The first substrate to bind is arbitrarily denoted as A, the second as B, the first product to leave as P, the last as Q. But in a real situation one does not know which substrate is A and which is B; that has to be determined and can present a problem that will be met in due course.

These ordered mechanisms, together with the random rapid equilibrium (R.R.E.) Bi Bi mechanism described at the end of the last section (Equation 83), all generate first degree rate equations of the general form:

$$
v_0 = \frac{n_{AB}[A]_0[B]_0 - n_{PQ}[P]_0[Q]_0}{D} \qquad (89)
$$

where n_{AB} and n_{PQ} are numerator coefficients consisting of the products of e_0 and all the forward or reverse rate constants respectively. The denominator D is different for each mechanism, but always consists of a series of positive terms. These are given in Table 5 for the different mechanisms that have been considered. These rate equations can be arrived at by the use of the King and Altman method and have been listed by Cleland.[36]

Table 5 shows that each term in the denominator is made up of a coefficient d_A, d_B, etc. multiplied by the concentrations of one or more of the substrates or products. In addition there is a constant d in all cases except the Ping Pong Bi Bi mechanism.

Table 5

Values of the denominator **D** in Equation 89 for different first degree mechanisms. The denominator coefficients d, d_A, d_B, etc., are composed of rate constants and are not necessarily the same for each mechanism. The structures of these coefficients are given by Cleland.[36]

Mechanism	Equation	Denominator, **D**, in Equation 89
Random Rapid Equilibrium Bi Bi	83	$d + d_A[A]_0 + d_B[B]_0 + d_{AB}[A]_0[B]_0$ $+ d_P[P]_0 + d_Q[Q]_0 + d_{PQ}[P]_0[Q]_0$
Ordered Bi Bi	84	As for Equation 83, plus: $d_{AP}[A]_0[P]_0 + d_{BQ}[B]_0[Q]_0$ $+ d_{ABP}[A]_0[B]_0[P]_0 + d_{BPQ}[B]_0[P]_0[Q]_0$
Iso Ordered Bi Bi	85	As for Equation 84, plus: $d_{APQ}[A]_0[P]_0[Q]_0 + d_{ABQ}[A]_0[B]_0[Q]_0$ $+ d_{ABPQ}[A]_0[B]_0[P]_0[Q]_0$
Theorell–Chance	86	As for Equation 83, plus: $d_{AP}[A]_0[P]_0 + d_{BQ}[B]_0[Q]_0$
Iso Theorell–Chance	87	As for Equation 83, plus: $d_{BQ}[B]_0[Q]_0 + d_{AP}[A]_0[P]_0$ $+ d_{ABQ}[A]_0[B]_0[Q]_0 + d_{APQ}[A]_0[P]_0[Q]_0$
Ping Pong Bi Bi	88	$d_A[A]_0 + d_B[B]_0 + d_{AB}[A]_0[B]_0 + d_Q[Q]_0$ $+ d_{PQ}[P]_0[Q]_0 + d_{AP}[A]_0[P]_0$ $+ d_{BQ}[B]_0[Q]_0$ (i.e. as for Equation 86 without the constant term d)

A general equation covering all the mechanisms in Table 5 would be as follows:

$$v_0 = \frac{n_{AB}[A]_0[B]_0 - n_{PQ}[P]_0[Q]_0}{d + d_A[A]_0 + d_B[B]_0 + d_{AB}[A]_0[B]_0 + d_P[P]_0 + d_Q[Q]_0} \tag{90}$$
$$+ d_{PQ}[P]_0[Q]_0 + d_{AP}[A]_0[P]_0 + d_{BQ}[B]_0[Q]_0$$
$$+ d_{ABP}[A]_0[B]_0[P]_0 + d_{ABQ}[A]_0[B]_0[Q]_0$$
$$+ d_{APQ}[A]_0[P]_0[Q]_0 + d_{BPQ}[B]_0[P]_0[Q]_0$$
$$+ d_{ABPQ}[A]_0[B]_0[P]_0[Q]_0$$

[N.B. $d_{AQ}[A]_0[Q]_0$ and $d_{BP}[B]_0[P]_0$ do not occur in any of these mechanisms.]

Note that, just as found earlier for the Uni Uni rate equation, when both products are present, this is never the equation of a hyperbola. No mechanism contains all these terms so, in order to distinguish between the various possible mechanisms, it is necessary to discover which denominator terms are missing in the rate equation for a given enzyme. For example the only mechanism for which $d = 0$ is the Ping Pong Bi Bi, whereas the only one for which d_{ABPQ} is not zero is the Iso Ordered Bi Bi. In addition to distinguishing the mechanism for a real enzyme, one would also wish to estimate the numerical values of all the coefficients in its rate equation, so as to be able to predict reaction rates in the presence of substrates and products together. It will become apparent that individual coefficients such as d, d_A, n_{AB}, etc., cannot be found, but ratios of each denominator coefficient to one or other of the two numerator coefficients can be determined by kinetic methods. The ratio of the two numerator co-efficients is always equal to the equilibrium constant for the overall reaction $A + B \rightleftharpoons P + Q$ because at equilibrium $v_0 = 0$ and therefore $n_{AB}[A]_{eq}[B]_{eq} = n_{PQ}[P]_{eq}[Q]_{eq}$, from which it follows that:

$$\frac{n_{AB}}{n_{PQ}} = \frac{[P]_{eq}[Q]_{eq}}{[A]_{eq}[B]_{eq}} = K_{eq} \tag{91}$$

(This is a Haldane relationship. There are other additional Haldanes for some of the mechanisms.)

The rate equation (Equation 90) can be re-cast in terms of the kinetically determined ratios of coefficients and the equilibrium constant. Dividing numerator and denominator of Equation 90 by n_{PQ} one obtains Equation 92:

$$v_0 = \frac{K_{eq}[A]_0[B]_0 - [P]_0[Q]_0}{\dfrac{d}{n_{PQ}} + \dfrac{d_A}{n_{PQ}}[A]_0 + \dfrac{d_B}{n_{PQ}}[B]_0 + \dfrac{d_{AB}}{n_{PQ}}[A]_0[B]_0 + \dfrac{d_P}{n_{PQ}}[P]_0 + \dfrac{d_Q}{n_{PQ}}[Q]_0 + \text{etc.}} \tag{92}$$

Provided that K_{eq} is known and the ratio of each denominator coefficient to *either* numerator coefficient can be found, the full rate equation can be constructed. If the ratio d_P/n_{PQ} is determined, for example, it can go straight into Equation 92, whereas if d_P/n_{AB} is found it must be multiplied by $K_{eq} = n_{AB}/n_{PQ}$ in order to convert it to d_P/n_{PQ}.

The replotting procedure for estimating the kinetic constants of first degree rate equations

A method which allows one to estimate the necessary ratios and hence decides which mechanism applies and its full rate equation, is the so-called re-plotting procedure originated by Dalziel.[37] The experimental procedure used is already familiar. Initial rates are measured in the absence of products, at a fixed concentration of one substrate and varying concentrations of the other. For the moment let us call the substrates X, the *fixed* one, and Y, the *varying* one, since it is not known which binds first to the enzyme i.e. which one is A. A plot of $1/v_0$ against $1/[Y]_0$ is constructed and the slope S and intercept I are noted. The procedure is repeated at several values of $[X]_0$, so that a series of slopes and intercepts $S_1, S_2, S_3 \ldots$ and $I_1, I_2, I_3 \ldots$ are obtained as shown in Figure 53(a).

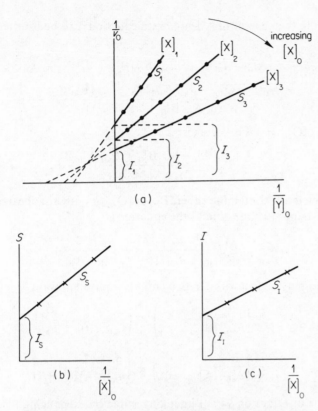

Figure 53 The replotting procedure of Dalziel. The significance of S_S, I_S, S_I and I_I are discussed in the text.
(a) Lineweaver–Burk plot for a two-substrate enzyme obeying
a first degree rate equation (Equation 90). The initial rate (v_0)
is measured in the absence of products at a fixed concentration ($[X]_0$) of one substrate, and varying concentrations
($[Y]_0$) of the other. Slopes (S) and intercepts (I) are measured
and replotted as shown below. The lines obtained at different
values of $[X]_0$ all intersect at the same point which can lie
above or below the abscissa at a value of $1/[Y]_0 = -d_Y/d$,
and $1/v_0 = (d_{AB} - d_X d_Y/d)/n_{AB}$, where d, d_X, d_Y, d_{AB} and n_{AB}
are coefficients in Equation 90. (b) Replotting of the slopes
S_1, S_2, S_3, from (a) against $1/[X]_0$ to obtain a new slope S_S
and intercept I_S. (c) Replotting of the intercepts I_1, I_2, I_3
from (a) against $1/[X]_0$ to obtain a slope S_I and intercept I_I

The slopes are then re-plotted against the value of $1/[X]_0$ for which each was
obtained, and a new slope S_S (slope of slopes) and intercept I_S (intercept of
slopes) are found (Figure 53(b).) The same is done for the original intercepts
(I) and again a slope S_I (slope of intercepts) and intercept I_I (intercept of intercepts) are found (Figure 53(c)). The values of these final slopes and intercepts
are related to coefficients of the rate equation in the following way.

Equation 90, the general first degree rate equation, can be inverted to give Equation 93:

$$\frac{1}{v_0} = \frac{\begin{array}{l} d + d_A[A]_0 + d_B[B]_0 + d_{AB}[A]_0[B]_0 + d_P[P]_0 + d_Q[Q]_0 \\ \quad + d_{PQ}[P]_0[Q]_0 + d_{AP}[A]_0[P]_0 + d_{BQ}[B]_0[Q]_0 \quad \ldots \text{etc.} \end{array}}{n_{AB}[A]_0[B]_0 - n_{PQ}[P]_0[Q]_0} \tag{93}$$

When $[P]_0 = [Q]_0 = 0$ this becomes:

$$\frac{1}{v_0} = \frac{d + d_A[A]_0 + d_B[B]_0 + d_{AB}[A]_0[B]_0}{n_{AB}[A]_0[B]_0}$$

because all the terms containing either $[P]_0$ or $[Q]_0$ or both disappear. Dividing out the right-hand side one obtains the equation:

$$\frac{1}{v_0} = \frac{d}{n_{AB}} \cdot \frac{1}{[A]_0[B]_0} + \frac{d_A}{n_{AB}} \cdot \frac{1}{[B]_0} + \frac{d_B}{n_{AB}} \cdot \frac{1}{[A]_0} + \frac{d_{AB}}{n_{AB}}$$

This equation may be factorized in two entirely equivalent ways:

$$\frac{1}{v_0} = \frac{1}{[A]_0}\left(\frac{d}{n_{AB}} \cdot \frac{1}{[B]_0} + \frac{d_B}{n_{AB}}\right) + \left(\frac{d_A}{n_{AB}} \cdot \frac{1}{[B]_0} + \frac{d_{AB}}{n_{AB}}\right)$$

$$= \frac{1}{[B]_0}\left(\frac{d}{n_{AB}} \cdot \frac{1}{[A]_0} + \frac{d_A}{n_{AB}}\right) + \left(\frac{d_B}{n_{AB}} \cdot \frac{1}{[A]_0} + \frac{d_{AB}}{n_{AB}}\right) \tag{94}$$

Both versions of Equation 94 are interconvertible by exchanging A for B and $[A]_0$ for $[B]_0$ whenever they occur (d_{AB} becomes d_{BA} which is the same as d_{AB}; n_{AB} becomes n_{BA} which is the same as n_{AB}). Thus, whichever substrate concentration is varied, the other being held constant, the equation has the form:

$$\frac{1}{v_0} = \frac{1}{[Y]_0}\left(\frac{d}{n_{AB}} \cdot \frac{1}{[X]_0} + \frac{d_X}{n_{AB}}\right) + \left(\frac{d_Y}{n_{AB}} \cdot \frac{1}{[X]_0} + \frac{d_{AB}}{n_{AB}}\right) \tag{95}$$

Thus the Lineweaver–Burk plot (Figure 53(a)) is a straight line whose slope and intercept are given by:

$$S = \frac{d}{n_{AB}} \cdot \frac{1}{[X]_0} + \frac{d_X}{n_{AB}} \tag{96}$$

$$I = \frac{d_Y}{n_{AB}} \cdot \frac{1}{[X]_0} + \frac{d_{AB}}{n_{AB}} \tag{97}$$

Further when S is re-plotted against $1/[X]_0$ (Figure 53(b)) the slope and intercept are given by:

$$S_S = \frac{d}{n_{AB}} \quad \text{and} \quad I_S = \frac{d_X}{n_{AB}}$$

And when I is re-plotted against $1/[X]_0$ (Figure 53(c)) the slope and intercept are given by:

$$S_I = \frac{d_Y}{n_{AB}} \quad \text{and} \quad I_I = \frac{d_{AB}}{n_{AB}}$$

(The various slopes and intercepts were originally defined differently by Dalziel[37] who used Φ_0 instead of I_I, Φ_{12} instead of S_S, Φ_1 and Φ_2 instead of S_I and I_S.)

At this point in the experimental procedure it becomes immediately apparent if the mechanism is Ping Pong Bi Bi because that is the only mechanism in which $d = 0$ (see Table 5) and hence for which $S_S = 0$. Indeed, one will have had a previous indication if this is the case because, as inspection of Equation 96 shows, the slopes of the initial plots will all be identical when $d = 0$, since $S = d_X/n_{AB}$ and is independent of $[X]_0$. (This, of course, is what leads to $S_S = 0$.) For the Ping Pong Bi Bi mechanism, then, a set of *parallel* Lineweaver–Burk plots is obtained when either substrate is held constant. The assessment, however, of parallel lines by eye is notoriously subjective and it is wise to establish by re-plotting the data, that S_S is actually zero before assuming that a Ping Pong Bi Bi mechanism operates.

If the mechanism is *not* Ping Pong Bi Bi on the other hand, one cannot distinguish between the remaining possibilities without further experiment. Note that the values of the individual coefficients n_{AB}, d_A, d_B, etc. cannot be calculated, only their ratios. The next experimental approach to the problem is to repeat the whole set of experiments in the absence of the substrates A and B, measuring the rate in the reverse direction with one of the products at fixed concentration and the other varied. This allows one to determine the ratios $S_S' = d/n_{PQ}$, $I_S' = d_{X'}/n_{PQ}$, $S_I' = d_{Y'}/n_{PQ}$, and $I_I' = d_{PQ}/n_{PQ}$ where X' and Y' are the fixed and variable products. Once again this should confirm the Ping Pong mechanism if it applies ($S_S' = 0$), and in addition allows one to decide whether the Theorell–Chance mechanism applies. If it does, Dalziel has shown that, because of the structure of the coefficients, $I_I' = S_I I_S/S_S$ and $I_I = S_I' I_S'/S_S'$.

Product inhibition studies

Next one measures the effect of *one* of the products, P or Q, on the rate in the forward direction. Provided only one of the products, say P, is allowed to be present the reverse reaction cannot occur and the numerator term $n_{PQ}[P]_0[Q]_0$ in Equation 90 remains zero, together with any denominator terms containing $[Q]_0$. Inversion of the rate equation now leads to an equation such as:

$$\frac{e_0}{v_0} = \frac{1}{[A]_0}\left\{\frac{1}{[B]_0}\left[[P]_0\left(\frac{d_P}{n_{AB}} + \frac{d_{ABP}}{n_{AB}}\right) + \frac{d}{n_{AB}}\right] + \left[\frac{d_B}{n_{AB}} + \frac{d_{AP}}{n_{AB}}[P]_0\right]\right\}$$
$$+ \left\{\frac{1}{[B]_0}\left[\frac{d_A}{n_{AB}} + \frac{d_{AP}}{n_{AB}}[P]_0\right] + \frac{d_{AB}}{n_{AB}}\right\} \tag{98}$$

Inspection of Equation 98 reveals that the product, P in this case, behaves rather like an inhibitor in the one-substrate situations discussed previously. Depending on the mechanism P (or Q) may affect only the slope of the Lineweaver–Burk plot without affecting the vertical intercept, in which case it is said to compete with Y the varied substrate, or it may affect both the slope and the intercept in which case it acts non-competitively. By investigating the product inhibition patterns of a given enzyme further light is shed on its kinetic mechanism. Table 6 lists the inhibition behaviour of each substrate–product pair for each mechanism.

Table 6

Product inhibition patterns for two-substrate enzymes that obey a first degree rate equation (Equation 90). Lineweaver–Burk plots are compared for the enzyme in the presence of a constant non-saturating concentration of one substrate, with and without the product. Competitive inhibition (COMP) between the product and the varied substrate is indicated when the two lines intersect on the ordinate, and non-competitive inhibition (N–C) occurs when the point of intersection lies to the left of the ordinate (N.B. not necessarily on the abscissa in this case, unlike the case of non-competitive inhibition of a one-substrate enzyme dealt with previously)

| Mechanism | Equation | Product | Varied substrate | |
			A	B
Random Rapid Equilibrium	83	P	COMP	COMP
Bi Bi		Q	COMP	COMP
Ordered Bi Bi	84	P	N-C	N-C
		Q	COMP	N-C
Iso Ordered Bi Bi	85	P	N-C	N-C
		Q	N-C	N-C
Theorell–Chance	86	P	N-C	COMP
		Q	COMP	N-C
Iso Theorell–Chance	87	P	N-C	COMP
		Q	N-C	N-C
Ping Pong Bi Bi	88	P	N-C	COMP
		Q	COMP	N-C

Table 6 shows that the Random Rapid Equilibrium Bi Bi and the Iso Ordered Bi Bi mechanisms can be distinguished from the others by this test. Theorell–Chance and Ping Pong Bi Bi mechanisms are distinguished from the remaining ones but not from each other; this does not matter because they will have been distinguished by the re-plotting procedure in the absence of products. The remaining two mechanisms, Ordered Bi Bi and Iso Theorell–Chance, present an interesting case in that they both show a pattern in which three of the four pairs are non-competitive and one is competitive, A/Q for the former and B/P for the latter. These two mechanisms are said to be complementary mechanisms. Unfortunately they cannot be distinguished by kinetic studies unless one knows which substrate is A and which product is Q. Steady state kinetics provide no insight into this problem, as pointed out earlier. The only way to distinguish these two mechanisms is to carry out independent binding studies. If the 'competitive' substrate binds to the free enzyme then the Ordered Bi Bi mechanism is indicated, but if the reverse is found, the Iso Theorell–Chance mechanism is correct. Of course one does not know, without binding studies, which substrate and product are which for any of the other mechanisms either; but it is not necessary to know in order to distinguish the others. In the Random Rapid Equilibrium Bi Bi, of course, A and B have no priority, either can bind to the free enzyme; and the same applies to P and Q. Similarly, in the Ping Pong Bi Bi mechanism either E or EG could be regarded as the 'free enzyme' since the mechanism can be written in two equivalent ways, so that once again the labels A and B or P and Q can have no 'priority' meaning:

$$
\begin{array}{cccccccc}
\text{A} & \text{P} & \text{B} & \text{Q} & & \text{B} & \text{Q} & \text{A} & \text{P} \\
\downarrow & \uparrow & \downarrow & \uparrow & \equiv & \downarrow & \uparrow & \downarrow & \uparrow \\
\hline
\text{E} & \text{EA} & \text{EG} & \text{EQ} & \text{E} & \text{EG} & \text{EQ} & \text{E} & \text{EA} & \text{EG}
\end{array}
$$

Having distinguished, by the various tests described, which mechanism applies to a given enzyme, it now only remains to find the other ratios of coefficients not yet determined. Lineweaver–Burk plots are constructed by measuring the rate at a constant concentration of one substrate and one product, with varying concentrations of the other substrate. This is repeated at a series of different concentrations of the fixed substrate, the product concentration being held constant, and finally the whole series of experiments is repeated with a series of product concentrations. A *double* replotting procedure for the slopes and intercepts now leads to the ratios d_{AP}/n_{AB}, d_{BP}/n_{AB}, d_{ABP}/n_{AB}, etc., which were not known before, together with re-estimates of some of the other ratios. The first part of the double re-plot is the same as before, but in the second part the values of S_S, S_I, I_S and I_I are plotted against the concentration of product (not its reciprocal) at which each was obtained. The other product can be used in an exactly similar manner to find other ratios and finally the effects of individual *substrates* on the *reverse* reaction leads to an evaluation of any remaining ratios such as d_{APQ}/n_{PQ}. Some of the possible ratios will be zero of course for certain mechanisms and this provides a check that the correct mechanism has been chosen.

Needless to say, the labour involved in a complete kinetic analysis of this sort is considerable and has been undertaken in only a few cases. Other possibilities may have to be taken into account that have not been discussed here, such as the possibility of the formation of abortive or 'dead-end' complexes, particularly important, for example, in the Random Rapid Equilibrium Bi Bi mechanism, where the pattern of inhibition shown in Table 6 would no longer apply. These represent enzyme–substrate or enzyme–product complexes that are catalytically inactive and therefore lock up enzyme in a useless form. These and other aspects of the kinetics of two-substrate enzymes are discussed at greater length in the references listed at the end of this chapter.

Michaelis constants and maximum velocities for two-substrate enzymes

Throughout the discussion of two-substrate enzymes the terms Michaelis constant (K_m) and maximum velocity (V_m) have been avoided, although this is not always done in other places. One has to define very carefully what is meant by the K_m of an enzyme which has more than one substrate and, whatever definition is used, the symbol K_m is not such a useful kinetic constant as in the case of Uni Uni reactions. The usual meaning given to the Michaelis constant for a given substrate, is the concentration of that substrate which in the absence of products leads to half the maximum possible rate when the enzyme is *saturated* with the other substrate. Clearly this is a highly artificial situation and may be quite unrelated to the conditions existing inside a living cell where products will be present and neither substrate may be at anything like a saturating concentration. It is better to leave it to the individual to work out apparent K_m s for the particular conditions applying to particular cases, rather than to label an enzyme with 'a K_m' which has an air of immutability. The same sort of restrictions apply to the term maximum velocity, which once again will depend on the concentrations of products and the other substrate.

Having made this point, however, the values of K_m, as defined above, and V_m the maximum velocity in the forward direction when saturated with both substrates, in the absence of products, can be found from Figure 53. The maximum velocity is clearly given by:

$$V_m = 1/I_1 = \frac{n_{AB}}{d_{AB}} \quad \text{and} \quad K_m^Y = I_S/I_1 = \frac{d_X}{n_{AB}} \cdot \frac{n_{AB}}{d_{AB}} = \frac{d_X}{d_{AB}}$$

from which it can be seen that $K_m^X = d_Y/d_{AB}$. Therefore

$$K_m^A = \frac{d_B}{d_{AB}} \quad \text{and} \quad K_m^B = \frac{d_A}{d_{AB}}$$

For the reverse direction, by analogy:

$$V_m' = \frac{n_{PQ}}{d_{PQ}}; \quad K_m^P = \frac{d_Q}{d_{PQ}} \quad \text{and} \quad K_m^Q = \frac{d_P}{d_{PQ}}$$

By using these definitions of V_m, K_m^A and K_m^B the rate equation for the Ping Pong Bi Pi mechanism, in the absence of products, can be written quite simply as:

$$v_0 = \frac{V_m}{1 + \dfrac{K_m^A}{[A]_0} + \dfrac{K_m^B}{[B]_0}}$$

The other first degree mechanisms all share a common rate equation in the absence of products:

$$v_0 = \frac{V_m}{1 + \dfrac{K_m^A}{[A]_0} + \dfrac{K_m^B}{[B]_0} + \dfrac{K_s^A K_m^B}{[A]_0[B]_0}}$$

where K_s^A is the dissociation constant for the reaction $EA \rightleftharpoons E + A$.

As soon as one or more products is present it becomes impossible to write a rate equation for any of the ordered mechanisms in terms of V_m s, K_m s and K_s s. Extra constants have to be defined which have no obvious kinetic meaning and little is gained in the attempt to simplify the rate equations. That is why the rate equations have been left with numerator and denominator coefficients in the foregoing treatment.

In some cases individual rate constants can be determined for some of the steps in a mechanism but not for all of them. Such cases have been discussed by Cleland[36] and Dalziel.[37]

ENZYMES WITH MORE THAN TWO SUBSTRATES OR PRODUCTS

A discussion of the kinetics of multisubstrate enzymes is beyond the scope of this book. Rules for dealing with these systems have been devised and are dealt with by Cleland in Reference 30 and by Dalziel in Reference 38.

Chapter 6

Enzyme Kinetics II: The Kinetics of Interacting Sites

HOMOTROPIC AND HETEROTROPIC INTERACTION

Up to now the kinetics of enzyme action have been dealt with as if enzymes had only one active site in each enzyme molecule. Many enzymes, however, are aggregates in which the oligomeric enzyme molecule contains a number of identical protein subunits or protomers each with its own active site. The kinetic behaviour of such enzymes will not differ from the behaviour of monomeric enzymes provided that the active site in each protomer behaves independently, in other words, provided that there is no interaction between the sites. This does indeed happen in some oligomeric enzymes. In such cases the individual active site binds substrates and catalyses reaction just as if the protomer on which it is sited were not aggregated with others. In other oligomeric enzymes, on the other hand, events at one active site affect what happens at the others in the oligomer. This is not surprising in view of the flexible nature of the protomer and the alterations in its structure that accompany binding and catalysis. Such conformational changes can be transmitted from one protomer to another and can give rise to the phenomenon of interaction or cooperativity. This type of interaction, where the sites which interact are identical ones, has been called *homotropic interaction*.

Another kind of interaction has already been met under the heading of allosteric interaction. In that case the events at an active site are affected by binding of an effector at an allosteric site. This second type of interaction has been called *heterotropic interaction* in order to distinguish it from the first type.

In both homotropic and heterotropic interaction, one binding site is able to interact with another by means of conformational changes in the protein structure that links them. In that sense there is no real distinction between the two types. It must be stressed, however, that homotropic interaction between active sites can exist in the absence of any allosteric site whatsoever. The failure to make this fundamental distinction can lead to misunderstanding about the various kinds of interaction.

186

LIGAND-LINKED-CONFORMATIONAL-CHANGE

Since interaction is mediated by conformational changes in the protein structure it is convenient to start by considering how conformational change is associated with the binding of ligands. When the 'induced fit' hypothesis was considered in Chapter 4 it was seen how the process of binding could be thought of as inducing the protein to change its shape, as if the protein wrapped itself around the ligand. This process could be represented schematically as follows, where the free protein is shown as a circle to which the ligand binds weakly. Conformational change then occurs in which the protein becomes represented by a square to which the ligand is tightly bound:

$$\text{C} + L \underset{}{\overset{K_1}{\rightleftharpoons}} \text{(L} \underset{}{\overset{K_2}{\rightleftharpoons}} \text{[L]} \; ; \; K_1 < 1 < K_2$$

where K_1 and K_2 are equilibrium constants.

In this scheme the ligand clearly *causes* the conformational change in the protein.

Another way of thinking about the process assumes that the protein can exist in two conformations say circle and square which are in equilibrium, one form, the circle, being favoured. The ligand is assumed to bind more tightly to the less favoured (square) form:

$$\text{C} \underset{}{\overset{K_3}{\rightleftharpoons}} \text{[} \; ; \quad K_3 < 1$$

$$\text{C} + L \underset{}{\overset{K_1}{\rightleftharpoons}} \text{(L} \; ; \quad \text{[} + L \underset{}{\overset{K_4}{\rightleftharpoons}} \text{[L]}$$

$$; \quad K_1 < 1 < K_4$$

In this scheme the ligand does not *cause* the alteration in conformation, it merely stabilizes it by binding to the altered form and pulling an otherwise unfavourable equilibrium over to the right.

Whichever way one thinks about the process, the result is the same: the ligand ends up bound predominantly to the conformationally altered (square) protein. One might link both processes together by the familiar thermodynamic box:

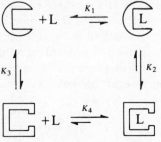

Going from top left to bottom right the process leads from free protein to conformationally altered, liganded protein whichever path is followed and so the product $K_1 K_2$ must equal $K_3 K_4$. In other words the condition that $K_1 < K_4$ which was inherent in the second of the two ways of viewing the process, implies that $K_2 > K_3$.

The fact that one can *envisage* both processes, and link them thermodynamically so as to relate the four equilibrium constants, does not mean that both actually occur. In a particular real case it may be that the process described by the bottom line of the box above (with equilibrium constant K_4) cannot occur because it is physically impossible for the ligand to get into the 'closed' binding site in the 'square' form of the protein. The process with equilibrium constant K_2, on the other hand, might occur in spite of this because the ligand is already partly in the binding site before the structure closes round it. In that case the route along the bottom of the box would be an imaginary one, whose rate constants in either direction would be zero. This, however, does not mean that K_4 is zero. The equilibrium constant, K_4, is the ratio of the two rate constants and 0/0 is indeterminate. One could assign a value to K_4 by measuring K_1, K_2 and K_3 and calculating $K_1 K_2 / K_3 = K_4$.

In another real situation the process with equilibrium constant K_2 might not occur, or might occur at an infinitesimally small rate so that to all intents and purposes it was insignificant. This might happen if the presence of L bound to either the 'circular' or the 'square' form of the protein reduced the rate constants for equilibration of the two conformations to zero. As before though, K_2 could still be calculated for the imaginary process provided K_1, K_3 and K_4 could be measured.

Of course it is important to know what processes actually do occur in the case of any real enzyme, and experiments can be devised to answer the question in some cases. On the other hand, both these ways of thinking about the causal relationships between binding and conformational change have served as starting points for theoretical descriptions of homotropic and heterotropic interactions. These theories have been developed and their predictions tested against real experimental results.

There are a number of ways in which binding sites could interact within an oligomeric structure. When the conformation of one protomer in an oligomeric assembly becomes altered, either before or as a result of binding a ligand, it must affect the conformations of other protomers if interaction is to result. There is no *a priori* reason why *all* the other protomers should have to change shape as a result of a change in one of them; nor is there any reason why a particular change in one should lead to the identical change in another. All sorts of possibilities come to mind. Consider first a dimeric protein made up of identical protomers represented as circles in one conformational isomer in which the ligand binds only weakly ((a) below).

(a)

weak/weak

(b)

strong/weaker

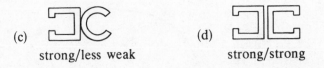

(c) strong/less weak   (d) strong/strong

When the conformation of one protomer alters to a form that binds ligand strongly, the other protomer may be constrained to take up a structure with a lower affinity than before ((b) above), or with a greater affinity than before ((c) above), or it may even be forced into the same strongly binding conformation as the first protomer ((d) above).

When we are considering an enzyme and the ligand is the substrate, in addition to alterations in the binding *affinity* for the substrate it must be borne in mind that similar changes may also occur in the ability of the active site to *catalyse* reaction once the substrate is bound. The ways in which conformational change can be transmitted from one protomer to another, and the effect this will have on the kinetics of the affected active sites, clearly depends on the actual structure of the protomers and their spatial relationships in the oligomer for any individual oligomeric enzyme.

When the kinetics of oligomeric enzymes began to be studied it was noticed that many of them gave rise to sigmoid plots of reaction velocity as a function of substrate concentration, rather than the 'classical' hyperbolic plot (Figure 54). In these cases it was clear that some kind of *positive interaction or cooperativity* was being observed. As the concentration of substrate is increased

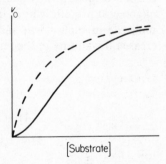

Figure 54   The dependence of reaction rate upon substrate concentration for a 'classical' enzyme (hyperboloid, pecked line) and an enzyme showing 'sigmoid' kinetics (solid line)

from low values the rate increases slowly at first but then it increases faster than predicted by the usual models of enzyme mechanism, suggesting that the catalysis of substrate at the first active site occupied in the oligomer makes catalysis at the other sites more efficient. This autocatalytic effect accounts for the upward turn in the lower half of the sigmoid curve, but as the substrate concentration is increased to the point where all the active sites are becoming filled, a saturation effect naturally leads to a levelling off in rate.

At first sight the sigmoid rate curve of Figure 54 bears a strong resemblance to the saturation curve of haemoglobin with oxygen (Figure 55). In the latter case there is no catalysis, but the *binding* of oxygen exhibits positive cooperativity.

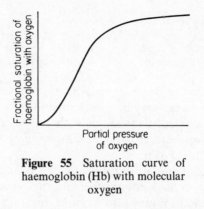

Figure 55 Saturation curve of haemoglobin (Hb) with molecular oxygen

HILL PLOTS AND THE HILL COEFFICIENT

One of the earliest attempts to explain the sigmoid oxygen binding curve of haemoglobin was put forward by Hill in the early years of this century. He first pointed out that an empirical equation of the form:

$$\overline{Y} = \frac{[O_2]^n}{K + [O_2]^n} \tag{99}$$

would fit the sigmoid shape of the binding curve, where $\overline{Y}$ represents the fractional saturation with oxygen, $[O_2]$ is the concentration, K is a constant and n is a small positive number. Later he suggested that haemoglobin was an oligomeric protein, whose individual protomers could each bind one oxygen molecule in such a way that the binding of successive oxygen molecules increased the affinity of the remaining protomers.

$$Hb + O_2 \rightleftharpoons HbO_2; \quad K_1$$

$$HbO_2 + O_2 \rightleftharpoons Hb(O_2)_2; \quad K_2$$

$$Hb(O_2)_2 + O_2 \rightleftharpoons Hb(O_2)_3; \quad K_3$$

$$Hb(O_2)_{n-1} + O_2 \rightleftharpoons Hb(O_2)_n; \quad K_n$$

where the equilibrium constants $K_1 \ldots K_n$ were progressively bigger, $K_1 \ll K_2 \ll K_3 \ll K_n$. In that case, he argued, the oxygen molecules would *effectively* bind n at a time (where n was the number of protomers in the haemoglobin oligomer), because the concentrations of HbO_2, $Hb(O_2)_2$ etc. would be vanishingly small. The binding process could then be summarized as:

$$Hb + nO_2 \rightleftharpoons Hb(O_2)_n; \quad K' = \frac{[Hb(O_2)_n]}{[Hb][O_2]^n} \tag{100}$$

where the equilibrium constant K' was an overall constant for the hypothetical process of n oxygen molecules binding simultaneously. This is a hypothetical process of course, because the likelihood of several molecules of oxygen binding to a single haemoglobin molecule at the same instant is exceedingly remote. Nevertheless rearrangement of Equation (100) leads to the empirical Equation (99) as shown below:

From Equation (100)

$$[Hb(O_2)_n] = K'[Hb][O_2]^n$$

The fractional saturation $\bar{Y}$ is the fraction of the total number of binding sites which are occupied at a given concentration of oxygen. Therefore

$$\bar{Y} = \frac{n[Hb(O_2)_n]}{n([Hb] + [Hb(O_2)_n])} = \frac{K'[Hb][O_2]^n}{[Hb] + K'[Hb][O_2]^n} = \frac{K'[O_2]^n}{(1 + K'[O_2]^n)}$$

$$\therefore \quad \bar{Y} = \frac{[O_2]^n}{\dfrac{1}{K'} + [O_2]^n} = \frac{[O_2]^n}{K + [O_2]^n} \qquad \text{i.e. Equation (99)}$$

where $\dfrac{1}{K'} = K$.

It is possible to rearrange Equation (99) to give a more useful form as follows:

$$(1 - \bar{Y}) = 1 - \frac{[O_2]^n}{K + [O_2]^n} = \frac{K + [O_2]^n - [O_2]^n}{K + [O_2]^n} = \frac{K}{K + [O_2]^n}$$

$$\therefore \quad \frac{\bar{Y}}{1 - \bar{Y}} = \frac{[O_2]^n}{K + [O_2]^n} \bigg/ \frac{K}{K + [O_2]^n} = \frac{[O_2]^n}{K}$$

$$\therefore \quad \log\left(\frac{\bar{Y}}{1 - \bar{Y}}\right) = n \log[O_2] - \log K \tag{101}$$

In this way a plot of $\log(\bar{Y}/1 - \bar{Y})$ against $\log[O_2]$ should give a straight line of slope n and intercept $-\log K$ if Hill's explanation is correct. The experimental data for haemoglobin, when plotted in this way, give maximum values for the slope, in the region of 2·5–3·0, although the plot is not a straight line. As we now know that haemoglobin is a tetrameric molecule the value of the slope should have been 4 if Hill's idea had been valid. Although his explanation turned out to be incorrect, his method of plotting the data has been widely used.

When one considers binding in more detail, even if the process is highly cooperative as Hill assumed, it is clear that at extremely low concentrations of oxygen, the molecules of oxygen must bind only one at a time. Similarly at very high oxygen concentrations, the last few molecules of oxygen to bind must do so one at a time. At the extremes of the Hill plot, the slope is therefore expected to be only 1. This is indeed found to be the case for haemoglobin as well as for other cooperative binding systems, as shown in Figure 56. At intermediate concentrations where the cooperativity is greatest, the molecules of ligand do indeed appear to bind more than one at a time, but the slope of the Hill plot never reaches the value of n, the number of protomers in the oligomer. The *maximum* slope of the Hill plot is denoted as h, the *Hill coefficient*. It is an

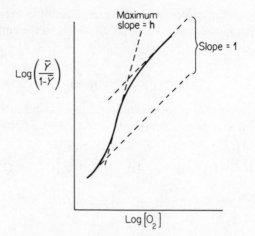

$$\text{Log}\left(\frac{\overline{Y}}{1-\overline{Y}}\right)$$

Maximum slope = h

Slope = 1

Log [O₂]

Figure 56 A Hill plot for the binding of oxygen to haemoglobin. The plot of the function $\log(\overline{Y}/(1 - \overline{Y}))$ against $\log[O_2]$ approaches a slope of one at both low and high partial pressures of oxygen. Its maximum slope h, the Hill coefficient, is a measure of the cooperativity of the binding process

indication of the cooperativity of the system and measures the apparent stoichiometry of the binding process at its most cooperative point.

The Hill plot can be applied to enzymes·that exhibit cooperative kinetics if the rate v is assumed to be proportional to the fractional saturation $\overline{Y}$. This will be strictly true only for enzymes that obey the rapid equilibrium assumption. Nevertheless a plot of $\log(v/V_m - v)$ against the log of substrate concentration has often been used to demonstrate cooperative enzyme kinetics. Enzymes with sigmoid kinetics give Hill plots that resemble the one shown in Figure 56.

Although such treatment of the data has its uses, it does not lead one to a deeper understanding of the process of cooperativity. The Hill coefficient merely puts a figure to the degree of cooperativity apparent in the kinetics.

THE ADAIR HYPOTHESIS

The next attempt to rationalize the haemoglobin saturation curve was made by Adair, who made similar assumptions about the model for oxygen binding as Hill had done but extended the mathematical treatment to take account of the fact that the oxygen molecules cannot bind four at a time. The fractional saturation is properly given as:

$$\overline{Y} = \frac{[Hb\,O_2] + 2[Hb(O_2)_2] + 3[Hb(O_2)_3] + 4[Hb(O_2)_4]}{4([Hb] + [Hb\,O_2] + [Hb(O_2)_2] + [Hb(O_2)_3] + [Hb(O_2)_4])}$$

Using the equilibrium constants K_1, K_2, K_3 and K_4 defined earlier for the successive binding steps and applying the weighting factors in exactly the same way as was done in arriving at the equation for the Scatchard plot in Chapter 4 (see p. 95) the equation becomes:

$$\bar{Y} = \frac{4K_1[\text{Hb}][O_2] + 2.\dfrac{4.3}{2}K_1K_2[\text{Hb}][O_2]^2 + 3.\dfrac{4.3.2}{2.3}K_1K_2K_3[\text{Hb}][O_2]^3 + 4.\dfrac{4.3.2.1}{2.3.4}K_1K_2K_3K_4[\text{Hb}][O_2]^4}{4\left([\text{Hb}] + 4K_1[\text{Hb}][O_2] + \dfrac{4.3}{2}K_1K_2[\text{Hb}][O_2]^2 + \dfrac{4.3.2}{2.3}K_1K_2K_3[\text{Hb}][O_2]^3 + \dfrac{4.3.2.1}{2.3.4}K_1K_2K_3K_4[\text{Hb}][O_2]^4\right)}$$

$$= \frac{K_1[O_2](1 + 3K_2[O_2] + 3K_2K_3[O_2]^2 + K_2K_3K_4[O_2]^3)}{1 + 4K_1[O_2] + 6K_1K_2[O_2]^2 + 4K_1K_2K_3[O_2]^3 + K_1K_2K_3K_4[O_2]^4} \quad (102)$$

(The equation actually given by Adair in a classic paper in 1925 was as follows:

$$= \frac{0.25\,K'_1[O_2] + 0.5\,K'_2[O_2]^2 + 0.75\,K'_3[O_2]^3 + K'_4[O_2]^4}{1 + K'_1[O_2] + K'_2[O_2]^2 + K'_3[O_2]^3 + K'_4[O_2]^4}$$

It is obtained by defining the equilibrium constants in a different manner. The weighting factors used above are absorbed into the value of the equilibrium constant and the constants for individual steps are merged so that, for example, K_1 used above is equal to $K'/4$ and $K_1K_2K_3K_4 = K'_4$.)

The values of $K_1–K_4$ can, of course, take any values and if they become progressively bigger, so that after one oxygen molecule has bound the binding constant for the second oxygen is greater than for the first and so on, then co-operativity occurs and a sigmoid plot of $\bar{Y}$ against $[O_2]$ will result. Conversely, if $K_1 = K_2 = K_3 = K_4$, then there is no cooperativity at all and the four binding sites are identical and *independent*. In that case the binding equation reduces to:

$$\bar{Y} = \frac{K_1[O_2](1 + K_1[O_2])^3}{(1 + K_1[O_2])^4} = \frac{K_1[O_2]}{1 + K_1[O_2]}$$

This is, of course, the classical hyperbolic binding equation, leading to the linear Scatchard plot (p. 96).

In principle the Adair equation, Equation 102, ought to be applicable to the binding of any ligand to any tetrameric protein, and can be extended to deal with higher oligomers. The values of the constants $K_1–K_n$ can be varied at will, but the initial assumptions of the model make no predictions as to *what* values of the constants can be expected, nor *why* they should be different. The model merely accepts the fact of interaction without adding to our understanding of why it should occur.

THE MONOD, WYMAN AND CHANGEUX (M.W.C.) MODEL

The first serious attempt to provide a molecular rationale for both homotropic and heterotropic interactions was proposed in 1965 by Monod, Wyman and Changeux.[39] Their model is elegant, relatively simple and in many ways is still the most intellectually satisfying. Their paper, see Reference 39, is a classic and should be obligatory reading for all students of enzymology.

They set out to explain a number of related facts which, at that time, were common to a large proportion of proteins that showed interaction effects in binding or enzymic catalysis. All such proteins were oligomeric and many gave sigmoid binding or rate curves, that is they showed positive homotropic effects. In addition, those that did often exhibited heterotropic effects of the type shown in Figure 57 where an activator shifts the saturation curve to the left and depresses the degree of cooperativity, while an inhibitor shifts the curve to

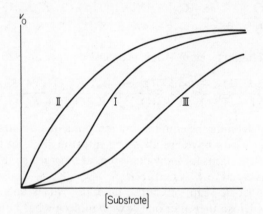

Figure 57 Typical rate curves for an 'allosteric' enzyme. I. The sigmoid curve of enzyme alone. II. The hyperbolic curve of enzyme in the presence of a saturating concentration of an allosteric activator, or of *desensitized* enzyme (either alone or with allosteric effectors). III. The enhanced sigmoid curve produced by the presence of an allosteric inhibitor

the right and increases cooperativity. Further, in many cases it had proved possible to modify the enzyme either by heat treatment or chemical modification so that although it was still active, and V_m was unaffected, the homotropic cooperativity (sigmoid curve) and the heterotropic effect (allosteric activation or inhibition) were *both* lost. This *desensitization* phenomenon convinced them that the explanation was to be found in interactions between the subunits of the oligomer, mediated by conformational changes in the protein structure.

They argued that oligomeric proteins of this 'allosteric' type were likely to be able to exist in at least two conformations, which they called the relaxed, or R, form and the tight, or T, form. In the absence of any ligand they assumed that both these forms would exist in solution in equilibrium. A basic assumption of their hypothesis was that all the protomers in the oligomeric assembly must have the same conformation. In other words, for a tetrameric enzyme, it might exist in the all-R form shown in Figure 58(a) or the all-T form shown in Figure 58(b); but hybrids with some protomers in one form and some in the other, as shown in Figure 58(c), could not exist. This model is thus a *concerted* model;

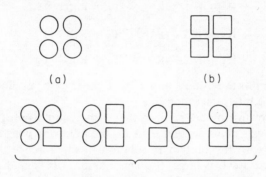

(a) (b)

(c)

Figure 58 Diagrammatic representation of the symmetry requirements of the M.W.C. model for a tetrameric protein. The protomers can exist in R form (circles) or T form (squares) but all the protomers in the tetramer must have the same form either all-R as in (a) or all-T as in (b). Hybrid forms shown in (c) are not allowed in this model

if one protomer changes conformation all must do so in the same way. Persuasive arguments were put forward by Monod, Wyman and Changeux to support the assumption of concerted conformational change in the absence of ligand. The equilibrium constant for the interconversion of the two forms, $R_0 \rightleftharpoons T_0$, was defined as:

$$L = \frac{[T_0]}{[R_0]} \tag{103}$$

where the small subscript 0 refers to the fact that no ligand is bound to these forms.

In retrospect it is unfortunate that Monod and his colleagues called the equilibrium constant, L, the 'allosteric constant'. In what follows it will be referred to as the 'concerted transformation constant' to avoid confusion. Both forms, T_0 and R_0, were assumed to be able to bind up to four molecules of a ligand F, that is one molecule of F to each protomer (in a tetramer). Each molecule of F that binds to the relaxed form is assumed to do so with the same intrinsic dissociation constant K_R no matter how many molecules of F are already bound. Similarly, each molecule of F binding to the tight form was assumed to bind with the intrinsic dissociation constant K_T. In this way the following equilibria would exist in the presence of F:

$$R_0 \overset{L}{\rightleftharpoons} T_0$$

$$R_0 + F \overset{K_R}{\rightleftharpoons} R_1 \qquad T_0 + F \overset{K_T}{\rightleftharpoons} T_1$$

$$R_1 + F \overset{K_R}{\rightleftharpoons} R_2 \qquad T_1 + F \overset{K_T}{\rightleftharpoons} T_2$$

$$R_2 + F \overset{K_R}{\rightleftharpoons} R_3 \qquad T_2 + F \overset{K_T}{\rightleftharpoons} T_3 \tag{104}$$

$$R_3 + F \overset{K_R}{\rightleftharpoons} R_4 \qquad T_3 + F \overset{K_T}{\rightleftharpoons} T_4$$

The small subscripts in R_1 to R_4 and T_1 to T_4 refer to the number of molecules of F bound to each tetramer molecule.

The authors of this model realized that if the equilibrium constant L was large and if K_R was much smaller than K_T the following consequences would result. Starting with the protein in solution in the absence of ligand, T_0 is predominant because L is large. If a small amount of ligand is added, since $K_R \ll K_T$, a proportion of the ligand molecules will bind singly to the low concentration of R_0 present, thereby forming R_1 and removing some R_0 molecules. This results in a displacement of the $R_0 \rightleftharpoons T_0$ equilibrium, pulling over some molecules from the T_0 to the R_0 form. In effect this means that *for every molecule of* F *that binds to* R_0, *three new high affinity R-type sites appear in solution.* If now more ligand is added it will find an increased concentration of high affinity sites and thus will bind more readily than the first batch of ligand, giving rise to the upward lift of the sigmoid curve at low ligand concentrations. In doing so the $R_0 \rightleftharpoons T_0$ equilibrium is displaced still further and so on until, eventually, all ligand binding sites become filled, on both relaxed and tight forms, and no more ligand can bind. As this saturating situation is approached every extra ligand molecule added will find fewer and fewer binding sites available and the curve will flatten out, thus accounting for the upper part of the sigmoid curve.

This qualitative explanation for sigmoid binding curves is fully borne out by a rigorous mathematical treatment as described below.

The binding curve for the M.W.C. model

The fractional saturation, $\overline{Y}$, at any concentration of F is defined as usual as:

$$\overline{Y} = \frac{\text{concentration of sites combined with F}}{\text{total concentration of sites available}}$$

$$= \frac{([R_1] + 2[R_2] + 3[R_3] + 4[R_4]) + ([T_1] + 2[T_2] + 3[T_3] + 4[T_4])}{4([R_0] + [R_1] + [R_2] + [R_3] + [R_4] + [T_0] + [T_1] + [T_2] + [T_3] + [T_4])} \tag{105}$$

The values of $[R_1], [R_2] \ldots [T_1], [T_2] \ldots$ can be expressed in terms of $[R_0]$ and F by making use of the dissociation constants K_R and K_T (Equation 104) and the concerted transformation constant L (Equation 103). For example the binding of F to R_0 is related to K_R as follows, bearing in mind that K_R is the *intrinsic* dissociation constant and that weighting factors must be applied as described in p. 95:

$$K_R = \frac{4[R_0][F]}{[R_1]}$$

$$\therefore \quad [R_1] = \frac{4[R_0][F]}{K_R} = 4[R_0]\alpha,$$

where $\quad \alpha = \dfrac{[F]}{K_R}$ (106)

Similarly $[R_2] = 6[R_0]\alpha^2$, $[R_3] = 4[R_0]\alpha^3$ and $[R_4] = [R_0]\alpha^4$. For the tight form it is convenient to introduce a new constant $c = K_R/K_T$ so that from Equation 106 it follows that $c\alpha = [F]/K_T$. From Equation 104:

$$[T_1] = \frac{4[T_0][F]}{K_T} = 4[T_0]c\alpha$$ (107)

But from Equation 103 we find that $[T_0] = L[R_0]$ and so Equation 107 becomes:

$$[T_1] = 4[R_0]Lc\alpha$$

Likewise $[T_2] = 6[R_0]Lc^2\alpha^2$, $[T_3] = 4[R_0]Lc^3\alpha^3$ and $[T_4] = [R_0]Lc^4\alpha^4$.
Equation 105 can now be re-written as follows:

$$\bar{Y} = \frac{\begin{aligned}[R_0](4\alpha + 12\alpha^2 + 12\alpha^3 + 4\alpha^4) \\ + [R_0](4Lc\alpha + 12Lc^2\alpha^2 + 4Lc^4\alpha^4)\end{aligned}}{\begin{aligned}4([R_0]\{1 + 4\alpha + 6\alpha^2 + 4\alpha^3 + \alpha^4\} \\ + [R_0]\{L + 4Lc\alpha + 6Lc^2\alpha^2 + 4Lc^3\alpha^3 + Lc^4\alpha^4\})\end{aligned}}$$

$$= \frac{\begin{aligned}(\alpha + 3\alpha^2 + 3\alpha^3 + \alpha^4) + L(c\alpha + 3c^2\alpha^2 + 3c^3\alpha^3 + c^4\alpha^4)\end{aligned}}{\begin{aligned}(1 + 4\alpha + 6\alpha^2 + 4\alpha^3 + \alpha^4) \\ + (1 + 4c\alpha + 6c^2\alpha^2 + 4c^3\alpha^3 + c^4\alpha^4)\end{aligned}}$$

$$= \frac{\begin{aligned}\alpha(1 + 3\alpha + 3\alpha^2 + \alpha^3) + Lc\alpha(1 + 3c\alpha + 3c^2\alpha^2 + c^3\alpha^3)\end{aligned}}{\begin{aligned}(1 + 4\alpha + 6\alpha^2 + 4\alpha^3 + \alpha^4) \\ + L(1 + 4c\alpha + 6c^2\alpha^2 + 4c^3\alpha^3 + c^4\alpha^4)\end{aligned}}$$

$$\therefore \quad \bar{Y} = \frac{\alpha(1 + \alpha)^3 + Lc\alpha(1 + c\alpha)^3}{(1 + \alpha)^4 + L(1 + c\alpha)^4}$$ (108)

As it stands Equation 108 is not very revealing. It should, however, be remembered that α is a measure of the ligand concentration [F] since it is defined

198

as $[F]/K_R$ from Equation 106, and that c is simply the ratio K_R/K_T. Of course L is the concerted transformation constant defined previously in Equation 103. Monod and his colleagues plotted curves of $\bar{Y}$ as a function of α for various values of L and c, as shown in Figure 59. As the figure shows, when L is large and c is small, the saturation curves are markedly sigmoidal. In this situation the ligand binds preferentially to R_0, (c is small therefore $K_R < K_T$) whereas T_0 is the favoured form in the $R_0 \rightleftharpoons T_0$ equilibrium (L is large) as predicted

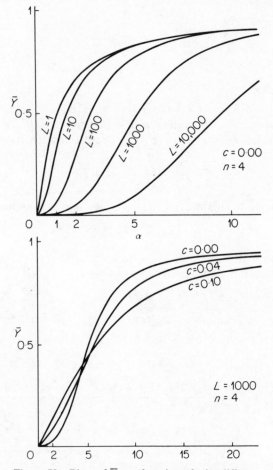

Figure 59 Plots of $\bar{Y}$ as a function of α for different values of L and c for a hypothetical protein obeying the M.W.C. model (according to Monod, Wyman and Changeux).[39] $\bar{Y}$ is the fractional saturation of the protein with ligand F; α is $[F]/K_R$ where K_R is the binding constant of F to R; c is K_R/K_T where K_T is the binding constant of F to T (reproduced, from Reference 16, by permission of Academic Press Inc. (London) Ltd)

qualitatively in the earlier discussion. Conversely, of course, exactly the same result is achieved when L is small and c is large; sigmoid binding curves result, but now because F binds preferentially to T_0 whereas R_0 is the favoured form.

The M.W.C. treatment can be generalized for an oligomer containing any number of protomers, n, in which case the binding equation (cf. Equation 108) becomes:

$$\bar{Y} = \frac{\alpha(1 + \alpha)^{n-1} + Lc\alpha(1 + c\alpha)^{n-1}}{(1 + \alpha)^n + L(1 + c\alpha)^n} \tag{109}$$

or if c is effectively zero,

$$\bar{Y} = \frac{\alpha(1 + \alpha)^{n-1}}{L + (1 + \alpha)^n} \tag{110}$$

Provided that $n > 2$, sigmoid curves will result when L is large and c small or vice versa. The larger the value of n, the more obvious the sigmoid nature of the curve becomes for any given values of L and c. When $n = 1$, Equation 109 reduces to:

$$\bar{Y} = \frac{\alpha(1 + \alpha)^0 + Lc\alpha(1 + c\alpha)^0}{1 + \alpha + L + Lc\alpha} = \frac{\alpha + Lc\alpha}{1 + L + \alpha + Lc\alpha}$$

$$= \frac{\alpha(1 + Lc)}{(1 + L) + \alpha(1 + Lc)}$$

$$\therefore \quad \bar{Y} = \frac{\alpha}{\alpha + (1 + L)/(1 + Lc)} = \frac{\alpha}{\alpha + \text{const.}} \tag{111}$$

This equation (111) is the equation of a hyperbola, as would be expected. In a similar way it can be easily shown that if $c = 1$, the equation also reduces to that of a hyperbola.

It is clear that the M.W.C. model is capable of explaining sigmoid binding curves, like those of haemoglobin for example. Further, if one can assume that the rate of an enzyme-catalysed reaction is proportional to the fractional saturation of the enzyme with substrate, then sigmoid enzymic rate curves can also be explained. This latter assumption implies that the enzyme obeys rapid equilibrium kinetics, that is to say the rate limiting process is the breakdown of the enzyme substrate complex, while the binding and dissociation of substrate are rapid processes. The M.W.C. model has been extended to the steady state case and once again it can explain sigmoid rate curves.

Heterotropic effects are encompassed within the model in a satisfactory manner as well. Consider first the effect of an allosteric ligand, A, that binds only to the R form of the enzyme, that is to the form that preferentially binds substrate, and which is not the favoured form in the $R_0 \rightleftharpoons T_0$ equilibrium (see Equation 104). Such a ligand will displace that equilibrium, providing a greater concentration of R_0 for the substrate to bind to and increasing the fractional saturation with substrate at low substrate concentrations. It will thus function as an activator, as shown in Figure 57. It is tacitly assumed that when A binds it does not alter either the affinity of the R_0 form for substrate (K_R), nor does it alter the catalytic efficiency of that form.

We may define the equilibrium constant for binding A to the R form as K_A. The form of the binding equation for substrate when c is very small now becomes altered from:

$$\overline{Y} = \frac{\alpha(1 + \alpha)^{n-1}}{L + (1 + \alpha)^n} \qquad \text{(Equation 110)}$$

to:

$$\overline{Y} = \frac{\alpha(1 + \alpha)^{n-1}}{L/(1 + \gamma)^n + (1 + \alpha)^n} \qquad \left(\text{where } \gamma \text{ is } \frac{[A]}{K_A}\right)$$

When [A] is very large relative to K_A then γ becomes very large. This results in an effective reduction of L to zero and the equation reduces to:

$$\overline{Y} = \frac{\alpha(1 + \alpha)^{n-1}}{(1 + \alpha)^n} = \frac{\alpha}{1 + \alpha} \qquad \text{(hyperbolic)}$$

So the effect of an activator of this sort is to reduce the homotropic cooperativity of substrate binding and to make the saturation curve more and more hyperbolic as the activator concentration increases, as shown in Figure 57.

An allosteric ligand I, on the other hand, that binds only to the T form, displaces the $R_0 \rightleftharpoons T_0$ equilibrium to the right and makes it harder for the substrate to bind. It acts as an inhibitor and, further, it increases the homotropic cooperativity of substrate binding. The saturation curve for substrate binding becomes more sigmoid in the presence of I as shown in Figure 57. The binding equation now becomes;

$$\overline{Y} = \frac{\alpha(1 + \alpha)^{n-1}}{L(1 + \beta)^n + (1 + \alpha)^n}$$

where $\beta = [I]/K_I$ and K_I is the equilibrium constant for binding I to T. When β is large, at high inhibitor concentrations, it has the effect of *enhancing L*, the concerted transformation constant, and thus increasing the homotropic cooperativity of substrate binding.

Ligands such as these therefore displace the substrate binding curve to the left or right, without altering the value of V_m. They act as if they were altering the 'K_m' of the enzyme, although strictly speaking, of course, the term Michaelis constant has no meaning for enzymes such as this and it is better to refer to the substrate concentration that gives half-maximal velocity as $S_{0.5}$. Nevertheless such systems of enzyme and allosteric effector are described as 'K systems', and a number of them are known to occur. In some cases the simple theory described above must be extended by allowing the activator or inhibitor to bind to *both* relaxed and tight forms of the enzyme. In these cases the binding equation becomes more complex, but the ratio of the affinities of the allosteric effector for the two forms is defined by the constant d, by analogy with c the ratio for substrate.[40]

Another type of system is also known in which the allosteric effector does not affect the half-saturation point for substrate binding ($S_{0.5}$), but does affect the velocity of the enzymic reaction at any substrate concentration. In particular the value of V_m is affected. Such systems were called 'V systems' in the M.W.C. terminology. Positive V systems are those in which the allosteric effector is an activator, while in negative V systems the effector is an inhibitor. An explanation for their behaviour supposes that there are, as before, two forms of the enzyme in equilibrium, $R_0 \rightleftharpoons T_0$, but now both forms bind the substrate *equally* tightly. One of the forms, say R_0, is assumed to have a greater catalytic efficiency than the other. The allosteric effector is assumed to bind more tightly to one form than the other. If it binds preferentially to the more efficient catalytic form, R, it will act as an activator, because the $R_0 \rightleftharpoons T_0$ equilibrium will be pulled over to favour R_0 in its presence. Conversely if it binds preferentially to the less active form T, it will behave as an inhibitor. Since both R and T bind the substrate equally well ($K_R/K_T = c = 1$), homotropic cooperativity in binding the substrate is absent and the binding curve or rate curve will be hyperbolic.

It is worth noting that in both K systems and V systems the binding of the *effector* itself can exhibit *homotropic* cooperativity. In other words, treating the effector simply as a ligand, and plotting a saturation curve for effector binding against effector concentration can give rise to a sigmoid curve because the effector, in both types of system, has a different affinity for the two forms of the enzyme. In a K system an activator will show homotropic effects in the absence of substrate as it binds preferentially to the unfavoured form. In the presence of a saturating concentration of substrate, the activator no longer exhibits homotropic binding because the $R_0 \rightleftharpoons T_0$ equilibrium will have been displaced already by the substrate. Conversely an inhibitor will not show homotropic binding in the absence of substrate but will do so in its presence.

The basic treatment of heterotropic interaction in the M.W.C. model is clearly capable of further sophistication. It assumes that in K systems the effector binds only to one form of the enzyme and that it affects neither the binding constant for substrate nor the catalytic efficiency of that form. If those restrictions were removed so that, for example, it could bind to both forms, or could affect the binding or catalytic efficiency towards substrate, the resulting mathematical equations would become more complex although they might represent a closer approach to reality in some cases. Similarly in V systems, if the substrate was allowed to bind preferentially to one form, then a mixed K and V system would result. The beauty of the M.W.C. model lies in its simplicity and it has contributed greatly to understanding the processes of cooperativity, both homotropic and heterotropic, from the simple assumption of two forms of the enzyme and the other simplifying assumptions mentioned.

It is worth noting that in the M.W.C. model, in contrast to the Hill or Adair models, the binding of ligand does not *cause* the structural change, and that successive molecules of ligand bind with the *same* binding constant. In fairness it should be pointed out that just as the Hill or Adair models did not rule out the possibility of binding following a change in structure, neither does the M.W.C. model rule out the possibility that binding is followed by conformational change. It is just a matter of emphasis. The M.W.C. model assumes that the only important species in solution in the presence of ligand will be T_0 and R_1–R_4 and, for convenience, it considers that binding occurs after conformational change. The important feature is the concerted transition.

Although this model has been successful in explaining the behaviour of a number of enzymes, it has certain limitations. Perhaps the most important of these is the fact that it makes no provision for *negative cooperativity*. What this apparently paradoxical term means is that the binding of successive molecules of a ligand is accompanied by a reduction in the affinity for subsequent ones instead of an increase, or alternatively a reduction in the catalytic efficiency of the other sites in the oligomer rather than an increase. Enzymes displaying only negative cooperativity do not have sigmoid rate curves, and the Hill coefficient, h, falls to a value of less than one at some point in the binding process. To complicate matters further, enzymes may display a mixture of positive and negative cooperativity where, for example, the first molecules of substrate to bind to the oligomer appear to *increase* the affinity of the other protomers, while later molecules *decrease* it. In order to provide a means of explaining negative, as well as positive, cooperativity and also to avoid the stringent requirements for *concerted* structural change inherent in the M.W.C. model, Koshland and his co-workers have proposed a somewhat different explanation for the kinetics of oligomeric enzymes which in some senses marks a return towards the Adair model.

THE SEQUENTIAL MODEL OF KOSHLAND[41]

As before two conformations of the protomer are assumed to exist and will be represented by circles and squares. Now, however, both types can exist together in the oligomer, as shown in Figure 58(c). This immediately removes the requirement for concerted conformational change that was inherent in the M.W.C. model. The two forms are called the A (circle) and B (square) forms in this model, rather than R and T, and there is assumed to be an intrinsic equilibrium constant for the conversion of A to B which is denoted as K_t the transformation constant. Whereas in the M.W.C. model both R and T forms were able to bind ligand, now only the B form is able to bind and it does so with an equilibrium constant K_s. A further difference between the two models is that in the simplest form of the sequential model the *free* B form of a protomer does not exist at any appreciable concentration; the B form only exists in combination with the ligand S. Taken together with the restriction that the A form does not bind S, one may conclude either that ligand binds to the A form and rapidly converts it to B form, or that the equilibrium between A and B strongly favours A while the binding of S to B is very tight. In either case AS and B do not exist at appreciable concentrations and the only forms that are observed will be A and BS. For convenience it is assumed that the latter sequence of events occurs:

$$A \underset{}{\overset{K_t}{\rightleftharpoons}} B; \qquad B + S \underset{}{\overset{K_s}{\rightleftharpoons}} BS$$

so that the net result is:

$$A + S \underset{}{\overset{K_tK_s}{\rightleftharpoons}} BS$$

Now in the oligomer the transformation of an A protomer to a B protomer is assumed to occur with the same intrinsic equilibrium constant (K_t) as before, but in addition an interaction constant will apply depending on what neighbours are found next to the newly formed B. Consider the following case where a dimer of A protomers changes to an AB dimer. Ignoring the subsequent binding of ligand for the moment, the conformational change can be represented as:

$$\qquad ; \qquad K_{eq} = K_t K_{AB}$$

The equilibrium constant of this process is the intrinsic one K_t multiplied by the interaction constant K_{AB}. If K_{AB} is greater than 1 the interaction stabilizes the new arrangement, but if less than 1 it destabilizes the new arrangement. When $K_{AB} = 1$ there is no interaction. The interaction constant is actually defined as the equilibrium constant for the hypothetical reaction:

$$AA + B \rightleftharpoons AB + A; \qquad K_{AB} = \frac{[AB][A]}{[AA][B]}$$

In a similar way the conversion of the second A protomer to B conformation:

$$\qquad ; \qquad K_{eq} = \frac{K_t K_{BB}}{K_{AB}}$$

involves the loss of the AB interaction and its replacement by a BB interaction, so the process has an equilibrium constant equal to K_t multiplied by a new interaction constant K_{BB} and divided by K_{AB}. Once again K_{BB} is defined for the reaction:

$$AB + B \rightleftharpoons BB + A; \qquad K_{BB} = \frac{[BB][A]}{[AB][B]}$$

It should be noted that K_{AA} is assumed to be 1; the AA pair is used as a reference point with which to compare the relative stabilities of AB and BB pairs. When the number of protomers in the oligomer is greater than two, interesting possibilities arise depending upon the geometrical arrangement of the protomers. For four protomers in a tetramer for example, we may imagine that they are arranged in a square pattern so that interactions occur between neighbours at adjacent corners but not between those at opposite corners:

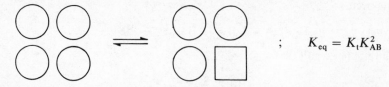 ; $K_{eq} = K_t K_{AB}^2$

In that case the B protomer has two adjacent A protomers with which it can interact and the interaction constant K_{AB} appears twice in the overall equilibrium constant.

For the second A to B conversion in a square tetramer, two alternatives exist:

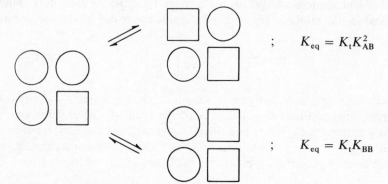

; $K_{eq} = K_t K_{AB}^2$

; $K_{eq} = K_t K_{BB}$

In the upper of these, both the original AB interactions are preserved and two new AB interactions occur so K_{AB} appears twice in the equilibrium constant. In the lower alternative there are two AB interactions originally and two in the product (although one of them is between a different AB pair) so K_{AB} does not appear in the equilibrium constant as there is no change in the number of AB interactions. A new BB interaction appears and is represented by K_{BB} in the equilibrium constant.

There is only one type of AB_3 tetramer, so the conversion of a third protomer to the B form will be as follows:

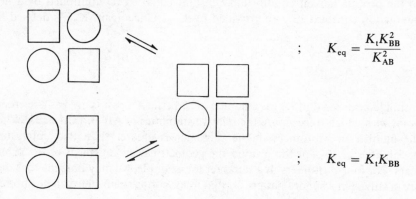

; $K_{eq} = \dfrac{K_t K_{BB}^2}{K_{AB}^2}$

; $K_{eq} = K_t K_{BB}$

In the upper process two AB interactions are lost and replaced by two BB interactions whereas in the lower process the two AB interactions are preserved (although one of them is between different partners in the product) and a new BB interaction appears. The conversion of the fourth protomer to B form has an equilibrium constant:

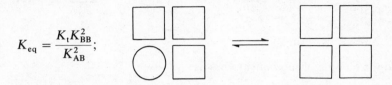

$$K_{eq} = \frac{K_t K_{BB}^2}{K_{AB}^2};$$

A different geometrical arrangement of the protomers may lead to different possibilities for interaction. For example in a tetrahedral array each protomer can interact with all the other three instead of only two of them. The conformational changes are as follows:

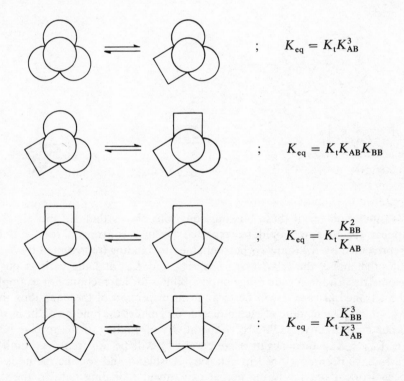

$$K_{eq} = K_t K_{AB}^3$$

$$K_{eq} = K_t K_{AB} K_{BB}$$

$$K_{eq} = K_t \frac{K_{BB}^2}{K_{AB}}$$

$$K_{eq} = K_t \frac{K_{BB}^3}{K_{AB}^3}$$

In all the above examples we have considered only the conformational change, but in its simplest form the model requires that as soon as a B protomer has been formed it must bind the ligand S with an equilibrium constant K_s. In every case, therefore, we should multiply the equilibrium constant given above by K_s to obtain the overall equilibrium constant for the binding process. For example in the tetrahedral case considered, the binding of the first ligand molecule to the tetramer is represented as:

$$K_{eq} = K_s K_t K_{AB}^3$$

and the binding of subsequent ones as:

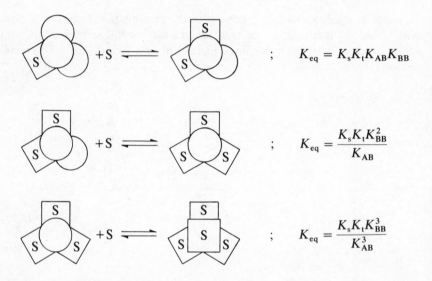

$$K_{eq} = K_s K_t K_{AB} K_{BB}$$

$$K_{eq} = \frac{K_s K_t K_{BB}^2}{K_{AB}}$$

$$K_{eq} = \frac{K_s K_t K_{BB}^3}{K_{AB}^3}$$

Consideration of all these binding equations shows that K_s and K_t always appear together as would be expected because conversion to the B form (represented by K_t) is always accompanied by binding (represented by K_s). On the other hand, the relative sizes of K_{AB} and K_{BB} and the various possible geometries lead to a wide range of possibilities for interaction. For example, in the tetrahedral case it will be clear from inspection of the equations that if $K_{BB} > K_{AB}^2$ the binding of each molecule of S makes the binding of the next one occur more tightly, leading to cooperativity of binding. If the reverse is true, i.e. $K_{BB} < K_{AB}^2$ then negative cooperativity will be apparent. A number of cases have been developed in detail by Koshland and routine methods have been worked out for fitting actual experimental binding data to the theoretical equations.[42] It should be added that the model also allows a situation where non-liganded B protomers do exist at appreciable concentrations, extra terms then appearing in the binding equations. Similarly the model could be extended to allow the ligand to bind to both A and B forms with different affinities.

The sequential model can be seen to be a more versatile one than the M.W.C. model. Indeed the latter may be considered as a special case of the former. This versatility is achieved by the use of extra constants, that is three in the sequential model, $(K_s K_t)$, K_{AB} and K_{BB}, as compared with two in the M.W.C. model, namely L and c. Of course if one builds a complex enough model with sufficient constants it becomes easier to fit almost any experimental set of results, as has often been pointed out. Conversely it becomes more and more difficult to distinguish between complex models if they involve a large number of constants. The sequential model appears to be complex enough to explain the behaviour of most oligomeric enzymes and yet it embodies relatively simple and realistic concepts of conformational change and interaction. At any rate it is an extremely useful working hypothesis at the present time.

Heterotropic effects of a variety of kinds are predicted by the sequential model, depending upon how the allosteric effector is assumed to bind to the A and B forms of the protomers. Three basic types of effector binding have been considered by Koshland and his colleagues.[42,43] In the first, which they call competitive binding, either substrate or effector can bind to a given protomer but both cannot be bound to the same protomer at one time. In the second type, independent binding, either can bind to a given protomer as before, but both can be present together. The third type, ordered binding, refers to a situation in which one ligand must bind before the other can bind to any given protomer. In all three types it is assumed that the two conformational forms of the protein, A and B, exist as before, but variations upon the three types are possible according to whether the substrate binds to the A form or to the B form. Similarly the effector may bind to either A or B forms. When, in addition, the possibility is considered that the binding of effector may either increase or decrease the catalytic efficiency of the protomer to which it is bound, it is clear that a large number of possibilities exist. Nevertheless, by considering the effect of the allosteric ligand upon the values of V_m, $S_{0.5}$ and R_s (a measure of the cooperativity of binding similar to the Hill coefficient h) it was shown how different possibilities could be distinguished using experimental criteria. In this way, by making many experimental observations and determining how the presence of effectors alters the enzymatic rate curve, one may slowly work towards an understanding of the molecular mechanism of heterotropic interactions.

ASPARTATE TRANSCARBAMOYLASE. AN EXAMPLE OF AN ALLOSTERIC ENZYME

Aspartate transcarbamoylase (ATCase) from the bacterium E. coli is an enzyme that has been associated with the development of ideas about regulatory enzymes and allostery since their beginning. As it is one of the most intensively studied of allosteric enzymes it is valuable to compare how the theoretical approaches to homotropic and heterotropic effects actually match up to experimental observations in the real situation of this enzyme, with its particular kinetic and structural properties.

The enzyme catalyses the reaction between aspartate and carbamoyl phosphate to give the products N-carbamoyl aspartate and orthophosphate:

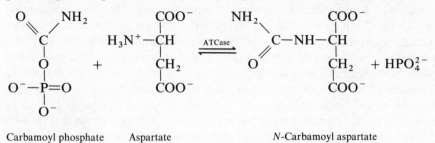

Carbamoyl phosphate   Aspartate N-Carbamoyl aspartate

This is the first step in the sequence of reactions that leads ultimately to the synthesis of the pyrimidine nucleotides, UMP, UDP, UTP and CTP, and of these CTP is known to be a feedback inhibitor of the enzyme *in vivo* (see Figure 63, Chapter 7).

When the initial velocity of the enzymic reaction is plotted as a function of aspartate concentration, with a saturating concentration of the second substrate carbamoyl phosphate throughout, the result is a sigmoid curve as shown in Figure 60. If CTP is present during the measurements, the rate curve becomes even more sigmoid, and $S_{0.5}$ is shifted to higher aspartate concentrations. In the presence of ATP, on the other hand, which functions *in vivo* as an activator, the rate curve becomes less sigmoid and at saturating ATP concentrations it finally becomes hyperbolic. The value of V_m achieved in the presence of saturating concentrations of aspartate is unaffected by the presence of either ATP or CTP. The data shown in Figure 60 can be replotted as a Hill plot in which case a value for h, the Hill coefficient, of 1·6 is found. If aspartate is maintained

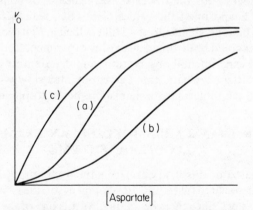

Figure 60   Rate curves for the enzyme aspartate transcarbamoylase (ATCase) in the presence of saturating carbamoyl phosphate. (a) Enzyme alone. (b) Enzyme plus the inhibitor CTP. (c) Enzyme plus the activator ATP

at a constant concentration and v_0 is plotted against the concentration of carbamoyl phosphate, a sigmoid plot is again found. All these findings suggest that ATCase fits the criteria for a typical K system in the M.W.C. terminology.

Further kinetic evidence supports this idea. Two substrate analogues are known, succinate and maleate, that closely resemble the substrate aspartate but, not having an amino group, cannot function as substrates in the enzymic reaction. At high concentrations both succinate and maleate are competitive inhibitors of the enzyme. On the other hand, in the presence of low concentrations of aspartate, low concentrations of either analogue give rise to an increase in activity. This odd situation is readily understood in terms of the M.W.C. model because the analogues could act by pulling the $R_0 \rightleftharpoons T_0$ equilibrium over in favour of the R_0 form thereby making available more of this high-affinity form for the substrate to bind to. In other words, low concentrations of analogue may act just like low concentrations of substrate in the region of the binding curve where it is swinging upwards. Effectively, a molecule of analogue makes available $(n - 1)$ high affinity sites in an oligomeric enzyme composed of n protomers. Even though it is not a substrate itself, so that the site it occupies is blocked and inactive, it can cause an activation effect by helping the low concentration of substrate to pull the $R_0 \rightleftharpoons T_0$ equilibrium. At high substrate concentrations, of course, it could not function in that way and would merely compete for the available active sites. Similarly at high analogue concentration its inhibitory effect would override its ability to mimic substrate in displacing the equilibrium between the high and low affinity forms. A similar explanation could be enunciated in terms of the sequential model. The analogue could mimic the substrate at low concentrations of both, insofar as it might stabilize tight binding conformations in the protomers neighbouring the one it binds to.

The binding curve for succinate in the presence of a saturating concentration of carbamoyl phosphate is found to be sigmoid, as required by the hypothesis that it mimics aspartate.

The binding of CTP, the allosteric inhibitor has also been studied. In the absence of both substrates, CTP binds in an essentially non-homotropic fashion. The Scatchard plot, however, shows good evidence for two classes of CTP binding sites, one class with high affinity and the other with much lower (about 43-fold) affinity. There are almost exactly three high affinity sites and three low affinity sites per molecule of ATCase as shown in Figure 61. This evidence can be interpreted as showing negative cooperativity in the binding of CTP, but other explanations are possible as will be discussed below.

In the presence of high aspartate concentrations, CTP binding follows a sigmoid saturation curve. The Hill plot for the binding data gives a value of $h = 3$. If the M.W.C. model applied to ATCase one would indeed expect the allosteric inhibitor to bind non-cooperatively in the absence of substrate, but cooperatively in its presence as discussed earlier, because the model assumes that the inhibitor binds preferentially to the T_0 form, the predominant form in the absence of substrate. Therefore, with no substrate present the inhibitor

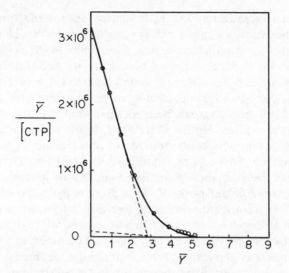

Figure 61 Binding of CTP to ATCase determined by equilibrium dialysis at 4°C in 10 mM potassium phosphate pH 7·0, 2 mM 2-mercaptoethanol, and 0·2 mM disodium EDTA. Radioactive (^{14}C) CTP was used. The data are presented as a Scatchard plot where $\overline{Y}$ symbolizes the moles CTP bound per mole ATCase, and [CTP] is the molar concentration of unbound CTP. The solid line is a calculated construction based on the assumption of the existence of two classes of CTP binding sites on ATCase, each class containing 2·90 sites, with K_d for one class = 9·30 × 10^{-7} M and for the other 4·0 × 10^{-5} M. The pecked lines represent Scatchard plots expected for each class separately and their summation yields the solid line (redrawn from C. C. Winlund and M. J. Chamberlin in *Biochemical and Biophysical Research Communications*, **40**, 45 (1970)

finds effectively only T_0 in solution, to which it binds tightly. Once substrate is present, displacing the equilibrium towards R_0, the inhibitor must pull the equilibrium back again to T_0 before binding, hence giving rise to the sigmoid binding curve.

A number of analogues of CTP also bind to the enzyme and act as inhibitors, although they function less effectively in the order CDP > CMP > cytidine. For each phosphate group removed from CTP, the affinity decreases by a factor of about four. Cytosine itself does not inhibit, nor does any cytidine nucleotide in which the amino group at position 4 in the pyrimidine ring has been methylated. Uridine and its phosphates are very poor inhibitors or else non-inhibitory. These facts indicate that the phosphate groups and the ribose moiety of CTP are important for its binding to ATCase, as is the 4-NH$_2$ group.

The ratio of the affinities of CTP for the T_0 and R_0 forms, that is the value of d in the M.W.C. model, is only small, about 1·5, and this ratio does not vary greatly among the four analogues of CTP already mentioned.

When one turns to the activator ATP, which now has adenine (i.e. 6-amino purine) in place of cytosine (4-amino pyrimidine) as the heterocyclic base, one finds a similar picture. ATP and ADP, together with adenosine tetraphosphate, are the only activators, and the value of d is of course less than 1 as required of an activator, but only a little less than 1. Both CTP and ATP appear to bind at the same site because they act competitively. It seems that small differences in the structure of the heterocyclic base make all the difference between an inhibitor and an activator, whereas the rest of the nucleotide provides the correct groups for binding to the regulatory site of the enzyme. It is worth pointing out that CTP, the most potent of the known allosteric inhibitors, never inhibits the enzyme to a greater extent than about 90%, even at saturating CTP concentrations and very low aspartate concentrations. This is reflected in the value of d close to 1.

All the features discussed so far can be explained by the M.W.C. model, with the possible exception of the negative cooperativity of binding CTP to the enzyme in the absence of its substrates. The sequential model, of course, can also explain all the data, so one is left with the task of trying to make a decision between the two hypothetical models. To attempt this one must turn to the study of the structural and physical properties of the ATCase molecule.

As described briefly in Chapter 3 (Figure 25), the oligomeric molecule can be dissociated by treatment with PCMB into two types of subunits. One of these, comprising 68% of the weight of the original ATCase molecule, can be separated from the other (32% of the original by weight) and, on removal of the PCMB by dialysis against mercaptoethanol, regains activity. This, the so-called *catalytic subunit*, is made up of three identical polypeptide chains, or *C-chains*, and has a molecular weight of $1·03 \times 10^5$ (each C-chain has a molecular weight of $3·4 \times 10^4$). The other subunit separated by PCMB treatment does not regain catalytic activity on dialysis against mercaptoethanol. It does, however, bind CTP and is therefore called the *regulatory subunit*. It has a molecular weight of $3·4 \times 10^4$ and consists of two identical polypeptide chains, the *R-chains*, each of molecular weight $1·7 \times 10^4$.

The native ATCase is found to contain six zinc ions (Zn^{2+}) per molecule and these are so tightly bound that they cannot be removed by extensive dialysis against solutions of chelating agents. After treatment with PCMB to separate the catalytic and regulatory subunits, the zinc is freely dialysable and can be removed with chelating agents. It is not necessary to add zinc ions back to the separated catalytic subunits after removing the PCMB in order to regain catalytic activity, so the zinc is not involved in the active site of the enzyme. Nor is it necessary to add it back to the separated regulatory subunits after removing PCMB, for them to bind CTP, so it appears to be unnecessary for the function of the allosteric site. The possible role of the zinc will be returned to later.

It is clear that **PCMB** breaks down the oligomeric ATCase into trimeric catalytic subunits and dimeric regulatory subunits and, since the molecular weight of ATCase is 3.1×10^5, it must be composed of two catalytic subunits and three regulatory subunits, that is to say six C-chains and six R-chains in all, the figure of six agreeing with the number of zinc ions found in native ATCase. This picture is substantiated by the X-ray crystallographic work so far completed which shows ATCase to be composed of two roughly triangular catalytic subunits held apart by three regulatory subunits as shown in Figure 62. X-ray crystallographic studies on the complex of ATCase with its allosteric inhibitor CTP show a very similar picture with a molecule of CTP bound to the outside of each of the three regulatory subunits as indicated in Figure 62. Presumably these three CTP molecules are bound at the high affinity sites mentioned earlier.

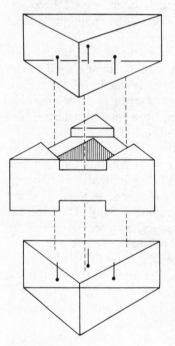

Figure 62 A diagrammatic representation of the structure of the ATCase molecule (cf. Figure 25). The triangular units at top and bottom represent the two catalytic trimers. The three black points in each trimer show the location of the 'active site' sulphydryl groups. The central region of the molecule, into which the catalytic trimers fit snugly, represents the three pairs of regulatory subunits. The main bulk of each pair lies towards the apices of this region but they merge into each other along the sides (reproduced, by permission of Alan R. Liss, Inc., from W. N. Lipscomb *et al.* in *Journal of Supramolecular Structure*, **2**, 89 (1974))

Before trying to construct a picture of how the oligomeric ATCase functions it is convenient to consider what is known about the separated catalytic and regulatory subunits.

The catalytic subunit

Although the catalytic subunit is itself an oligomer, in that it is composed of three apparently identical polypeptide chains, and although it is enzymically active, it shows neither homotropic nor heterotropic effects. The plot of v_0 against either aspartate or carbamoyl phosphate concentration is hyperbolic.

Steady state kinetic studies have established that carbamoyl phosphate must bind first at the active site, followed by aspartate. Apparently, however, the three active sites are quite independent of each other. One slight problem arises in the determination of the number of binding sites because, although three binding sites can be demonstrated for carbamoyl phosphate as would be expected, only two can be shown by analysis of succinate binding data. The affinity of the catalytic subunit for succinate is about 27 times greater in the presence of carbamoyl phosphate than in its absence. This evidence, plus the evidence for an ordered mechanism in which carbamoyl phosphate must bind before aspartate, suggests that the binding of carbamoyl phosphate is associated with a change in conformation that leads to greater affinity for aspartate or succinate. There is additional evidence to support this view, in that changes in the gross structure can be demonstrated by a variety of physical and chemical techniques when carbamoyl phosphate binds. These experiments show in addition that further structural changes occur when succinate subsequently binds.

The conclusion from all these studies is that when substrates bind to each of the three active sites in the catalytic subunit, local changes in tertiary structure occur, but these changes cannot be transmitted from one active site to another. The lack of heterotropic effects in the catalytic subunit is easily explained because, of course, it does not possess the binding site for CTP or ATP, which resides in the regulatory subunit. The failure to exhibit homotropic effects must be due to the lack of any effective structural mechanism for transmitting changes in tertiary structure across the interfaces between the three C-chains. Further evidence bearing on this point is that the C-chains can be separated and then recombined to reconstitute active catalytic subunit. If the separated C-chains are chemically modified by succinylation of amino groups, the reconstituted trimer is inactive. But if hybrid trimers are reconstituted, containing one or two modified chains and two or one unmodified chains, then 66 % or 33 % of full activity respectively is regained. This evidence clearly shows that inactivation of one or two of the component C-chains does not affect the active sites in the remaining chains in the catalytic subunit.

The regulatory subunit

The amino acid sequence of the R-chain has been established and it is clear that both R-chains in the regulatory subunit are identical in primary structure. They are held together in the dimeric regulatory subunit by a fairly weak interaction because dissociation to individual chains can be detected. Twenty-four of the thirty sulphydryl (SH) groups in native ATCase are located in the R-chains, which possess four each. (The remaining six are located one in each of the C-chains.) The four SH groups per R-chain are probably not involved in the interface between the two R-chains in the regulatory subunit because they react readily with thiol reagents such as PCMB and their modification does not lead to increased dissociation of the dimer. They may be located in the region that forms the contact area with the catalytic subunits in the ATCase oligomer, which might explain why PCMB dissociates the catalytic and regulatory subunits from each other. The zinc ions may also reside in this interface, possibly associated with some of the SH groups. Addition of zinc, however, is not required for reconstitution of ATCase from regulatory and catalytic subunits although there is some evidence that mercuric ions derived from PCMB replace the lost zinc ions. This evidence stems from the observation that, even after extensive dialysis of regulatory subunits against mercaptoethanol to remove PCMB, some Hg^{2+} remains tightly bound. About 0·7 mercuric ions are bound per R-chain. Further, in reconstitution experiments where catalytic and regulatory subunits are mixed to re-form ATCase, between 12% and 33% of the regulatory subunits are unable to recombine. This suggests that a metal ion, either Zn^{2+} or Hg^{2+}, is necessary in the interface between each R-chain and the catalytic subunit. The X-ray crystallographic studies carried out so far, while not establishing the position of the Zn ions with certainty, do show that they are indeed near this interface if not in it.

The regulatory dimer is able to bind two molecules of CTP, but the first binds about 20 times more tightly than the second, a situation reminiscent of that found in native ATCase. ATP also binds and can displace CTP, at least from the high affinity sites, but the affinity for ATP is about 30-fold lower than for CTP. The question as to why there should be such a discrepancy between the affinities of the two sites for CTP when the two R-chains are apparently identical has not yet been resolved. It may be that a negative homotropic effect of the Koshland type is at work, so that once one of the two sites in the dimer is occupied by CTP the second site suffers a distortion that reduces its affinity. Alternatively the effect may be due to a more direct form of interaction if the two binding sites lie close to the interface between the R-chains. In that case the physical presence of a CTP molecule in one binding site may block access to the second site. Very little is known about possible structural changes that may accompany CTP binding beyond the fact that the dissociation of the dimer is reduced when CTP binds.

The ATCase oligomer

The major features of the complete molecule are shown in diagrammatic form in Figure 62. The two catalytic subunits are at the top and bottom of the molecule with an equatorial region between them largely occupied by the regulatory subunits. The central cavity is an intriguing feature of the molecule because the six active sites appear to lie on the *inside* faces of the catalytic subunits, facing into the cavity. There is no access to the cavity from above or below, through the catalytic subunits, but six channels in the equatorial region do allow access to it. The evidence for the active sites being on the inner faces of the catalytic subunits is not conclusive, but hinges on the observation that the single SH group of each C-chain is located there. Modification of this SH group by reaction with any one of a number of mercurial derivatives leads to loss of activity and loss of ability to bind succinate, suggesting that it is close to the active site; but if it is converted to an –SCN group, activity is retained. This latter finding shows that the SH group is not necessary for activity. If the active sites are indeed 'inside' the cage constructed by the protein, then substrates would have to diffuse through the equatorial channels to reach them.

To date no three-dimensional structure for the enzyme with bound substrates or analogues has been worked out. It is, however, known that the conformation changes to a looser and more expanded one when the substrates are bound. Carbamoyl phosphate alone causes a small decrease in the sedimentation co-efficient, small increases in the rate of reaction with PCMB and in the rate of proteolytic digestion by trypsin, and a change in the optical rotation. Carbamoyl phosphate plus succinate cause larger changes in all these. It is interesting that in all these cases except trypsin digestion the changes are in the opposite direction to those found when carbamoyl phosphate plus succinate bind to the isolated catalytic subunit.

It is becoming clear that when substrates or analogues bind to the catalytic subunit, local changes in the tertiary structure of the C-chains occur. In isolated catalytic subunits these structural changes remain localized and do not lead to changes in quaternary structure, nor do they affect adjacent active sites. In the complete ATCase molecule, on the other hand, the local changes are transmitted to other active sites and do cause changes in quaternary structure. Presumably this interaction is made possible via the regulatory subunits. Moreover, it is very likely that the quaternary structure changes in a concerted manner because, when a hybrid ATCase molecule is prepared in which one trimeric catalytic subunit is completely inactivated by modification of lysyl residues, the enzyme still shows homotropic and heterotropic cooperativity. Further, when only one of the C-chains in each of the two catalytic subunits is modified, the same result is found. One hypothesis suggested to explain these properties is that the enzyme exists in two forms, one of which corresponds to the diagram shown in Figure 62. This is a 'tight' form, predominating normally in the absence of substrates, in which the two catalytic subunits are eclipsed when viewed from above. This form would have low affinity for substrates but a higher affinity for CTP. The other form, a 'relaxed' one, would have one catalytic subunit rotated through 120° relative to the other, so as to appear staggered when viewed from above. The rotation would necessarily require a change in the relative orientations of the regulatory subunits and might involve a new interfacial contact between R- and C-chains. This relaxed form is assumed to bind the substrates more tightly. The hypothesis is essentially one of concerted change in quaternary structure and, as such, is close to the M.W.C. model. Another possibility is that, in the 'relaxed' form, the channels through which substrates must diffuse if the active sites are inside are widened relative to the channels in the 'tight' form. This would provide freer access for substrates in the 'relaxed' form and might account for the homotropic effects.

Until the X-ray crystallographic studies produce a structure for the enzyme in the presence of substrates or analogues, all such suggestions must remain conjecture. It appears, however, that neither the simple M.W.C. model nor the simple sequential model will be adequate to explain all the subtleties of this intriguing enzyme. Perhaps that should have been expected in any case!

Although ATCase is a fine example of an oligomeric, interacting enzyme it should not be taken as typical of all such enzymes. It is, for example, still the only one where the active and allosteric sites have been clearly shown to lie on different subunits. That has made it easier in some respects to disentangle what goes on when the various ligands bind, but it is probably an extreme case of allostery.

Chapter 7

Enzymes and the Control of Metabolism I: Fine Control of Enzyme Activity

INTRODUCTION

The question of metabolic self-regulation in the cellular milieu is an extremely complex one. Some of the problems were outlined in Chapter 1 and some of the metabolic solutions were mentioned briefly, including the role of negative feedback in maintaining homeostasis. In a multicellular organism various differentiated tissues and organs may mutually regulate each other via the humoral fluids, such as the blood of animals. The problems are therefore more complex than, but of the same general type as, those encountered in the regulation of unicellular metabolism. Many insights into these problems have been gained by biochemists in recent years but the overall goal of a *quantitative* description of metabolism has not yet been reached. By a quantitative description one means a formulation of the metabolic pathways, the enzymes they contain and the distributions of both of these throughout the cell in such a way that allows one to predict the concentrations of enzymes and intermediary metabolites, together with the rates of interconversion of the intermediates under a variety of externally imposed conditions. Such a quantitative model of cellular metabolism must of necessity rely heavily on mathematical description. It would, however, allow us to predict rates of growth, respiration, utilization of growth substrates and excretion of waste products. It is certain that if such a predictive model of even a simple unicell is ever realized, it will depend upon computer simulation because the complexity of the many interactions is too great to be comprehended by mental arithmetic. Nevertheless, fortunately for most biochemists, one does not have to be a computer expert to understand in a qualitative way, or even a semi-quantitative way, the insights into metabolic control that have emerged.

STEADY STATE FLUXES

Let us begin by imagining a very simple set of enzymes that collectively can catalyse the conversion of a substance A to a product Z, the overall conversion

$A \rightleftharpoons Z$ being reversible:

$$A \overset{E_1}{\rightleftharpoons} B \overset{E_2}{\rightleftharpoons} C \overset{E_3}{\rightleftharpoons} \cdots \overset{E_n}{\rightleftharpoons} Z$$

This series of enzymatic reactions can be thought of as a simple linear metabolic pathway.

Now imagine that a solution of all the enzymes (E_1, E_2, $E_3 \ldots E_n$) is placed in a bag of semipermeable membrane that retains them but does not bind them, so that they remain freely in solution inside the bag. At this stage in the thought experiment none of the substrates A, B, C, etc. is present. What will happen if the bag of enzymes is placed into a solution of A at a fixed concentration $[A]_0$ that can be maintained constant indefinitely? At first A will diffuse into the bag and its internal concentration $[A]_i$ will tend to rise with time. The enzyme E_1 will then start catalysing the conversion of A to B at a rate depending on $[A]_i$. As $[A]_i$ increases the enzyme will work faster and the concentration of B will rise, thus allowing E_2 to catalyse its conversion to C and so on down the line until Z is produced. It will simplify matters if we imagine that the intermediates B, C, etc. cannot diffuse out of the bag although Z can. As time passes the concentration of Z inside the bag, $[Z]_i$, will increase and it will start to diffuse out.

Eventually the external concentration, $[Z]_0$ of Z will reach a value such that $[Z]_0/[A]_0 = K_{eq}$, the equilibrium constant for the overall reaction $A \rightleftharpoons Z$. Long before this happens the build-up of the concentrations of the intermediates B, C, etc. inside the bag will mean that the reverse reactions at each step will also be catalysed by the enzyme for that step so that the *net* rate of conversion of A to B, B to C, and so on, will start to decrease. When overall equilibrium is reached, the net rate of conversion of A to Z must be zero and, consequently, the net rate of each step will also be zero and no further net change can occur. Now if we had arranged a 'sink' for Z so that its external concentration $[Z]_0$ was maintained at a constant value such that $[Z]_0/[A]_0$ is less than K_{eq}, then equilibrium could never be achieved. Instead a steady state would be reached. The concentrations of the intermediates would reach constant values such that their net rates of synthesis were balanced by their net rates of removal. Take the intermediate B for example. Its net rate of production from A, catalysed by E_1, will be v_1 and the value of v_1 will depend upon the concentrations of A and B and E_1 together with the kinetic constants of E_1 (K_m^A, K_m^B, V_m^A and V_m^B). The net rate of conversion of B to C by E_2 will be v_2, again determined by [B], [C], $[E_2]$ and the kinetic constants for E_2. When $v_1 = v_2$ the concentration of B will remain constant. Similarly when $v_2 = v_3$ the concentration of C will remain constant and so on down the line. In other words the steady state will be characterized by fixed concentrations of the intermediates and by *a constant flux through the pathway*. This flux, **F**, will be the rate of conversion of A to Z and will be the same as the net rate from left to right for each of the enzymic steps in the pathway. Of course A must be supplied and Z removed externally at the rate **F** also, otherwise the concentrations $[A]_0$

and $[Z]_0$ could not be maintained at constant values. What determines the value of the flux through the pathway? Clearly the fixed external concentrations of A and Z will be involved, as will the concentrations of the enzymes E_1, E_2 etc., and also the kinetic constants of all the enzymes. In addition it must not be forgotten that the rates at which A can enter the bag and Z can leave it will be involved in determining the steady state flux and these rates may depend upon the nature of the membrane.

If any one of these factors is altered one may expect the flux and the concentrations of the intermediates to alter in response. For example increasing $[A]_0$ may lead to an increase in both the flux and all the intermediate concentrations. A decrease in the concentration of any one of the enzymes may lead to a decrease in the flux and to an increase in the concentrations of intermediates that precede it in the pathway accompanied by a decrease in the concentrations of those that follow it. Such possibilities can be comprehended intuitively but it is difficult to estimate the effects quantitatively.

The imaginary 'cell' we have just been considering is a very poor model for a real living cell, but it does illustrate the principle that any collection of enzymes constituting a linear reversible pathway for the conversion of one substance to another will, at fixed concentrations of the two substances, adopt a steady state characterized by a flux and constant concentrations of intermediates, and that the values of these are determined by the whole system.

The model cell considered above is a poor model for a number of reasons which are worth delineating. In the first place we considered only a single starting substance A outside the cell whereas in real life cells require a number of such substances. For example, a minimum requirement for a carbon source such as glucose, a nitrogen source such as ammonium ion, inorganic phosphate, oxygen, and trace metal ions is exhibited by many micro-organisms. In order to make use of these substrates a real cell needs several metabolic pathways, and produces a number of products instead of just one. Further, in the model we considered, the single product Z was excreted, so there was no possibility for growth. Real cells do excrete unwanted waste products but they retain a proportion, and often a high proportion, of what they consume in order to provide for growth.

All the enzymes considered in the model acted on one substrate producing one product, so they would be classed as Uni Uni enzymes in Clelands' nomenclature; whereas many, if not most, real enzymes catalyse multisubstrate reactions. The cell contains coenzymes which act as shuttles between one enzyme and another and between one metabolic pathway and another. In the model, any one intermediate such as B was produced by only one enzyme and utilized by only one other; and there was no possibility for interaction of one part of the pathway with remote parts, no possibility for feedback or feedforward.

The enzymes E_1, E_2, etc. in our model were assumed to be present at fixed concentrations whereas enzyme levels in real cells may change due to synthesis or degradation. They were all considered to be freely in solution although of course some enzymes are membrane bound and others are more or less sealed off in cellular compartments or organelles in living systems. The model cell and its reactions were all considered to be readily reversible. In principle there is nothing wrong with that, although in reality many cellular reactions are effectively irreversible and irreversible reactions or pathways play important roles in cellular metabolism.

Perhaps the greatest weakness of the model system is the fact that it was totally at the mercy of the environment in which it was situated. There was no way in which it could buffer its internal concentrations or flux against change in the external concentrations of A and Z. It showed no propensity for self-regulation.

FLUX REGULATION BY FEEDBACK

Although the simple linear pathway is a poor model for the complex branching, interlinked system of a real cell, nevertheless one can often identify *parts* of the total cellular metabolism that can be considered as subsystems with a beginning and an end in a meaningful and useful way. Before attempting to isolate such subsystems for analytical purposes it is necessary to know the 'metabolic map' of the whole cellular system in some detail. This knowledge of the major pathways, the enzymes and intermediates they contain and the points at which they interact, is a prerequisite for attempting to understand cellular regulation, but in many cases this knowledge has been collected by the patient work of biochemists. For example, in certain situations the glycolytic pathway can be considered as a subsystem with a beginning at glucose and an end at pyruvate, with all the reactions taking place in the cytoplasm.

Returning to the model we considered before, and turning a blind eye to many of its inadequacies for the moment, one can make it a more meaningful model of a *subsystem* by a simple modification. Instead of allowing the end-product of the pathway, Z, to be excreted, let us imagine that it is retained in the cell and utilized for some purpose such as growth or movement. We can represent this utilization, whatever its real nature, by the device of a 'dummy' enzyme which will act at a variable rate to remove Z as required. The role of the subsystem can then be considered as supplying Z and maintaining it at a suitable concentration, or within a limited range of concentration, whatever demands are made on it as shown below:

$$A_0 \longrightarrow A_i \xrightarrow{E_1} B \xrightarrow{E_2} C \xrightarrow{E_3} \cdots \xrightarrow{E_n} Z \xrightarrow{E_z}$$

This diagrammatic representation shows E_z as the 'dummy' enzyme representing a variable demand and also includes a provision for feedback. The end product Z is allowed to inhibit the first enzyme of the pathway. Now when the demand for Z is high (i.e. E_z is active) a slight drop in the concentration of Z will occur; this will result in a release of inhibition at the first step in the pathway. E_1 will then work faster and Z will be synthesized at an increased rate to supply the demand. As soon as the demand for Z falls (E_z becomes less active) the concentration of Z will tend to rise and this will inhibit E_1 and thus reduce the flux through the pathway and Z will no longer be made as rapidly.

Many examples of this type of subsystem are documented. One of the best studied is the pathway of pyrimidine biosynthesis shown in Figure 63. Starting with carbamoyl phosphate and aspartic acid a linear sequence of eight enzymic reactions leads to the end-product cytidine triphosphate (CTP) which acts as an allosteric inhibitor of the first enzyme of the pathway, ATCase, as described in Chapter 6. Other examples are found in the biosynthetic pathways leading to various amino acids such as isoleucine or histidine. These and other examples are described elsewhere[44] and will not be elaborated here. In all cases the principle of negative feedback operates to control the rate of synthesis of the end-product. In some cases a slight complication arises in that the pathway may branch at some point to produce two or more final products:

$$A \xrightarrow{E_1} B \xrightarrow{E_2} C \xrightarrow{E_3} D \begin{array}{c} \xrightarrow{E_4} E \rightarrow \cdots \rightarrow X \\ \xrightarrow{E_5} F \rightarrow \cdots \rightarrow Y \end{array}$$

In this case if the end-product X were to inhibit the first enzyme of the overall pathway (E_1) the rate of synthesis of X would certainly be capable of self-regulation but that of Y would not. In circumstances where Y was required, but X was not, there would be no way of achieving the synthesis of Y. Conversely when X was being made, Y would be also, whether or not it was required. One solution is that shown above, where X inhibits the first enzyme after the branch point leading to its own synthesis. This enzyme (E_4) is the first enzyme that is *committed* to synthesis of X. Similarly Y inhibits the first committed enzyme (E_5) on its branch of the pathway. That solution, however, leaves the early part of the pathway from A to D uncontrolled. This could be uneconomical, especially in circumstances where neither X nor Y was required, because the substrate A would be converted unnecessarily into B, C and D when perhaps it might be used by the cell for some other purpose. Further, the concentrations of B, C and D might rise to dangerously high levels in the cell. Two alternatives present themselves. The intermediate D could act as a feedback inhibitor of E_1. Alternatively X and Y *in combination* could secure the inhibition of E_1 whereas neither alone would have much effect. The latter possibility, called synergistic inhibition, is a subtle adaptation that is found in some cases. An enzyme that is subject to regulation by more than one effector is called a polyvalent regulatory enzyme.

Negative feedback is an obvious way of controlling biosynthetic pathways but is of less value in a catabolic situation where cellular activity must be geared to the supply of the initial substrate. In that case the principle of positive feedforward may be found to operate. A good example is found in the glycolytic pathway where fructose diphosphate, an early intermediate in the pathway, activates pyruvate kinase, the last enzyme in the pathway.

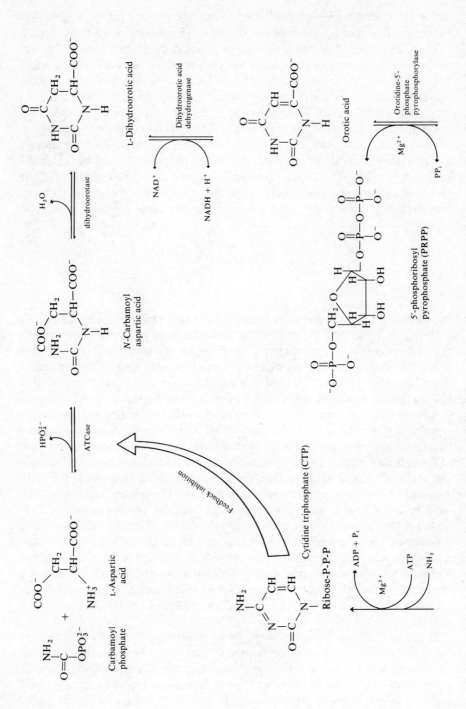

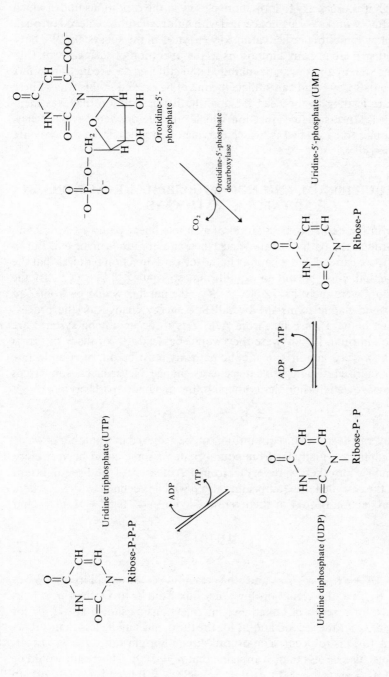

Figure 63 The pathway of pyrimidine biosynthesis, showing end-product inhibition of the first enzyme of the pathway (ATCase) by CTP

Although the principles of homeostasis and flux control in subsystems are clear enough it is necessary to look more closely at the conditions under which they can actually work. We are concerned with autoregulation achieved through flux control by inhibiting or activating key enzymes in the subsystem. We have to ask whether there are any limitations to the operation of such control. Can any enzymic step in a pathway be inhibited and still lead to a reduction in flux through the pathway? And can sufficient control be exerted in this way? These questions are leading to the need for quantitative description that was mentioned earlier. An associated question which must be answered when a new situation comes to be studied is 'which enzyme or enzymes in the pathway are actually controlled?'.

NEAR-EQUILIBRIUM AND NON-EQUILIBRIUM REACTIONS IN METABOLIC PATHWAYS

As we saw in the earlier consideration of a simple linear pathway A...Z, no net flux through the pathway can occur when equilibrium is achieved. Put in another way we can say that a net flux in the forward direction requires that the overall reaction must be out of equilibrium so that $[Z]/[A] < K_{eq}$. (If the disequilibrium were such that $[Z]/[A] > K_{eq}$, the net flux would be from Z to A.) In thermodynamic terms the overall free energy change for the process $A \rightarrow Z$ must be negative if flux from A to Z is to occur. Living systems are never in equilibrium. If they were they would be dead! A corollary of this is that each and every step in a metabolic pathway must be out of equilibrium, and is characterized by a negative free energy change. Nevertheless some steps may be closer to equilibrium than others. In the imaginary pathway:

$$A \underset{}{\overset{E_1}{\rightleftharpoons}} B \underset{}{\overset{E_2}{\rightleftharpoons}} C \underset{}{\overset{E_3}{\rightleftharpoons}} D$$

any given overall degree of disequilibrium can be achieved in a number of ways. Each step could be equally far from equilibrium, or some could be very close to equilibrium while others were very far from it. A measure of the disequilibrium is given by the so-called disequilibrium ratio ρ which is defined as Γ/K_{eq} where Γ is the mass action ratio. For the overall reaction $A \rightleftharpoons D$, $\Gamma = [D]/[A]$, and therefore:

$$\rho = \frac{[D]/[A]}{K_{eq}} \tag{112}$$

When $[D]/[A] = K_{eq}$, $\rho = 1$ and the reaction is at equilibrium. When $[D]/[A] < K_{eq}$, i.e. the condition for a net flux from A to D, then $\rho < 1$. In other words, for a series of linked enzymic reactions as shown above, ρ_1 for the first step, ρ_2 for the second and ρ_3 for the third, must all be less than 1 if net flux from A to D is to occur. The overall disequilibrium ratio $\rho = \rho_1 \cdot \rho_2 \cdot \rho_3$ and this must also be less than 1. Imagine that $\rho = 0.001$. This result would be achieved if $\rho_1 = \rho_2 = \rho_3 = 0.1$ or if $\rho_1 = 0.99$, $\rho_2 = 0.1$, and $\rho_3 = 0.0101$. In the first case all three steps are equally far from equilibrium, but in the second case the first step is very close to equilibrium while the other two are far from it. An infinite number of other possibilities also exist, giving the same overall disequilibrium ratio.

Consider the first case where $\rho_1 = 0.99$ and the reaction $A \rightleftharpoons B$ catalysed by the enzyme E_1 is virtually at equilibrium. This implies that a high concentration of the enzyme E_1 is present so as to catalyse both directions of the reaction, $A \rightarrow B$ and $A \leftarrow B$, at almost equal rates. There is clearly a great excess of E_1 present, as far as its ability to maintain the net flux F through the pathway is concerned. The flux is given by the difference between the rate in the forward direction, which will be defined as v_f for this step, and the rate, v_b, in the reverse direction:

$$F = v_f - v_b$$

If v_f and v_b are almost equal, then both must be much larger than the flux F. It follows that, even if the concentration or specific activity of E_1 is reduced very considerably, the flux will not be greatly altered. The reason for this lies in the fact that any reduction in the activity of the enzyme catalysing a 'near-equilibrium' reaction will be offset by a small reduction in the concentration of the product of that reaction, thereby effectively restoring the flux through that step. The following example illustrates the point.

In general, the ratio of the rate of the reverse reaction v_b to the rate of the forward reaction v_f, can be shown to equal the disequilibrium ratio for the step in question:

$$\rho_1 = v_b/v_f \ .$$

This is proved as follows. The net rate catalysed by the enzyme E_1 will be given by an equation of the form:

$$v = v_f - v_b = \frac{V_m^A K_m^B [A] - V_m^B K_m^A [B]}{K_m^A K_m^B + K_m^B [A] + K_m^A [B]} \quad \text{(cf. Equation 50 in Chapter 5)}$$

where V_m^A, V_m^B, K_m^A and K_m^B are the appropriate kinetic constants. Thus:

$$v_f = \frac{V_m^A K_m^B [A]}{D} \quad \text{where } D = (K_m^A K_m^B + K_m^B [A] + K_m^B [B]) \quad (113)$$

and:

$$v_b = \frac{V_m^B K_m^A [B]}{D}$$

$$(114)$$

$$\therefore \quad \frac{v_b}{v_f} = \frac{V_m^B K_m^A [B]}{V_m^A K_m^B [A]}$$

Using the Haldane relationship $V_m^A K_m^B / V_m^B K_m^A = K_{eq}$ (cf. Equation 52 in Chapter 5):

$$\therefore \quad \frac{v_b}{v_f} = \frac{[B]}{K_{eq}[A]}$$

But from Equation 112 we know that $[B]/K_{eq}[A] = \rho_1$

$$\therefore \quad \frac{v_b}{v_f} = \rho_1 \tag{115}$$

The exact form of the rate equation applicable to E_1 does not matter. In all cases the result of Equation 115 is obtained.

In the particular case in question, where E_1 catalyses a 'near-equilibrium' reaction, $\rho_1 = 0.99$ and therefore $v_b = 0.99v_f$ from Equation 115. It follows that

$$v = v_f - v_b = v_f - 0.99v_f = 0.01v_f$$

Of course v is equal to $\mathbf{F}$ the flux through the pathway in the steady state.

If we now imagine that the concentration of E_1 is reduced to half of its initial value, the immediate result will be that the rates v, v_f and v_b will all fall to half of their original values because V_m^A and V_m^B contain e_1, the concentration of E_1. If nothing else in the imaginary pathway

$$A \underset{}{\overset{E_1}{\rightleftharpoons}} B \underset{}{\overset{E_2}{\rightleftharpoons}} C \underset{}{\overset{E_3}{\rightleftharpoons}} D$$

is altered and $[A]$ remains at a constant value, the enzymes E_2 and E_3 will continue at their original rates with the result that the concentration of B will start to fall. When, however, it has fallen to 99 % of its original value, the new value of ρ_1 will be 0.98 instead of 0.99. This is because $\rho_1 = [B]/[A]K_{eq}$ and since neither $[A]$ nor K_{eq} alters, a reduction in $[B]$ to 99 % of its original value reduces ρ_1 from 0.99 to $0.99 \times 0.99 = 0.98$. This in turn means that the new ratio of the reverse and forward rates is given by:

$$\frac{v_b'}{v_f'} = 0.98$$

The value of v_f is virtually unaffected by the change in $[B]$ since it depends mainly on $[A]$ and a small change in $[B]$ affects only one term in $\mathbf{D}$ the denominator of Equation 113. It will fall to half of its original value due to the fall in e_1. The value of v_b is affected by the change in $[B]$ as well as by the fall in e_1 as seen from Equation 114, and is now given by:

$$v_b' = 0.98v_f' = 0.98 \times 0.5v_f = 0.49v_f$$

The net rate catalysed by E_1 will now be given by:

$$v' = v_f' - v_b' = 0.5v_f - 0.49v_f$$

Thus $v' = 0.01v_f$ which is equal to the original value of v. This result clearly shows how the net rate catalysed by E_1 returns rapidly to its original value when the concentration of E_1 is cut to half, and requires only a 1 % fall in the product concentration $[B]$ to achieve this. The same result applies if, instead of reducing the concentration of E_1 its activity is reduced by an inhibitor. For this reason enzyme reactions that are close to equilibrium are not expected to be good sites for metabolic regulation. A feedback inhibitor would have to alter the activity of a 'near-equilibrium' enzyme so drastically to achieve a measurable effect upon the flux through the pathway that it would be an uneconomical approach to regulation. Of course a 99 % inhibition of a 'near-equilibrium' enzyme would affect the flux, but would also remove so much of the enzyme activity that the step would no longer be 'near-equilibrium'.

One of the first experimental objectives in pinpointing the steps involved in flux regulation in a metabolic pathway is now clear. The values of ρ for each step must be estimated, by measuring the metabolite concentrations and the equilibrium constants for each step. Values of ρ close to unity indicate 'near-equilibrium' steps and these can then be eliminated as possible regulatory steps. As a purely empirical guide it has been suggested by Rolleston[45] that values of ρ greater than 0·2 should be taken to indicate 'near-equilibrium' steps, while values less than 0·05 indicate 'non-equilibrium' steps. Values of ρ lying between 0·05 and 0·2 are ambiguous and may or may not indicate regulatory possibilities.

Non-equilibrium steps characterized by low values of ρ are possible sites of regulation because v_b is negligible in comparison with v_f. The flux through such steps is largely determined by v_f and so changes in the amount or activity of the enzymes catalysing them can lead to substantial changes in rate. This is not to say that such non-equilibrium steps *must* be regulatory; only that they could be. In practice it makes good metabolic sense to regulate a non-equilibrium step for the following reason. Being far from equilibrium the free energy change associated with such a step will be large and negative. By failing to regulate the step by, for example, feedback inhibition, an unnecessary build up of all intermediates following it will result and this may be uneconomical in both energetic terms and in view of the loss of substrate material. For this reason the first step in a metabolic pathway is often found to be both irreversible, that is 'non-equilibrium', and regulated.

SATURATED OR SUBSTRATE INDEPENDENT REACTIONS: THE IMPORTANCE OF V_m

Another factor of importance in deciding whether an enzymic step in a pathway is a possible candidate for flux regulation is its V_m. Consider a particular step in a metabolic pathway, let us say $\cdots \rightarrow B \xrightarrow{E} C \cdots \rightarrow$ catalysed by the enzyme E, where the reaction is far from equilibrium, that is ρ is small. First imagine that the steady state concentration of B is of the same order as its K_m, so that E is not saturated by B and the flux is much less than V_m for the enzyme. If the activity or concentration of E is reduced to half its original value, the rate catalysed by the enzyme will fall to half its original value immediately. Any subsequent drop in the concentration of C, the product, will be unable to restore the rate because the reaction is far from equilibrium, but the concentration of B, the substrate, may tend to rise because E is no longer removing it as rapidly as before. In that case the flux can be partially restored since E will now catalyse the reaction more rapidly in response to the increased concentration of B. In such a situation the flux will be sensitive to changes in the activity of E, but not as sensitive as one might imagine, owing to the buffering or compensating effect that the substrate concentration imposes.

Now imagine a different steady state situation in which [B] is sufficient to saturate E, that is [B] > $10K_m$, and the flux is close to V_m for the enzyme. When the activity or concentration of E is now reduced to half its original value, the rate catalysed by E falls to half and no subsequent increase in the substrate concentration can restore it. This step is said to be substrate independent and the flux is directly proportional to the activity of E. Such a step would be quite a good candidate for flux regulation, either via changes in the concentration of the enzyme, or by feedback inhibition.

THE PIPE ANALOGY: PACEMAKERS AND BOTTLENECKS

When considering the flux of metabolites through a metabolic pathway it is tempting to draw an analogy with a flow of liquid through a pipe. At a constant pressure differential over the length of the pipe, the flow of water will be determined by the flow that can be accommodated by the narrowest part of the pipe. If that bottleneck is enlarged the flow will increase and vice versa. It is often suggested that every metabolic pathway has one step that acts as a bottleneck. The enzyme catalysing this step is referred to as the pacemaker enzyme and of course it is the obvious candidate for regulation of the flux. A substrate independent and, by inference, non-equilibrium enzyme is a good case where such a concept might be justified. One must, however, beware of oversimplification of what are after all highly complex systems. Somewhat paradoxically a pathway may have more than one pacemaker, or it may have none. The first situation can arise when one enzyme will act as pacemaker under one set of conditions but another may take over under different conditions. The second situation may be nearer to reality in many cases in that all the enzymes and substrates together determine the flux over a wide range of conditions. Indeed Kacser and Burns in an excellent paper on flux control[46] have stressed the importance of considering the system as a whole. Moreover the system should be analysed under the particular conditions at which it operates physiologically and *small* perturbations of those conditions used to gain insight into the controlling elements that operate in those circumstances. If a large change to a new set of physiological conditions sometimes occurs, for example in altered dietary or developmental states, then the whole analytical process should be repeated since new regulatory factors may come into prominence. In short a cell is not a pipe. A metabolic pathway is much more flexible than a water main.

THE THEORY OF KACSER AND BURNS

An outline of the theoretical framework set up in 1973 by Kacser and Burns[46] may help to clarify some of the concepts that have been touched upon in the previous sections.

First a clear distinction must be drawn between factors external to a metabolic system or subsystem which in a sense are under the investigator's control, and internal factors which are not under this direct control. External substrate concentrations which can be set *in vivo* by the environment or *in vitro* by the experimenter come into the former category, as do the predetermined gene doses which set the levels of enzymes. Such factors are called *parameters* in this theory. A parameter may be altered; the system then responds to the alteration by various changes in the levels of intermediary metabolites, leading to a new steady state characterized by a new flux or fluxes. The responding factors over which the investigator has no direct control are called *variables*.

Next the theory defines the concept of *sensitivity* (sometimes called 'control strength' by other workers). The flux through any given subsystem will be more dependent upon the activity of some of the enzymes in the pathway than upon others, as was demonstrated earlier in the discussion of near-equilibrium and non-equilibrium enzymes and in the consideration of the importance of V_m. The sensitivity, Z, of a particular step is defined as the fractional change in flux produced by a given fractional change in the concentration of the enzyme that catalyses it:

$$\frac{\delta F}{F} = Z\frac{\delta e}{e} \tag{116}$$

where δF is a small change in the flux F produced by a small change δe in enzyme concentration e. The sensitivity of a given step need not be constant. Indeed it is not expected to be so when large changes are considered, but over a small range of e and F it may be considered as a constant.

An important result emerging from the theory is that the sensitivities of all the enzymic steps in a subsystem are interrelated. For example, under rather simplified circumstances where all the enzymes in the subsystem are not saturated, the sensitivities must add up to 1. Since the maximum value for any given sensitivity is 1, and all must be positive, this result shows clearly that, under any particular set of circumstances, not more than one of the steps can be totally controlling the flux ($Z = 1$), but it may happen that no single one is controlling, they may all contribute equally.

The second important concept is that of *controllability*, K. The controllability of an enzyme is defined as the fractional change in rate produced by a given fractional change in a parameter, which may for example be the concentration of an externally applied inhibitor I:

$$\frac{\delta v}{v} = K_1\frac{\delta[I]}{[I]} \tag{117}$$

It can be shown that the fractional change in flux through a pathway resulting from the fractional change in [I] acting upon the step with sensitivity Z and controllability K_1, is given by:

$$\frac{\delta F}{F} = ZK_1\frac{\delta[I]}{[I]}$$

In other words both the sensitivity and the controllability of a particular step are involved in determining the flux change.

Another important concept is the *elasticity* of a step, ε. This is defined as the fractional change in enzyme rate produced by a given fractional change in a variable, such as substrate, product, or allosteric effector concentration. For example:

$$\frac{\delta v}{v} = \varepsilon_S \frac{\delta[S]}{[S]} \tag{118}$$

where ε_S is the elasticity of the step to changes in the concentration of S. Note that since we are dealing now with a variable, it is changes in the rate v that are considered, not changes in flux. Once again the elasticity of a step towards any particular variable is not constant, but over a small range may be considered as a constant.

One of the greatest achievements of the theory is the way in which it shows that the sensitivities of the various steps in a pathway are linked via the elasticities of successive steps towards common intermediates. Take for example two successive steps in a pathway:

$$\cdots \rightarrow B \xrightarrow{E_1} C \xrightarrow{E_2} D \cdots \rightarrow$$

Let the steady state flux be F, and let the rates catalysed by E_1 and E_2 be v_1 and v_2 respectively ($F = v_1 = v_2$). The sensitivities of E_1 and E_2 are Z_1 and Z_2 respectively, and their elasticities towards the intermediate C are $^1\varepsilon_C$ and $^2\varepsilon_C$ respectively. Now imagine that the concentration of E_1 is reduced from e_1 to $(e_1 - \delta e_1)$ at the same time as the concentration of E_2 is increased from e_2 to $(e_2 + \delta e_2)$ in such a way that the two changes exactly balance each other in their effects on the flux through the pathway, so that $\delta F = 0$. The only result will be a fall in the steady state concentration of C from [C] to ([C] $- \delta[C]$). If this seems an unreasonable conclusion, it can be approached from a different point of view.

Consider the enzyme E_1 in isolation and let the concentration of C fall by $\delta[C]$. The rate will increase by an amount δv_1 such that, from equation 118:

$$\frac{\delta v_1}{v_1} = {}^1\varepsilon_C \frac{\delta[C]}{[C]}$$

This can be offset by decreasing the concentration of E_1 by an amount δe_1 so that no net change in rate at this step occurs. Then:

$$\frac{\delta e_1}{e_1} + {}^1\varepsilon_C \frac{\delta[C]}{[C]} = 0 \tag{119}$$

Similarly for E_2, the drop in [C] will lead to a reduction in rate which can be offset by an increase in the concentration of E_2 such that:

$$\frac{\delta e_2}{e_2} + {}^2\varepsilon_C \frac{\delta[C]}{[C]} = 0 \tag{120}$$

For the pathway as a whole, the net change in flux is given by the sum of the effects that each change in enzyme concentration would have, that is, from Equation 116:

$$\frac{\delta F}{F} = 0 = Z_1 \frac{\delta e_1}{e_1} + Z_2 \frac{\delta e_2}{e_2} \tag{121}$$

Now from Equations 119 and 120 we can substitute for $\delta e_1/e_1$ and $\delta e_2/e_2$ so that Equation 121 becomes:

$$\frac{\delta F}{F} = 0 = Z_1 {}^1\varepsilon_C \frac{\delta[C]}{[C]} + Z_2 {}^2\varepsilon_C \frac{\delta[C]}{[C]} = \frac{\delta[C]}{[C]}(Z_1 {}^1\varepsilon_C + Z_2 {}^2\varepsilon_C)$$

and since $\delta[C]/[C]$ is not zero:

$$Z_1 {}^1\varepsilon_C + Z_2 {}^2\varepsilon_C = 0$$

$$\therefore \quad \frac{Z_1}{Z_2} = -\frac{{}^2\varepsilon_C}{{}^1\varepsilon_C} \tag{122}$$

In other words the sensitivities of E_1 and E_2 are related to their elasticities towards their common intermediate C by the inverse relationship given in Equation 122.

The sensitivity of an enzyme cannot be measured easily in a real system, unless one can actually bring about changes in its concentration and measure the change in flux. Its elasticities, on the other hand, can be estimated from a knowledge of its kinetic behaviour. Once the rate equation for an enzyme has been established by studies *in vitro* as described in Chapters 5 and 6, the elasticities towards substrates, products and effectors can be easily obtained. In the case we have just considered, for example,

$$\cdots \rightarrow B \xrightarrow{E_1} C \xrightarrow{E_2} D \cdots \rightarrow$$

the two enzymes E_1 and E_2 might be expected to have rate equations of the form:

$$v_1 = \frac{V_m^B K_m^C[B] - V_m^C K_m^B[C]}{K_m^B K_m^C + K_m^C[B] + K_m^B[C]} \quad \text{for } E_1$$

$$v_2 = \frac{V_m'^C K_m'^D[C] - V_m'^D K_m'^C[D]}{K_m'^C K_m'^D + K_m'^D[C] + K_m'^C[D]} \quad \text{for } E_2$$

By partial differentiation of these equations with respect to [C] it can be shown that the elasticity of E_1 with respect to C is given by:

$$^1\varepsilon_C = \frac{-[C]/K_1}{[B] - [C]/K_1}\left(1 + \frac{F}{V_m^C}\right) \tag{123}$$

where K_1 is the equilibrium constant for the reaction $B \rightleftharpoons C$ and F is the flux through the pathway. (For a proof of this equation the reader is referred to Kacser and Burns paper).[46] Since ρ_1, the disequilibrium ratio for this step, is given by $[C]/[B]K_1$ (cf. Equation 112) it follows that:

$$^1\varepsilon_C = \frac{-\rho_1}{1 - \rho_1}\left(1 + \frac{F}{V_m^C}\right) \qquad (124)$$

Similarly it can be shown that for enzyme E_2:

$$^2\varepsilon_C = \frac{[C]}{[C] - [D]/K_2}\left(1 - \frac{F}{V_m^{'C}}\right) \qquad (125)$$

where K_2 is the equilibrium constant for the step $C \rightleftharpoons D$. Once again, since $\rho_2 = [D]/[C]K_2$ it follows that:

$$^2\varepsilon_C = \frac{1}{1 - \rho_2}\left(1 - \frac{F}{V_m^{'C}}\right) \qquad (126)$$

We can now use Equations 124 and 126 to substitute for $^1\varepsilon_C$ and $^2\varepsilon_C$ in Equation 122 so that:

$$\frac{Z_1}{Z_2} = \frac{\dfrac{1}{1 - \rho_2}\left(1 - \dfrac{F}{V_m^{'C}}\right)}{\dfrac{\rho_1}{1 - \rho_1}\left(1 + \dfrac{F}{V_m^C}\right)}$$

$$= \frac{(1 - \rho_1)\left(1 - \dfrac{F}{V_m^{'C}}\right)}{\rho_1(1 - \rho_2)\left(1 + \dfrac{F}{V_m^C}\right)} \qquad (127)$$

In cases where the flux F through the pathway is much less than the V_m s of the enzymes in the pathway, Equation 127 simplifies to:

$$Z_1 : Z_2 = (1 - \rho_1) : \rho_1(1 - \rho_2)$$

By extending the logic for the rest of the pathway it can be seen that:

$$Z_1 : Z_2 : Z_3 : Z_4 \cdots = (1 - \rho_1) : \rho_1(1 - \rho_2) : \rho_1\rho_2(1 - \rho_3) : \rho_1\rho_2\rho_3(1 - \rho_4) \qquad (128)$$

In cases where F is comparable with the V_m s of the enzymes, extra terms appear in Equation 128, but it is now clear why the sensitivities of all the enzymes in a pathway are linked to the disequilibrium ratios and V_m s for the steps. Thus it is only necessary in principle to know the sensitivity of any *one* enzyme and all the others can be found from the disequilibrium ratios and V_m s of the enzymes, using a relationship of the form of Equation 128.

By approaches such as these it is now becoming possible to build quantitative models of metabolic pathways. Detailed knowledge of the rate equations for each enzyme can be incorporated into computer programs together with information on metabolite levels and suchlike in order to simulate the whole pathway and study the way in which it will respond to alterations in its environment and how it is regulated.

SWITCH MECHANISMS

Changes in *flux* through a pathway are, on the whole, expected to be less than the changes in *rate* that would be produced by some alteration in the activity of an isolated enzyme, because of the buffering effects described earlier. In view of this it is necessary to ask whether the observed properties of enzymes actually allow sufficient control to be exerted, for example in feedback regulated systems. In order to achieve the required degree of control over fluxes it may be necessary to have a very efficient 'switch mechanism' operating upon a regulatory enzyme. The regulatory enzyme should have a very high elasticity towards its allosteric regulator if the latter is to be able to 'turn off' or 'turn on' the flux through the pathway. The evolution of regulatory enzymes that show homotropic as well as heterotropic effects has probably resulted from selective pressures towards this kind of regulatory efficiency. If we consider the rate curves of a typical regulatory enzyme (Figure 64) it is clear that, in the range of

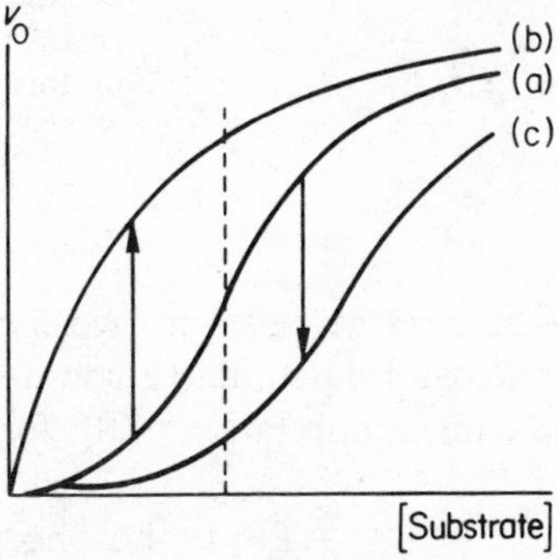

Figure 64 An illustration of the 'switch' mechanism inherent in the kinetic properties of a typical regulatory enzyme. Curve (a) shows a typical sigmoid rate curve in the absence of effectors. The dotted line indicates the substrate concentration at which the rate is most sensitive to changes in substrate concentration. Curve (b) shows the curve in the presence of an activator and curve (c) shows the rate curve in the presence of an inhibitor. The arrows indicate the maximum 'switch' effects that could be produced by activator and inhibitor

substrate concentrations around the point of inflection of the sigmoid curve found in the absence of effectors, the rate is very responsive to changes in substrate concentration. This, however, is probably not as important as the fact that a sigmoid curve allows an activator to exert a large 'switch' effect, especially at low substrate concentrations. Similarly an inhibitor can have a large effect, particularly at the higher substrate concentrations. Thus if the cellular range of concentrations lies around the midpoint of the sigmoid, as shown in Figure 64, a greater degree of control is possible than if no homotropic effect was present, that is to say if the curve was not sigmoid as shown in Figure 65. In some cases, however, it may be that even this degree of controllability is insufficient and it has been suggested that some of the more complex regulatory systems found in living cells may have arisen as a result of the need for even greater control.

In the glycolytic pathway two such examples of complex control mechanisms may illustrate the point. The first step in the breakdown of glycogen is promoted by glycogen phosphorylase which catalyses the production of glucose-1-phosphate:

$$\text{glycogen} + \text{inorganic phosphate} \xrightleftharpoons{\text{phosphorylase}} \text{glucose-1-phosphate}$$

234

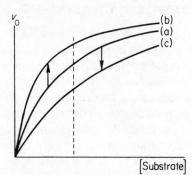

Figure 65 The lack of significant 'switch' effects in the kinetic properties of non-regulatory enzymes. The three curves are labelled (a), (b) and (c) as in Figure 64. Much smaller 'switch' effects are possible in this case by comparison with Figure 64

The enzyme exists in two forms, phosphorylase *b*, which is inactive, and its phosphorylated active counterpart, phosphorylase *a*. The latter is formed by the phosphorylation of the former, at the expense of ATP:

$$\text{ATP} + \text{phosphorylase } b \longrightarrow \text{ADP} + \text{phosphorylase } a$$

The enzyme which catalyses this reaction is called phosphorylase *b* kinase and it is also an enzyme existing in an inactive non-phosphorylated form and an active phosphorylated form. Its conversion from one form to the other is catalysed by another kinase which is activated by cyclic AMP.

By this means a cascade process of activation can be set in train by very small changes in the concentration of cyclic AMP that leads to a large increase in the rate of glycogen breakdown.

Other examples of interconvertible forms of regulatory enzymes are the mammalian pyruvate dehydrogenase system, which is activated by de-phosphorylation and inactivated by phosphorylation, and glutamine-synthetase from *E. coli* which is adenylated in the inactive form and requires de-adenylation for activation. Another example of a cascade process is found in the mechanism that controls blood-clotting.

Another type of switch mechanism has been suggested by Newsholme[47] in the form of an apparently wasteful cycle of enzyme action brought about by the two enzymes phosphofructokinase (PFK) and fructose diphosphatase (FDPase). These two enzymes catalyse respectively the phosphorylation of fructose-6-phosphate to fructose-diphosphate and the dephosphorylation of the latter back to fructose-6-phosphate:

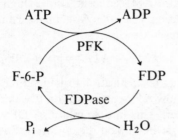

During glycolysis F-6-P must be converted to FDP which then acts as the substrate for aldolase in the next step of the pathway. The glycolytic flux is known to be regulated at the PFK step and phosphofructokinase is indeed an allosteric enzyme with interesting regulatory properties as shown in Figure 66.

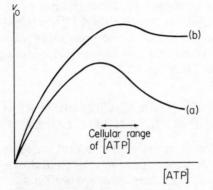

Figure 66 Rate curves for the enzyme phosphofructokinase (PFK). Curve (a) shows the rate curve as a function of ATP concentration at a fixed concentration of the second substrate, fructose-6-phosphate. Curve (b) shows the rate curve under the same conditions but in the presence of the activator AMP at a fixed concentration

In particular AMP acts as an activator of PFK while high ATP concentrations lead to its inhibition. FDPase, on the other hand, is also an allosteric enzyme, but now it is inhibited by AMP. If both enzymes were present and active at the same time and the metabolites F-6-P and FDP both had access to both enzymes, the net result would simply be the wasteful hydrolysis of ATP to ADP and inorganic phosphate. It has been suggested by Newsholme that a certain degree of wastefulness is tolerated by the metabolic economy in order to achieve a high degree of control at this step. The control is achieved by means of the intracellular concentration of AMP which is an inverse signal of the energy status of the cell. When the AMP concentration is high, the ATP concentration must be low, because the enzyme adenylate kinase catalyses a near equilibrium reaction that equilibrates both with ADP:

$$AMP + ATP \xrightleftharpoons[\text{kinase}]{\text{adenylate}} 2ADP$$

The sum of the cellular concentrations of the three adenine nucleotides is virtually constant so a low ATP concentration is very effectively signalled by a high AMP concentration. This in turn inhibits FDPase and at the same time activates PFK, thereby switching on the glycolytic flux which tends to restore the ATP level and diminish the AMP level. As the ATP concentration in the cell reaches its optimal level the AMP concentration will have fallen to such a value that FDPase is released from inhibition, whereas PFK will be de-activated. When the activities of these two enzymes are exactly equal, zero net glycolytic flux results.

Such a futile cycle represents a highly sensitive switch mechanism. Indeed the switch can be thrown into reverse if the ATP level climbs and the AMP level plunges to such values that FDPase becomes more active than PFK. In that case the glycolytic flux may actually be reversed, resulting in gluconeogenesis which of course soaks up excess ATP.

THE ROLE OF NEAR-EQUILIBRIUM REACTIONS IN MAINTAINING METABOLITE CONCENTRATIONS

So far in this discussion of metabolic control we have been concentrating attention on the maintenance of *end product* concentrations by means of flux regulation due to feedback effects. Similar considerations may be applied to feedforward effects. The metabolic pathways have been considered as involving only single-substrate enzymes whereas of course the true situation involves enzymes that utilize coenzymes and other second-substrates. One of the remarkable features of metabolism to emerge from the work of Krebs and others[48] is that the concentrations of coenzymes such as NAD and NADH remain within certain strictly defined limits under a wide variety of metabolic situations. In parts this aspect of homeostasis may result from the operation of feedback mechanisms such as the one mentioned for the maintenance of ATP levels in glycolytic tissues. Another very important contribution to the maintenance of coenzyme concentrations, and probably of other intermediary metabolite concentrations, comes from the existence of near-equilibrium enzymic steps. Although, as was seen earlier, near-equilibrium steps are poor places at which to attempt to control *fluxes*, nevertheless they are ideally suited to the control of the concentrations of the metabolites involved at such steps. It was shown earlier that a near-equilibrium step shows great capacity to buffer the flux against changes in the activity of the enzyme catalysing the step. A corollary of this is that it can absorb great changes in flux with little change in the metabolite concentrations. A near-equilibrium enzyme shows much greater capacity in this respect than a non-equilibrium enzyme, as the following example demonstrates. Consider a step in a pathway catalysed by the enzyme E and having disequilibrium ratio $\rho = [C]/[B]K_{eq}$

$$\cdots \to B \underset{}{\overset{E}{\rightleftharpoons}} C \cdots \to$$

Let us imagine what would happen to the flux through the pathway if the disequilibrium ratio is decreased by 2% owing to a regulatory effect on one of the other enzymes in the pathway. This implies that the ratio $[C]/[B]$ decreases by 2% and could be the result, for example, of a 1% increase in $[B]$ and a 1% decrease in $[C]$.

First take the near-equilibrium situation when $\rho_1 = 0.9$ initially, falling to $\rho_2 = 0.88$. The initial flux $F_1 = v_{f_1} - v_{b_1}$ and since $v_{b_1}/v_{f_1} = \rho_1 = 0.9$, then $F_1 = 0.1v_{f_1}$. The final flux $F_2 = v_{f_2} - v_{b_2}$ and since $v_{b_2}/v_{f_2} = \rho_2 = 0.88$, then $F_2 = 0.12v_{f_2}$. Now assuming that the value of v_f in each case is proportional to $[B]$, then since $[B]$ increases by 1%, $v_{f_2} = 1.01v_{f_1}$. Therefore $F_2 = 0.12 \times 1.01v_{f_1} = 0.121v_{f_1}$. So

$$\frac{F_2}{F_1} = \frac{0.121v_{f_1}}{0.1v_{f_1}} \quad \text{and} \quad F_2 = 1.21F_1.$$

The flux is increased by 21%.

Next consider what happens in the non-equilibrium situation when $\rho_1 = 0.050$ initially, falling to $\rho_2 = 0.049$. Proceeding as before, the initial flux is $F_1 = 0.95v_{f_1}$ and the final flux is $F_2 = 0.951v_{f_2} = 0.951 \times 1.01v_{f_1} = 0.9605v_{f_1}$. In this case

$$\frac{F_2}{F_1} = \frac{0.9605v_{f_1}}{0.95v_{f_1}} \quad \text{and} \quad F_2 = 1.01F_1.$$

The flux is increased by only 1%.

In other words the near-equilibrium reaction can accommodate a 21% increase in flux whereas the non-equilibrium step only allows a 1% increase in flux for the same small change in metabolite concentrations. This property may explain why many metabolic steps are found to be near-equilibrium and incidentally may provide an answer to the often-voiced question of why so many enzymes are present at concentrations greatly in excess of what is apparently 'needed' to catalyse observed fluxes.

COMPARTMENTATION AS A MEANS OF CONTROL

It is not difficult to imagine that, in all the complexity of metabolic interactions found within living cells, situations may arise in which conflicting demands upon a metabolite could occur. If a metabolite is used in two different pathways serving quite different purposes in the cell, it may be required at high concentration in one pathway but at a low concentration in another. One solution to this problem could be to separate the enzymes for the two pathways into different 'compartments' of the cell, so that the conditions prevailing in one pathway could be different from those affecting the other. This will only provide a solution as long as the barrier separating the compartments is not permeable to the metabolite in question, otherwise the concentrations will tend to equilibrate across the barrier, which is of course usually a membrane. In addition to impermeability to the metabolite in question, the membrane should not allow ready equilibration of other metabolites that are easily derived from, or coupled to, the metabolite, otherwise an indirect equilibration could occur. This latter difficulty actually makes compartmentation of rather limited use for the purposes of independent control because the metabolism of the whole cell is so fundamentally indivisible, as pointed out earlier. Nevertheless in some cases there seems to be a role for this kind of compartmentation.

Perhaps one example occurs in the metabolism of carbamoyl phosphate. This intermediate is required for two main purposes in mammalian liver. On the one hand excess ammonium ions are converted to carbamoyl phosphate as the first step in the synthesis of urea which is then excreted from the body. On the other hand carbamoyl phosphate is required for the first step in the synthesis of pyrimidine nucleotides, as shown in Figure 63. It is clear that these two metabolic functions are quite unrelated and the needs of the liver cell for one may be expected to be quite independent of its needs for the other. What is found is that the synthesis of carbamoyl phosphate from ammonium ions destined to enter the urea cycle takes place in the mitochondria catalysed by the mitrochondrial carbamoyl phosphate synthetase (CPS) according to the following reaction:

$$NH_4^+ + HCO_3^- + 2\,ATP^{4-} \xrightarrow{\text{CPS}}$$

$$NH_2{\cdot}CO{\cdot}OPO_3^{2-} + 2\,ADP^{3-} + HPO_3^{2-} + 2\,H^+$$

Magnesium ions are required for this enzyme to be active, as is N-acetylglutamic acid, which may play a regulatory role. Carbamoyl phosphate required for pyrimidine synthesis on the other hand is manufactured in the cytoplasm by a different reaction utilizing glutamine as the source of nitrogen instead of ammonia and catalysed by the cytoplasmic CPS:

$$\text{glutamine} + HCO_3^- + 2\,ATP^{4-} \longrightarrow$$

$$NH_2{\cdot}CO{\cdot}OPO_3^{2-} + 2\,ADP^{3-} + HPO_3^{2-} + \text{glutamic acid}$$

Although this enzyme also requires magnesium ions for activity, N-acetyl glutamate is not involved. Both reactions are effectively irreversible and carbamoyl phosphate is not easily able to cross the mitochondrial membrane, so the two reactions are effectively isolated for practical purposes.

In this chapter some of the ways in which control of enzymic activity may be brought about so as to achieve metabolic autoregulation have been touched upon. The complexity of the problem should be obvious, but it would be true to say that understanding it is one of the major challenges of modern bio-chemistry. Perhaps in conclusion two things should be stressed. The first is that individual enzymes cannot be considered only in isolation if the metabolism of whole cells is to be understood; and the second is that one should not expect to find that single 'regulatory' enzymes hold the key to control problems.

Chapter 8

Enzymes and the Control of Metabolism II: Coarse Control of Enzyme Activity

COARSE CONTROL

In the previous chapter metabolic regulation was considered from the point of view of fine control, that is to say, the mechanisms by which metabolite concentrations can be controlled through the interaction of effectors with regulatory enzymes. In passing it was noted that the amount or, rather, concentration of an enzyme in a cell would affect the fluxes to which it contributed; but, apart from considering the concept of sensitivity, it was passed over as being under the control of the genetic machinery and not readily altered from outside.

Nevertheless enzyme concentrations do change in an important way in most living systems and these changes contribute very significantly to the control of the cellular metabolism. For example, cultures of the bacterium *E. coli* grown on the sugar galactose as carbon source, elaborate a series of enzymes that convert galactose to glucose-1-phosphate which then enters the glycolytic pathway in the normal way. When such cultures are transferred to a medium in which glucose replaces the galactose as carbon source, they stop synthesizing the now useless enzymes of galactose metabolism, but they do synthesize hexokinase so as to be able to utilize the glucose. In a similar way, if the amino acid histidine is present in the culture medium, bacteria lose the enzymes needed to synthesize histidine; but, if grown in a medium without it, they are able to re-synthesize the enzymes necessary for its production. Changes in the enzymes present in mammalian liver follow changes in diet or developmental state and enzyme levels are even found to fluctuate on a diurnal basis. Similar changes occur in the enzyme complement of plant cells.

Quite clearly all forms of life from the simplest to the most complex are able to regulate the concentrations of enzymes present in the cell according to need; this type of control is called 'coarse control'. At its coarsest it represents the ability to turn on or off the synthesis of enzymes in an absolute manner when they are or are not needed. But, more than that, it allows the *rates* of synthesis and degradation of enzymes to be balanced according to their relative need in the cell.

THE OPERON THEORY OF JACOB AND MONOD[49]

The control of enzyme synthesis will be dealt with before turning to a consideration of enzyme degradation, primarily because more is known about it. There is no doubt that the synthesis of all cellular proteins is directed by the genetic information stored in the cellular DNA. Sometimes some of the information is not used as when, in the examples quoted earlier, bacterial cells do not elaborate a particular enzyme that they are competent to make under other circumstances. Two types of situation can be distinguished here. In some cases the enzyme or enzymes are only made when a particular substance is present in the growth medium. This is typified by the example of the galactose-utilizing enzymes mentioned before. In the other type of situation the enzymes are *not* made when a particular substance is present, as typified in the case of histidine. In the former case the enzymes are said to be inducible by the substance that causes them to be made. In the other case the enzymes are repressible by the substance that prevents their synthesis. Although sometimes only a single enzyme is induced or repressed in this way, in the majority of cases a series of enzymes, all concerned in the metabolism of the inducing or repressing substance, are all coordinately induced or repressed. The problem was investigated first in prokaryotic micro-organisms for a variety of reasons, but principally because their genetic structure is much simpler than that of the eukaryotes. By genetic mapping techniques it was found that in most of the cases where a series of enzymes are coordinately induced or repressed, the genes that carry the information for each of the enzymes all bunch very closely together on the single chromosome of the prokaryote. For this reason it became clear that, whatever mechanism converted the genetic information in the chromosomal DNA into proteins, it could be switched on in such a way that it 'read' a whole series of neighbouring genes in succession and, when switched off, none of those genes could be 'read'. Monod and Jacob proposed a scheme that served as a framework for understanding this mechanism for 'reading' genetic information and explained the coordinate nature of the switch. Their scheme, simple and elegant though it is, still stands up 15 years later, after an explosion of research on the subject. It was, however, proposed to explain the phenomena in prokaryotes. The mechanism in eukaryotes appears to be more complex and is still not understood.

In essence Monod and Jacob proposed that genes are grouped together into *operons*. An operon comprises a series of genes or cistrons as they are called separated by short stretches of intercistronic DNA which carries no significant information content, at least in terms of protein structure. At one end of the operon is a stretch of DNA that serves as a regulatory region and embodies the switch mechanism. They further proposed that the first act in the process of protein synthesis consists in transcribing the whole of the information in one operon into a complementary molecule of RNA, the so-called messenger RNA or mRNA. For this purpose an enzyme, DNA dependent RNA polymerase, is required to attach itself to the DNA at the regulatory end of an operon and then work its way along the operon catalysing the synthesis of a mRNA molecule exactly complementary in base sequence to one of the two strands of DNA. This mRNA molecule constitutes a polycistronic message, being a transcription or copy of the information contained in the original operon. When the end of the operon is reached the mRNA is released. When this mRNA molecule has been synthesized, or even before its synthesis is completed, ribosomes attach themselves to it near the beginning and proceed to travel along it, translating the information coded in the sequence of nucleotides into a polypeptide chain. When a ribosome reaches the end of a cistron on the mRNA the polypeptide chain is completed and is released into the cytoplasm. The intercistronic region of the mRNA is not translated, but the ribosome proceeds to the start of the next cistron and starts translating it into a second polypeptide and so on until all the polypeptides coded by the original operon have been synthesized. Many ribosomes may be attached to a single mRNA molecule, spaced out at intervals, each following the first and synthesizing proteins, so that a number of identical protein molecules can be synthesized from a given mRNA molecule.

OPERATORS AND REPRESSORS

To return to the initial event in this process, the attachment of the enzyme RNA polymerase to the DNA, the site of attachment is a short stretch of DNA of a particular nucleotide sequence that has a high affinity for the enzyme, a so-called RNA polymerase recognition site. This site is known as the *promoter* for the operon in question. It lies at the very beginning of the operon. Between the promoter and the start of the first cistron there lies a region of DNA that serves a most important function, for it is here that the switch mechanism actually operates. For this reason it is called the *operator* region. In the operator is a specific sequence of nucleotides that acts as another recognition site, this time recognizing not the RNA polymerase but a special protein molecule called the *repressor*. When the repressor is bound to the operator, the RNA polymerase cannot proceed past it and therefore cannot reach the first cistron of the operon in order to begin the process of transcription, that is mRNA synthesis is blocked. The operon is repressed. The interaction between operator (DNA) and repressor (protein) is exceedingly specific and the repressor is bound very tightly indeed. There is a specific repressor protein for each operator.

Although a given bacterial cell contains only a small number of molecules of each repressor, and although the bacterial chromosome may contain hundreds of different operators, the specificity of recognition is such that an operator spends most of its time in combination with its repressor, thus effectively preventing transcription of the operon. In order for the genetic information contained in an operon to be expressed, the repressor must be caused to come off the operator, thus allowing the RNA polymerase to commence its journey of transcription.

The repressor is an allosteric protein, though not an enzyme. In addition to a binding site that is complementary to, and therefore binds, the operator DNA sequence, it has another binding site for a small molecule which might be likened to an effector. There are two types of repressor proteins, corresponding to the two types of control system mentioned earlier, the inducible and repressible systems. In inducible systems it will be recalled that a particular substance, galactose in the example considered, causes the induction of the set of enzymes encoded in an operon, the enzymes of galactose metabolism or *gal* enzymes in the example. This substance is called the inducer of the particular operon and it functions by binding like an effector to the second site of the repressor molecule. When it binds, the repressor molecule undergoes a conformational change that weakens its affinity for the operator and it is no longer able to remain bound to the operator DNA. The opposite situation occurs in repressible systems. In that case, exemplified by the enzymes of histidine biosynthesis coded by the *his* operon, the presence of histidine brings about repression of the operon and it is clear that histidine, or a metabolite derived from it, must be bound in the second site of the repressor in order for it to recognize the *his* operator. Once the histidine concentration falls, the histidine is no longer able to bind to the repressor, holding it in its correct conformation for attachment to the operator, and de-repression of the operon occurs. The molecule that brings about repression is called the co-repressor of the operon.

In this way induction is caused by the *combination* of inducer and repressor, leading to release of the latter from the operator; whereas de-repression is caused by the *release* of the co-repressor from the repressor, leading to the same result. The two types of control system are illustrated in Figure 67.

Cellular proteins are not all inducible or repressible in the manner described. Some are synthesized all the time at a constant rate, more or less whatever the nutritional state of the organism. Repressor proteins are one obvious class of such proteins. They are synthesized under the direction of genes that do not belong to any operon and have no operator or other regulatory mechanisms. Such proteins are said to be *constitutive*. Inducible or repressible proteins may become constitutive as a result of mutation. For example if a bacterium loses the gene that directs the synthesis of the *gal* repressor, then no repressor for the *gal* operon can be made and the bacterium becomes constitutive for all the *gal* enzymes. Mutation in the operator region can also result in constitutive synthesis of all the proteins of an operon if the mutation leads to loss of the ability to bind the repressor.

From mutants of these types a number of interesting facts have emerged. It appears that in many cases the rate of protein synthesis is limited by the rate at which mRNA can be synthesized. This in turn may depend upon the efficiency of the promoter. Some promoters appear to be more efficient than others in 'attracting' RNA polymerase and so this can play a role in the determination of relative rates of synthesis of different proteins. The concentration of inducer or co-repressor will also play a part in determining the rate of synthesis of inducible or repressible proteins, via regulation of the rate of mRNA synthesis. Other factors may affect the rate of protein synthesis at this transcriptional level. For example the *gal* operon has another recognition site associated with its regulatory region, which is located close to the promoter. Here the cAMP

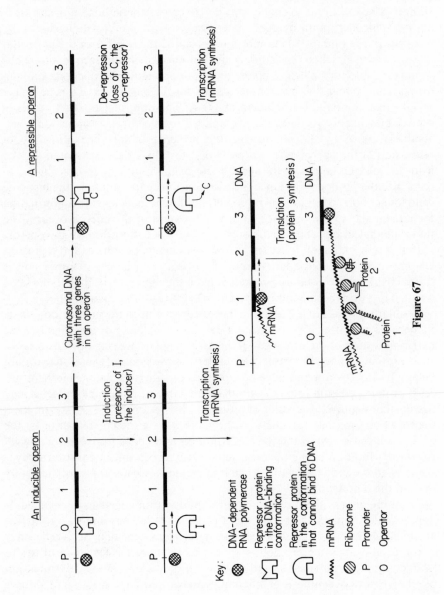

Figure 67

binding protein is recognized and bound in a manner that resembles the binding of repressor. In this way the concentration of cAMP in the cell has a regulatory effect on *gal* enzyme synthesis.

After messenger synthesis there are still many places at which the rate of protein synthesis can be regulated. Post-transcriptional control, or translational control to use its other name, can operate via the frequency with which ribosomes attach themselves to the starting point of a cistron, the so-called initiation site. In bacteria this always seems to be the nucleotide triplet AUG, although the neighbouring nucleotides in the mRNA must also play some part in determining initiation because AUG occurs within cistrons, coding for methionine, whereas when it marks the initiation site it causes *N*-formyl-methionine to be built into the polypeptide. The availability of amino acids, and of the transfer RNA molecules which carry them in an activated form to the ribosome, can also affect the rates of protein synthesis, along with many other factors.

In bacteria, however, transcriptional control seems at the present time, to play the major role in the coarse control of protein synthesis. Through control of the rate of mRNA synthesis, the varying needs of the cell in terms of the different enzymes it requires and their varying concentrations are regulated.

WHAT CONTROLS THE RATE OF ENZYME DEGRADATION?

Much less is known about the degradation of enzymes and other proteins within the living cell. In part this is due to the difficulty of measuring the turnover of protein *in vivo*. When a protein is broken down by proteolytic enzymes the resulting amino acids are returned to the cellular pool and may be re-incorporated into newly synthesized protein. To the extent that this happens, the true rate of protein degradation can be underestimated. Pulse and chase experiments with radioactive amino acids, or double-labelling experiments, can overcome some of the problems; but a variety of experimental approaches have been found to be necessary to establish the true rates of protein degradation.[50]

It is now clear that individual proteins are degraded at widely differing rates within the cell and that the individual rates can vary under different physiological conditions. A variety of proteolytic enzymes is usually present in a given cell and the total proteolytic capacity of a cell may change, as well as the distribution of the activity amongst the different proteases. For example, in exponentially growing cultures of *E. coli* in a minimal medium there is only a low total protease activity; but when the cells become starved of nutrients and enter a stationary phase, proteolytic enzymes are induced, presumably to mobilize the now unnecessary cellular proteins and make their amino acids available for a siege economy. In germinating seeds, on the other hand, proteolytic enzymes rapidly make their appearance in order to mobilize amino acids from the stored seed proteins and make them available for growth of the plant.

A picture of some complexity thus emerges in which the *average* rate of protein degradation within a cell may vary, broadly speaking, with the nutritional or developmental state of the cell, whereas within these general movements, the rates of degradation of individual proteins may change either in parallel with the average trend, or contrary to it.

FACTORS AFFECTING PROTEIN TURNOVER

Some generalizations can be made from the evidence available. Abnormal proteins which may from time to time appear in a cell, either because of mutation, or because of an error in protein synthesis, or because of a chemical modification of the protein, are usually degraded faster than their normal counterparts. This has been ascertained from many studies in which abnormalities are deliberately produced in particular proteins and the rates of degradation then compared with the normal situation. This phenomenon most likely represents an adaptation protecting cells from the effects of abnormal proteins.

In a number of cases it has been established that known regulatory enzymes in a given metabolic pathway are degraded more rapidly than the non-regulatory ones. Again this makes biochemical sense because a regulatory enzyme may have to change in concentration quite rapidly. Rapid degradation allows greater flexibility in the control of the enzyme's concentration.

Large proteins or, in the case of oliogmeric proteins, those with large sub-units, seem on the whole to be degraded more rapidly than small ones. A number of possible reasons can be suggested. The machinery of protein synthesis occasionally makes errors and inserts a wrong amino acid into the polypeptide chain. A long chain is thus more likely to contain an error than a short one and, since abnormal proteins are degraded more rapidly, would be expected to turn over more quickly. Alternatively a large protein may be intrinsically less stable in aqueous solution because of the difficulty of allowing all the hydrophilic residues to be at the surface of the molecule at once. Large globular molecules have a smaller surface to volume ratio than small ones, so unless they have a correspondingly smaller proportion of hydrophilic residues, some of the latter must get tucked away in the core. Of course other solutions to this problem do suggest themselves. For instance, a series of roughly spherical small 'blobs' of polypeptide could be joined up by stretches of exposed peptide. Nevertheless, a large protein may have many energetically equivalent or near-equivalent conformations and may move from one to another, thus exposing weak points to proteolytic attack. It is well known from studies *in vitro* that denatured proteins are more susceptible to proteolysis than native ones.

The presence of specific ligands in the cell may also affect the rate of degradation of certain proteins. Tryptophan pyrrolase in rat liver turns over more slowly when tryptophan is present, for example, and this kind of behaviour is mirrored in many studies on purified enzymes where substrate, or substrate analogue, or cofactor exerts a protective effect against attack by proteolytic enzymes. This is almost certainly due to a conformational change attendant on the binding of the ligand to the protein and causing a 'tightening' of the three-dimensional structure, a shielding of susceptible peptide bonds against proteolytic attack.

It is interesting to speculate on why some proteins are degraded rapidly in the cell and others more slowly. How is it that the assembly of proteins that constitutes a cell is more or less immune to proteolytic attack by its own proteases except when conditions demand that some of those proteins should be broken down? Teleologically one may imagine that evolution would strive after such a situation since it offers so many advantages in terms of flexible control, as discussed above.

The answers to such questions are still obscure but clearly they are to be found in the specific three-dimensional structures of the individual proteins and in the specificities of the proteolytic enzymes that attack them. The modulating effects of ligands and perhaps also of pH and temperature must be taken into account as well.

CONCLUSION

Clearly, in some respects, the study of the enzyme molecule has only just begun, especially insofar as it relates to the functioning of whole living systems. Many questions still remain to be asked and many others are only incompletely answered. Only by patient research into the structures, stabilities and kinetics of enzyme molecules and their interactions with each other and with the many other types of molecules present in cells shall we begin to comprehend the vast and elegant complexity of life.

Appendix 1

Nomenclature and Classification of Enzymes

In 1956 a special International Commission on Enzymes was set up by the International Union of Biochemistry (IUB) in consultation with the International Union of Pure and Applied Chemistry (IUPAC). The report of this *Enzyme Commission* was published in 1964, and was subsequently brought up to date in 1972 in the form of *Enzyme Nomenclature* (*1972*) published by Elsevier Scientific Publishing Co. The rules for naming and classifying enzymes are set out in detail in this latest publication, together with a complete list of all the enzymes known at the time. A brief summary of the rules on nomenclature is given here together with a note on classification and a key to the numbering of the different classes of enzymes.

NOMENCLATURE

1. Systematic names

An enzyme is named after the reaction it catalyses according to well defined rules set out in *Enzyme Nomenclature* (*1972*). For example any enzyme catalysing an oxidation–reduction reaction is called an *oxidoreductase* and the systematic name is constructed by taking the name of the substrate which is oxidized, i.e. the hydrogen *donor*, followed by the name of the substrate which is reduced, i.e. the hydrogen *acceptor*, so that a general name would be of the form 'donor : acceptor oxidoreductase'. Of course this immediately presents a problem because any such reaction:

$$DH_2 + A \rightleftharpoons D + AH_2$$

where DH_2 represents the donor and A the acceptor, can be written, and indeed must be catalysed, in either direction. The question then arises as to whether this enzyme should be called DH_2:A oxidoreductase of AH_2:D oxidoreductase. Either name would be equally acceptable but in practice certain conventions have become established. For example in all reactions involving the interconversion of NAD^+ and NADH, NAD^+ is taken to be the acceptor, rather than considering NADH as the donor. In other cases only one direction can be demonstrated or is biochemically important, so this is used. Once such a convention is established all similar enzymic reactions are named in the same reaction direction. In the case considered, all NAD^+ utilizing oxido-reductases are named according to the reaction:

$$DH_2 + NAD^+ \rightleftharpoons D + NADH + H^+$$

that is as $DH_2:NAD^+$ oxidoreductases. Similar considerations apply to other groups of enzymes insofar as choosing a uniform direction of reaction is con-concerned. The six major groups of enzymes and rules for naming within them are dealt with under *Classification* below.

2. Recommended or trivial names

Not surprisingly the systematic names arrived at by logical application of the rules can lead to some extraordinarily long names. Many enzymes were known before the rules were laid down, and most of them had been given names, although in some cases the names were confusing or even misleading. Nevertheless many of these 'working names' are more convenient for everyday use and, provided everyone knows what they mean, are recommended to be used rather than the systematic names. *Enzyme Nomenclature* (*1972*) lists the acceptable working names as *Recommended Names* together with the Systematic Names and reactions catalysed, so reference to the lists will always clarify any doubt. For example, the enzyme whose systematic name is D-glyceraldehyde-3-phosphate:NAD^+ oxidoreductase (phosphorylating) can be referred to by its Recommended Name of glyceraldehydephosphate dehydrogenase. An even greater simplification occurs by referring to ribonucleate 3'-pyrimidino-oligonucleotidohydrolase as, simply, ribonuclease I.

3. Naming of multienzyme complexes

Many instances are now known where several enzymes are physically aggregated to form a functionally discrete entity. For example, the oxidation of pyruvate to acetyl CoA is carried out by one such assembly, consisting of several molecules of each of three enzymes. The whole assembly, a *multienzyme complex*, should not be referred to as if it were one enzyme, a pyruvate dehydrogenase, but as a *system of enzymes* and the proper terminology in this case would be the *pyruvate dehydrogenase system*.

4. Species and tissue differences

Neither the systematic nor the recommended name of an enzyme take any account of the fact that the enzyme from one species that catalyses a particular reaction will often, and perhaps usually, differ in primary structure from the enzyme that catalyses the same reaction in another species. Although the enzymes from the two sources may be different molecules and indeed may function by different mechanisms, nevertheless they are given the same name because nomenclature is based upon the reaction catalysed. It is important to specify the *source* of an enzyme along with its *name* whenever it is being discussed. Where multiple forms of any enzyme occur within a single species, for example in different tissues or in mitochondria and cytoplasm of a single cell, the appropriate details should be given whenever possible.

CLASSIFICATION

Enzymes are grouped into six major classes according to the type of reaction catalysed. The first of these classes contains all the *oxidoreductases*, which have been mentioned already. The second class consists of all enzymes that catalyse the transfer of a group from one substrate molecule to another; these enzymes are called *transferases*. Class 3 contains all the *hydrolases*, enzymes that catalyse hydrolytic reactions, while Class 4 comprises the *lyases*, which catalyse elimination reactions that lead to the formation of double bonds. All enzymes that catalyse isomerization reactions are grouped together as *isomerases* in Class 5. The sixth and last class contains the enzymes that catalyse the joining together of two molecules while, at the same time, ATP (or some similar pyrophosphate compound) is converted to ADP and inorganic phosphate or AMP and inorganic pyrophosphate. These enzymes, all broadly synthetic and all using the free energy of hydrolysis of pyrophosphate bonds, are called *ligases*.

Within each of the six major classes there are a number of sub-classes and within each sub-class there are sub-sub-classes, until finally one comes down to individual enzymes. An enzyme is given a number that contains four separate parts. The first part is the number, one to six, of the major class to which it belongs, the second and third parts are the numbers of the sub-class and sub-sub-class and the fourth part is a sort of accession number, given to enzymes in the order in which they were discovered. In this way the enzyme alcohol: NAD^+ oxidoreductase, or alcohol dehydrogenase to use its Recommended Name, bears the number 1.1.1.1 as it clearly belongs to Class 1, the oxidoreductases, and within that major group it belongs to sub-class 1, sub-sub-class 1, and was the first enzyme in that sub-sub-class to be discovered.

The following list describes the sub-classes and sub-sub-classes of the six major groups.

KEY TO NUMBERING AND CLASSIFICATION OF ENZYMES

1. OXIDOREDUCTASES

1.1   Acting on the CH–OH group of donors

1.1.1   With NAD^+ or $NADP^+$ as acceptor
1.1.2   With a cytochrome as acceptor
1.1.3   With oxygen as acceptor
1.1.99   With other acceptors

1.2   Acting on the aldehyde or keto group of donors

1.2.1   With NAD^+ or $NADP^+$ as acceptor
1.2.2   With a cytochrome as acceptor
1.2.3   With oxygen as acceptor
1.2.4   With a disulphide compound as acceptor
1.2.7   With an iron–sulphur protein as acceptor
1.2.99 With other acceptors

1.3   Acting on the CH–CH group of donors

1.3.1   With NAD^+ or $NADP^+$ as acceptor
1.3.2   With a cytochrome as acceptor
1.3.3   With oxygen as acceptor
1.3.7   With an iron–sulphur protein as acceptor
1.3.99 With other acceptors

1.4   Acting on the $CH–NH_2$ group of donors

1.4.1   With NAD^+ or $NADP^+$ as acceptor
1.4.3   With oxygen as acceptor
1.4.4   With a disulphide compound as acceptor
1.4.99 With other acceptors

1.5   Acting on the CH–NH group of donors

1.5.1   With NAD^+ or $NADP^+$ as acceptor
1.5.3   With oxygen as acceptor
1.5.99 With other acceptors

1.6   Acting on NADH or NADPH

1.6.1   With NAD^+ or $NADP^+$ as acceptor
1.6.2   With a cytochrome as acceptor
1.6.4   With a disulphide compound as acceptor
1.6.5   With a quinone or related compound as acceptor
1.6.6   With a nitrogenous group as acceptor
1.6.7   With an iron–sulphur protein as acceptor
1.6.99 With other acceptors

1.7   Acting on other nitrogenous compounds as donors

1.7.2   With a cytochrome as acceptor
1.7.3   With oxygen as acceptor
1.7.7   With an iron–sulphur protein as acceptor
1.7.99 With other acceptors

1.8   Acting on a sulphur group of donors

1.8.1   With NAD^+ or $NADP^+$ as acceptor
1.8.2   With a cytochrome as acceptor
1.8.3   With oxygen as acceptor
1.8.4   With a disulphide compound as acceptor

1.8.5   With a quinone or related compound as acceptor
1.8.6   With a nitrogenous group as acceptor
1.8.7   With an iron–sulphur protein as acceptor
1.8.99   With other acceptors

1.9   Acting on a haem group of donors

1.9.3   With oxygen as acceptor
1.9.6   With a nitrogenous group as acceptor
1.9.99   With other acceptors

1.10   Acting on diphenols and related substances as donors

1.10.2   With a cytochrome as acceptor
1.10.3   With oxygen as acceptor

1.11   Acting on hydrogen peroxide as acceptor

1.12   Acting on hydrogen as donor

1.12.1   With NAD^+ or $NADP^+$ as acceptor
1.12.2   With a cytochrome as acceptor
1.12.7   With an iron–sulphur protein as acceptor

1.13   Acting on single donors with incorporation of molecular
oxygen (oxygenases)

1.13.11 With incorporation of two atoms of oxygen
1.13.12 With incorporation of one atom of oxygen (internal monooxygenases
or internal mixed function oxidases)
1.13.99 Miscellaneous (requires further characterization)

1.14   Acting on paired donors with incorporation of molecular oxygen

1.14.11 With 2-oxoglutarate as one donor, and incorporation of one atom
each of oxygen into both donors
1.14.12 With NADH or NADPH as one donor, and incorporation of two
atoms of oxygen into one donor
1.14.13 With NADH or NADPH as one donor, and incorporation of one
atom of oxygen
1.14.14 With reduced flavin or flavoprotein as one donor, and incorporation
of one atom of oxygen
1.14.15 With a reduced iron–sulphur protein as one donor, and incorporation
of one atom of oxygen
1.14.16 With reduced pteridine as one donor, and incorporation of one atom
of oxygen
1.14.17 With ascorbate as one donor, and incorporation of one atom of
oxygen
1.14.18 With another compound as one donor, and incorporation of one
atom of oxygen

1.14.99   Miscellaneous (requires further characterization)

1.15   Acting on superoxide radicals as acceptor

1.16   Oxidizing metal ions

1.16.3   With oxygen as acceptor

1.17   Acting on $-CH_2$ groups

1.17.1   With NAD^+ or $NADP^+$ as acceptor
1.17.4   With a disulphide compound as acceptor

2.   TRANSFERASES

2.1   Transferring one-carbon groups

2.1.1   Methyltransferases
2.1.2   Hydroxymethyl-, formyl- and related transferases
2.1.3   Carboxyl- and carbamoyl-transferases
2.1.4   Amidinotransferases

2.2   Transferring aldehyde or ketonic residues

2.3   Acyltransferases

2.3.1   Acyltransferases
2.3.2   Aminoacyltransferases

2.4   Glycosyltransferases

2.4.1   Hexosyltransferases
2.4.2   Pentosyltransferases
2.4.99   Transferring other glycosyl groups

2.5   Transferring alkyl or aryl groups, other than methyl groups

2.6   Transferring nitrogenous groups
2.6.1   Aminotransferases
2.6.3   Oximinotransferases

2.7   Transferring phosphorus-containing groups

2.7.1   Phosphotransferases with an alcohol group as acceptor
2.7.2   Phosphotransferases with a carboxyl group as acceptor
2.7.3   Phosphotransferases with a nitrogenous group as acceptor
2.7.4   Phosphotransferases with a phospho-group as acceptor
2.7.5   Phosphotransferases with regeneration of donors (apparently cata-lysing intramolecular transfers)
2.7.6   Diphosphotransferases
2.7.7   Nucleotidyltransferases
2.7.8   Transferases for other substituted phospho-groups
2.7.9   Phosphotransferases with paired acceptors

2.8   Transferring sulphur-containing groups

2.8.1   Sulphurtransferases
2.8.2   Sulphotransferases
2.8.3   CoA-transferases

3. HYDROLASES

3.1   Acting on ester bonds

3.1.1   Carboxylic ester hydrolases
3.1.2   Thiolester hydrolases
3.1.3   Phosphoric monoester hydrolases
3.1.4   Phosphoric diester hydrolases
3.1.5   Triphosphoric monoester hydrolases
3.1.6   Sulphuric ester hydrolases
3.1.7   Diphosphoric monoester hydrolases

3.2   Acting on glycosyl compounds

3.2.1   Hydrolysing O-glycosyl compounds
3.2.2   Hydrolysing N-glycosyl compounds
3.2.3   Hydrolysing S-glycosyl compounds

3.3   Acting on ether bonds

3.3.1   Thioether hydrolases
3.3.2   Ether hydrolases

3.4   Acting on peptide bonds (peptide hydrolases)

3.4.11 α-Aminoacylpeptide hydrolases
3.4.12   Peptidylamino-acid or acylamino-acid hydrolases
3.4.13   Dipeptide hydrolases
3.4.14   Dipeptidylpeptide hydrolases
3.4.15   Peptidyldipeptide hydrolases
3.4.21   Serine proteinases
3.4.22   SH-proteinases
3.4.23   Acid proteinases
3.4.24   Metalloproteinases
3.4.99   Proteinases of unknown catalytic mechanism

3.5   Acting on carbon–nitrogen bonds, other than peptide bonds

3.5.1   In linear amides
3.5.2   In cyclic amides
3.5.3   In linear amidines
3.5.4   In cyclic amidines
3.5.5   In nitriles
3.5.99   In other compounds

3.6   Acting on acid anhydrides

3.6.1   In phosphoryl-containing anhydrides
3.6.2   In sulphonyl-containing anhydrides

3.7   Acting on carbon–carbon bonds

3.7.1   In ketonic substances

3.8   Acting on halide bonds

3.8.1   In C-halide compounds
3.8.2   In P-halide compounds

3.9   Acting on phosphorus–nitrogen bonds

3.10   Acting on sulphur–nitrogen bonds

3.11   Acting on carbon–phosphorus bonds

4.   LYASES

4.1   Carbon–carbon lyases

4.1.1   Carboxy-lyases
4.1.2   Aldehyde-lyases
4.1.3   Oxo-acid-lyases
4.1.99   Other carbon–carbon lyases

4.2   Carbon–oxygen lyases

4.2.1   Hydro-lyases
4.2.2   Acting on polysaccharides
4.2.99   Other carbon–oxygen lyases

4.3   Carbon–nitrogen lyases

4.3.1   Ammonia-lyases
4.3.2   Amidine-lyases

4.4   Carbon–sulphur lyases

4.5   Carbon–halide lyases

4.6   Phosphorus–oxygen lyases

4.99   Other lyases

5.   ISOMERASES

5.1   Racemases and epimerases

5.1.1   Acting on amino acids and derivatives
5.1.2   Acting on hydroxy acids and derivatives
5.1.3   Acting on carbohydrates and derivatives
5.1.99   Acting on other compounds

5.2   *Cis–trans* isomerases

5.3   Intramolecular oxidoreductases

5.3.1   Interconverting aldoses and ketoses
5.3.2   Interconverting keto and enol groups
5.3.3   Transposing C=C bonds
5.3.4   Transposing S—S bonds
5.3.99   Other intramolecular oxidoreductases

5.4   Intramolecular transferases

5.4.1   Transferring acyl groups
5.4.2   Transferring phosphoryl groups
5.4.3   Transferring amino groups
5.4.99   Transferring other groups

5.5   Intramolecular lyases

5.99   Other isomerases

6. LIGASES (SYNTHETASES)

6.1   Forming carbon–oxygen bonds

6.1.1   Ligases forming aminoacyl–tRNA and related compounds

6.2   Forming carbon–sulphur bonds

6.2.1   Acid–thiol ligases

6.3   Forming carbon–nitrogen bonds

6.3.1   Acid–ammonia ligases (amide synthetases)
6.3.2   Acid–amino-acid ligases (peptide synthetases)
6.3.3   Cyclo–ligases
6.3.4   Other carbon–nitrogen ligases
6.3.5   Carbon–nitrogen ligases with glutamine as amido-N-donor

6.4   Forming carbon–carbon bonds

6.5   Forming phosphate ester bonds

SOME NOTES ON THE RULES OF NOMENCLATURE

It is not necessary to go into all the detailed rules for naming the enzymes within all the classes listed above, as these are laid out clearly in *Enzyme Nomenclature* (*1972*) which should always be referred to in any case of doubt. A few of the major points are worth bearing in mind. Among the oxidoreductases (Class 1) the recommended name usually contains the word dehydrogenase or, more rarely, reductase. The only cases where an enzyme may be called an *oxidase* are those where molecular oxygen (O_2) is the hydrogen acceptor. In

Class 2, the transferases, systematic names are formed according to the general reaction scheme $DX + A \rightleftharpoons D + AX$ where X is the group transferred, D is the donor molecule and A the acceptor. A general systematic name is thus donor:acceptor group transferase. A specific example from Sub-class 1, Sub-sub-class 1, is the enzyme methylamine:L-glutamate N-methyl transferase (2.1.1.21) which catalyses the reaction:

methylamine + L-glutamate $\rightleftharpoons$ ammonia + N-methyl-L-glutamate

The hydrolases (Class 3) are a group of widely different enzymes but the systematic name always includes the word hydrolase, whereas the recommended name is often merely the name of the substrate with '-ase' added to it. This is the only situation where the rules allow one to call an enzyme a 'substratase'. For example acetylcholine hydrolase (3.1.1.7) catalyses the hydrolysis of acetylcholine to choline and acetate and is usually known as acetylcholinesterase. Some confusion may arise between *hydrolases* which belong to Class 3 and *hydro-lyases* which belong to Class 4. The systematic names of lyases are constructed by considering the substrate and the group which is eliminated in forming the double bond and naming them as substrate group-lyases. Thus the systematic name for enolase is 2-phospho-D-glycerate hydro-lyase (4.2.1.11) because it catalyses elimination of water from 2-phosphoglycerate:

2-phospho-D-glycerate $\rightleftharpoons$ phosphoenolpyruvate + H_2O

Note that this is *not* a hydrolase as no hydrolytic cleavage occurs; the hyphen in the name hydro-lyase is present to help in making the distinction clear.

Carboxy-lyases may be confused with *carboxylases* in a similar way if the presence of the hyphen is overlooked. The carboxy-lyases (systematic names) catalyse elimination of CO_2 and the formation of a double bond whereas the term carboxylase is used in the recommended names of a number of enzymes to imply the involvement of CO_2 in the reaction, sometimes in true carboxy-lyase reactions but on other occasions in other types of reaction.

Another possible source of confusion in the enzymes of Class 4 concerns the use of the word *synthase* in recommended names. This is used where the biochemically important direction of reaction is the one where something is *added across a double bond* rather than being eliminated to form a double bond. An example occurs with the important enzyme citrate synthase, the first enzyme of the tricarboxylic acid cycle. This catalyses the condensation of oxaloacetate and acetyl CoA together with water to form citrate and CoA. Its systematic name is citrate oxaloacetate-lyase (4.1.3.7). Synthases also occur in other classes. For example in Class 2 there is a glutamate synthase which catalyses the reaction:

L-glutamine + 2-oxoglutarate + NADPH $\rightleftharpoons$ 2 L-glutamate + NADP$^+$

Its systematic name is L-glutamine:2-oxoglutarate aminotransferase (NADPH-oxidizing) (2.6.1.53). The important thing is not to confuse *synthases* with *synthetases.* The latter term is used in the recommended name of some *ligases* and always implies the involvement of ATP or a similar compound.

The isomerases of Class 5 contain a number of different types and the systematic names may or may not contain the word isomerase. They can be called *racemases, epimerases, tautomerases,* or *mutases* depending upon the type of isomerization catalysed, and there are also simple *isomerases, cis–trans-isomerases,* and *cyclo-isomerases.*

Lastly the *ligases* (Class 6) are named systematically according to the molecules X and Y which are joined together and the nucleotide that is formed in the process, to give a general name X:Y ligase (ADP-forming) for the cases where ATP gives ADP and inorganic phosphate.

ALPHABETICAL LIST OF ENZYMES REFERRED TO IN THE TEXT

Name referred to in the text	Reaction catalysed	E.C. number
Acetyl-CoA carboxylase	$ATP + acetyl-CoA + CO_2 + H_2O \rightleftharpoons$ $ADP + malonyl-CoA + orthophosphate$	6.4.1.2
Adenylate kinase	$ATP + AMP \rightleftharpoons ADP + ADP$	2.7.4.3
Aldolase	D-fructose-1,6-bisphosphate $\rightleftharpoons$ D-glyceraldehyde-3-phosphate + dihydroxyacetone phosphate	4.1.2.13
Aspartate transcarbamoylase (ATCase)	Carbamoylphosphate + L-aspartate $\rightleftharpoons$ N-carbamoyl-L-aspartate + orthophosphate	2.1.3.2
Carbamoyl-phosphate synthetase (CPS) (ammonia)	$2 ATP + NH_3 + CO_2 + H_2O \rightleftharpoons 2 ADP$ + carbamoyl phosphate + orthophosphate	2.7.2.5
Carbamoyl-phosphate synthetase (CPS) (glutamine)	$2 ATP + glutamine + CO_2 + H_2O \rightleftharpoons 2 ADP$ + glutamate + carbamoyl phosphate + orthophosphate	2.7.2.9
Carboxypeptidase A	Peptidyl-L-amino acid + $H_2O \rightleftharpoons$ peptide + L-amino acid	3.4.12.2
Chymotrypsin	Hydrolytic cleavage of peptide bonds within a polypeptide chain, preferentially following the residues Tyr, Trp, Phe, Leu	3.4.21.1
DNA-dependent RNA polymerase	n Nucleoside triphosphate $\rightleftharpoons n$ pyrophosphate + RNA_n (requires DNA as a template)	2.7.7.6
Elastase	Hydrolytic cleavage of peptide bonds within a polypeptide chain, preferentially following uncharged non-aromatic residues	3.4.21.11
Enolase	2-Phospho-D-glycerate $\rightleftharpoons$ phosphoenolpyruvate + H_2O	4.2.1.11
Fructose diphosphatase (FDPase)	D-fructose-1,6-bisphosphate + $H_2O \rightleftharpoons$ D-fructose-6-phosphate + orthophosphate	3.1.3.11
Fumarase	L-malate $\rightleftharpoons$ fumarate + H_2O	4.2.1.2
β-Galactosidase	Hydrolytic cleavage of terminal non-reducing β-D-galactose residues from β-galactosides	3.2.1.23

Name referred to in the text	Reaction catalysed	E.C. number
Glucokinase	ATP + D-glucose $\rightleftharpoons$ ADP + D-glucose-6-phosphate	2.7.1.2
Glutamine synthetase	ATP + L-glutamate + NH_3 $\rightleftharpoons$ ADP + L-glutamine + orthophosphate	6.3.1.2
Glyceraldehyde-phosphate dehydrogenase (GPD)	D-glyceraldehyde-3-phosphate + orthophosphate + NAD^+ $\rightleftharpoons$ 3-phospho-D-glycerol phosphate + NADH (i.e. 1,3-diphosphoglycerate)	1.2.1.12
Glycogen phosphorylase	$(1,4\text{-}\alpha\text{-D-glycosyl})_n$ + orthophosphate $\rightleftharpoons$ $(1,4\text{-}\alpha\text{-D-glucosyl})_{n-1}$ + α-D-glucose-1-phosphate	2.4.1.1
Hexokinase	ATP + D-hexose $\rightleftharpoons$ ADP + D-hexose-6-phosphate	2.7.1.1
Lactate dehydrogenase (LDH)	L-lactate + NAD^+ $\rightleftharpoons$ pyruvate + NADH	1.1.1.27
Lysozyme	Hydrolysis of 1,4-β-linkages between N-acetylmuramic acid and 2-actamido-2-deoxy-D-glucose residues in a mucopolysaccharide or mucopeptide	3.2.1.17
Papain	Hydrolytic cleavage of peptide bonds within a polypeptide chain, preferentially following the residues Arg or Lys, or the Phe . X sequence, where X can be anything	3.4.22.2
Pepsin	Hydrolytic cleavage of peptide bonds within a polypeptide chain, preferentially following the residues Phe or Leu	3.4.23.1
Phosphofructokinase (PFK)	ATP + D-fructose-6-phosphate $\rightleftharpoons$ ADP + D-fructose-1,6-bisphosphate	2.7.1.11
Phosphoglucomutase	α-D-glucose-1,6-bisphosphate + α-D-glucose-1-phosphate $\rightleftharpoons$ α-D-glucose-6-phosphate + α-D-glucose-1,6-bisphosphate	2.7.5.1
Phosphoglyceromutase	2,3-Bisphospho-D-glycerate + 2-phospho-D-glycerate = 3-phospho-D-glycerate + 2,3-bisphospho-D-glycerate	2.7.5.3
Phosphorylase (see glycogen phosphorylase)		
Phosphorylase b kinase	4 ATP + 2 phosphorylase b $\rightleftharpoons$ 4 ADP + phosphorylase a	2.7.1.38
Pyruvate kinase	ATP + pyruvate $\rightleftharpoons$ ADP + phosphoenolpyruvate	2.7.1.40
Ribonuclease A (RNase A)	Endonucleolytic cleavage of RNA at the 3′-position of a pyrimidine nucleotide residue	3.1.4.22
RNA polymerase	(see DNA-dependent RNA polymerase)	(2.7.7.6)
Streptococcal proteinase	Hydrolytic cleavage of peptide bonds within a polypeptide chain, preferentially following the sequence X:Y., where X has a bulky side chain and Y can be anything	3.4.22.10
Subtilisin	Hydrolytic cleavage of peptide bonds in polypeptide chains. No clear specificity	3.4.21.14
Thermolysin	Hydrolytic cleavage of peptide bonds in polypeptide chains, preferentially following the residues Leu or Phe	3.4.24.4

Name referred to in the text	Reaction catalysed	E.C. number
Triose phosphate isomerase	D-glyceraldehyde-3-phosphate $\rightleftharpoons$ dihydroxyacetone phosphate	5.3.1.1
Trypsin	Hydrolytic cleavage of peptide bonds within a polypeptide chain, preferentially following the residues Arg or Lys	3.4.21.4
Tryptophan pyrrolase	L-tryptophan + $O_2 \rightleftharpoons$ L-formylkynurenine	1.13.11.11
Urease	Urea + $H_2O \rightleftharpoons CO_2 + 2\,NH_3$	3.5.1.5
UDPgalactose-4-epimerase	UDPgalactose $\rightleftharpoons$ UDPglucose	5.1.3.2
UDPglucose dehydrogenase	UDPglucose + $2\,NAD^+ + H_2O \rightleftharpoons$ UDPglucuronate + 2 NADH	1.1.1.22

Appendix 2

The Purification of Proteins

Methods for purifying enzymes are no different in principle from those used to purify proteins in general. They fall into three categories. The first group of methods involve differential *precipitation* of proteins and are based upon the differences in solubility that exist between proteins. The second group consists of *chromatographic* methods, using differences in various physical properties of proteins, such as their size or charge, in order to separate them on columns packed with a variety of media. The methods in the third category are based upon differences in ionic mobility and are *electrophoretic* methods.

PRECIPITATION METHODS

1. Ammonium sulphate precipitation

The most common method in this group uses ammonium sulphate as the precipitating agent. Ammonium sulphate is very soluble in water and concentrations of up to about 4 M can be obtained, depending to some extent on temperature. It is a mild reagent in the sense that, when pure, its presence in solution even at high concentrations does not lead to any permanent alteration in biological activity of proteins. As the ammonium sulphate concentration is raised from zero the solubility of any given protein at first usually increases but then, at quite low concentrations of the salt, this 'salting-in' effect comes to an end and as the salt concentration is raised to higher values a 'salting-out' effect comes into operation and the protein becomes progressively less soluble. Salting-in and salting-out effects depend largely on ionic strength so, in principle, any one of a number of salts could be used to precipitate proteins. Ammonium sulphate is commonly used because of its very high solubility, allowing high ionic strengths to be achieved.

In practise the method is usually employed in the following way. A small sample of a solution of the impure protein is cooled to a temperature of 0–5 °C and finely powdered solid ammonium sulphate is added with stirring until visible precipitation of protein occurs. The amount of salt added is noted and the suspension is centrifuged to remove the precipitated proteins. The supernatant liquid is assayed in order to determine the concentration of the protein of interest. If it is all still in solution the precipitated material can be discarded and more ammonium sulphate added to the supernatant liquid until precipitation occurs again. This process is repeated until the required protein starts to precipitate. The addition of a little more ammonium sulphate then completes the precipitation of this protein and any others still remaining in solution are discarded. A typical result for an experiment of this sort is shown in Figure 68, where the concentration of the protein in question is plotted against the ammonium sulphate concentration. In this hypothetical example the protein remained in solution when the ammonium sulphate concentration reached 40 % of the fully saturated value, but was partly precipitated by 50 % saturated ammonium sulphate, mostly precipitated by 60 % saturated ammonium sulphate, and at 70 % saturated ammonium sulphate, or above, only a very small amount still remained in solution.

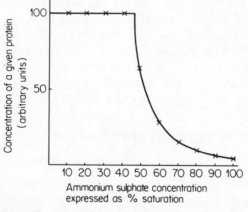

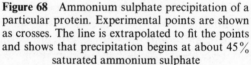

Figure 68 Ammonium sulphate precipitation of a particular protein. Experimental points are shown as crosses. The line is extrapolated to fit the points and shows that precipitation begins at about 45 % saturated ammonium sulphate

With this information the experimenter then extrapolates the precipitation curve to find the point at which precipitation of the protein begins, about 45 % in this case. He may then return to the bulk of his impure protein solution and calculate how much ammonium sulphate is needed to bring it to 45 % saturation. This quantity is added and, when fully dissolved, the unwanted protein is centrifuged out. The supernatant liquid is then raised to say 70 % of saturation by adding more ammonium sulphate and then centrifuged again. This time the solid pellet of precipitated protein contains the bulk of the required protein. Of course it will also contain other proteins which were precipitating in the range between 45 % and 70 % saturation. At the expense of a poorer yield, the experimenter may decide to take a narrower cut, say from 45 % to only 60 % saturation, in order to exclude some of these other unwanted proteins and thereby achieve a higher degree of purification for the protein of interest.

A number of points deserve note in connection with the practical use of this method. The first is that one must have a *specific assay* for the protein of interest, in order to be able to find the range in which it precipitates. With enzymes there is no problem as one uses the enzyme's catalytic activity in order to assay it, but with some other proteins problems may be encountered in this respect. Even with enzymes, however, it is advisable to remember to remove or dilute out the ammonium sulphate before attempting to assay because high ammonium sulphate concentrations may be inhibitory.

The second point to note is that the concentration of ammonium sulphate at which any given protein begins to precipitate will depend upon the initial *concentration of that protein*. If, for example, one had commenced the experiment shown in Figure 68, with a concentration of only 50 in arbitrary units, instead of 100 as shown, then precipitation would not have commenced until the solution was 52 % saturated with ammonium sulphate, rather than 45 %.

The third point concerns the mode of expressing ammonium sulphate concentrations. It is usual practice to express them as a percentage of the concentration of a saturated solution (4·1 M at 23 °C) rather than as actual molarities of the salt. This convention arises out of the fact that when solid ammonium sulphate is added to aqueous solutions, rather large and non-linear volume changes occur so that calculation of molarities becomes a tedious business. The problem was overcome by Kunitz who established by careful measurements the weights of solid ammonium sulphate that must be added to a litre of pure water to raise it to various concentrations, and the weights required to raise a litre from any one concentration to another. These valuable experimental data were expressed as percentages of the saturated concentration in tabular form as shown in Table 7 and they have been used in this form ever since. The table refers to solutions at 23 °C and although differences will occur by working at 0–5 °C, they are small and are often ignored for practical purposes. Table 7 also shows the volume of saturated ammonium sulphate solution that must be added to 100 ml of a solution to raise it from any one concentration to another for those cases where addition of solid ammonium sulphate is not desirable. The addition of solid ammonium sulphate usually leads to some 'frothing' of the solution and, although this is not usually a problem, some enzymes may be denatured readily by the presence of froth due to the phenomenon of surface denaturation. In those cases addition of saturated ammonium sulphate solution would be preferable in spite of the larger volumes thereby incurred.

The ammonium sulphate concentration at which a given protein starts to precipitate may depend upon what other proteins are present with it in solution. This can sometimes lead to confusion if further purification is attempted by re-precipitating the protein with ammonium sulphate at a later stage in its isolation.

Ammonium sulphate is a weak acid due to the ammonium ion ($pK_a = 9.25$) so that concentrated solutions, such as employed for purification, will have pH values below 7. If the enzyme of interest is liable to acid denaturation it is advisable to carry out the precipitation in a suitable buffer.

Table 7

Ammonium sulphate concentration conversion tables

Grams solid ammonium sulphate added to 1000 ml solution:

0																			
27.0	54.9	83.7	113.4	144.0	175.7	208.4	242.3	277.3	313.5	351.1	390.0	430.4	472.3	515.8	561.1	608.1	657.1	708.2	761.4
.05	27.5	55.8	85.1	115.2	146.4	178.7	212.0	246.5	282.2	319.2	357.5	397.3	438.5	481.4	526.0	572.4	620.6	670.9	723.4
0	.10	27.9	56.7	86.4	117.1	148.9	181.7	215.7	250.8	287.3	325.0	364.2	404.8	447.0	490.9	536.6	584.1	633.6	685.3
5.26	.05	.15	28.4	57.6	87.9	119.1	151.4	184.9	219.5	255.3	292.5	331.1	371.1	412.6	455.9	500.8	547.6	596.4	647.2
11.1	5.58	.10	.20	28.8	58.6	89.3	121.1	154.0	188.1	223.4	260.0	298.0	337.3	378.3	420.8	465.0	511.1	559.1	609.1
17.7	11.8	5.88	.15	.25	29.3	59.6	90.9	123.2	156.8	191.5	227.5	264.8	303.6	343.9	385.7	429.3	474.6	521.8	571.1
25.0	18.8	12.5	6.25	.20	.30	29.8	60.6	92.4	125.4	159.6	195.0	231.7	269.9	309.5	350.7	393.5	438.1	484.5	533.0
33.3	26.7	20.0	13.3	6.67	.25	.35	30.3	61.6	94.1	127.7	162.5	198.6	236.1	275.1	315.6	357.7	401.6	447.3	494.9
42.9	35.7	28.6	21.4	14.3	7.14	.30	.40	30.8	62.7	95.7	130.0	165.5	202.4	240.7	280.5	321.9	365.1	410.0	456.9
55.9	46.2	38.5	30.8	23.1	15.4	7.69	.35	.45	31.4	63.8	97.5	132.4	168.7	206.3	245.5	286.2	328.6	372.7	418.8
66.7	58.3	50.0	41.7	33.3	25.0	16.7	8.33	.40	.50	31.9	65.0	99.3	134.9	171.9	210.4	250.4	292.1	335.5	380.7
81.8	72.7	63.7	54.6	45.5	36.4	27.3	18.2	9.10	.45	.55	32.5	66.2	101.2	137.5	175.3	214.6	255.5	298.2	342.6
100.0	90.0	80.0	70.0	60.0	50.0	40.0	30.0	20.0	10.0	.50	.60	33.1	67.5	103.2	140.3	178.9	219.0	260.9	304.6
122.2	111.1	100.0	88.9	77.8	66.7	55.6	44.4	33.3	22.2	11.1	.55	.65	33.7	68.8	105.2	143.1	182.5	223.6	266.5
150.0	137.5	125.0	112.5	100.0	87.5	75.0	62.5	50.0	37.5	25.0	12.5	.60	.70	34.4	70.1	107.3	146.0	186.4	228.4
187.5	171.4	157.1	142.9	128.6	114.3	100.0	85.7	71.4	57.1	42.9	28.6	14.3	.65	.75	35.1	71.5	109.5	149.1	190.4
233.3	216.7	200.1	183.3	166.7	150.0	133.3	116.7	100.0	83.3	66.7	50.0	33.3	16.7	.70	.80	35.8	73.0	111.8	152.3
300.0	280.0	260.0	240.0	220.0	200.0	180.0	160.0	140.0	120.0	100.0	80.0	60.0	40.0	20.0	.75	.85	36.5	74.5	114.2
400.0	375.0	350.0	325.0	300.0	275.0	250.0	225.0	200.0	175.0	150.0	125.0	100.0	75.0	50.0	25.0	.80	.90	37.3	76.1
566.7	533.3	500.0	466.7	433.3	400.0	366.7	333.3	300.0	266.7	233.3	200.0	166.7	133.3	100.0	66.7	33.3	.85	.95	38.1
900	850	800	750	700	650	600	550	500	450	400	350	300	250	200	150	100	50	.90	1.0

Millilitres saturated ammonium sulphate (23 °C) added too 100 ml solution:

| 0 | 50 | 100 | 150 | 200 | 250 | 300 | 350 | 400 | 450 | 500 | 550 | 600 | 650 | 700 | 750 | 800 | 850 | 900 |

Finally it should be remembered that the concentrated solutions of ammonium sulphate used in this technique have densities of up to 1·24 at 20 °C, and that the density of precipitated protein is not usually much greater than this, being about 1·37 for most proteins. For this reason the precipitated material settles out of suspension only extremely slowly and powerful centrifuges are required to assist the process. Typically one uses a preparative ultracentrifuge capable of centrifuging two or three litres of suspension at speeds of up to 20,000 rev/min giving rise to centrifugal forces of about 50,000 times the force of gravity.

2. The use of other soluble salts for protein precipitation

A variation on the procedure outlined above is to use high concentrations of other salts, such as potassium phosphate, to achieve differential precipitation. Pure dipotassium hydrogen phosphate or potassium dihydrogen phosphate may be used but one should note that the former leads to pH values in the region of pH 8 while the latter leads to pH values below 7, so that once more care should be taken to ensure that the enzyme in question will survive these conditions. In some cases these approaches have been used with success but they are by no means as common as the use of ammonium sulphate.

3. The use of organic solvents

Acetone and ethanol, among other water-miscible organic solvents, have been used to precipitate proteins. They probably function by altering the dielectric constant of the solvent, thereby enhancing protein–protein interactions at the expense of those between protein and solvent. These organic solvents must be used with great care as they invariably lead to some irreversible loss of enzyme activity and may, if used unwisely, cause complete denaturation. The acetone or alcohol is usually cooled to 0 °C or below and added with good stirring, to the aqueous protein solution which is in turn maintained at a temperature near 0 °C. The precipitated protein is then removed as quickly as possible from the remaining solution. In other respects the procedure is exactly as described for precipitation by ammonium sulphate except that powerful centrifuges are not usually needed to remove precipitated proteins. Once precipitated, the required enzyme is usually re-dissolved in an aqueous buffer and dialysed to remove entrained organic solvent at the earliest opportunity.

CHROMATOGRAPHIC METHODS

1. Molecular sieve chromatography or gel filtration

The principle and an outline of the practice of this method are given in Chapter 3 (p. 48) where its use is described for the estimation of the molecular weight of a pure protein. One does not need to know the molecular weight of an enzyme to be able to use gel filtration in purifying it. The fact is that a crude extract will contain proteins of all sizes and shapes, so that passage of an extract through a column containing a molecular sieve such as Sephadex or Agarose will sort the proteins according to size, the larger molecules travelling most rapidly through the column and the smaller ones emerging later, as the column is eluted with a suitable buffer. The choice of molecular sieve depends very largely upon circumstances and is usually a matter of trial and error with a new enzyme. As most proteins have molecular weights in excess of 20,000 there is little point in using a molecular sieve of small pore size that excludes molecules of this size and above. No purification would result as all the proteins would pass straight through the column, passing between the beads of gel. On the other hand, few enzymes have molecular weights larger than 1,000,000 so the very large pore molecular sieves are equally unlikely to be of use. The method would be tested in the first place with a gel filtration column that could discriminate between molecules of molecular weight in the range of say 30,000 to 200,000, in the hope that a good purification could be obtained. In the light of the experience gained from this experiment, the choice of molecular sieve could be refined.

It is often useful to carry out a gel filtration quite early in the sequence of steps needed to purify an enzyme, because much unwanted protein can be separated in this way. On the other hand, efficient use of the method requires that the bed volume of the column (that is the total volume occupied by the packed column) should be as large as possible relative to the volume of protein solution applied to it. In any case this ratio should not be less than 10:1. This means that efficient gel filtration of 100 ml of protein solution requires a column occupying at least 1 litre, and more likely 5 litres if the proteins to be separated are rather close in size. A 5-litre column is beyond the capabilities of many research laboratories. It would have to be 8 cm in diameter and nearly 100 cm long. This places a severe restriction on the volume of protein solution that can be handled which is why gel filtration is not often used directly on crude cell-extracts, but is usually kept in reserve until the enzyme of interest has been concentrated into a smaller volume, perhaps by the use of a precipitation step.

Automatic fraction collectors

All the column-chromatographic methods rely on the use of a fraction collector to keep the different proteins apart once they have been separated by the column. A typical procedure is illustrated diagrammatically in Figure 69. The protein solution is loaded onto the column through a three-way tap and a pump. The latter is usually a peristaltic pump employing silicone rubber tubing so that the enzymes do not come into contact with metal parts. When all the protein solution has been pumped onto the column the three-way tap is turned to admit a buffer into the pump and this is then pumped through the column until at least one bed-volume of buffer has emerged from the bottom of the column. For example, 10 ml of protein solution might be loaded onto a molecular sieve column 5 cm in diameter and 25 cm long (bed volume = 491 ml) and then eluted with a 0·01 M potassium phosphate buffer at pH 7·0 until

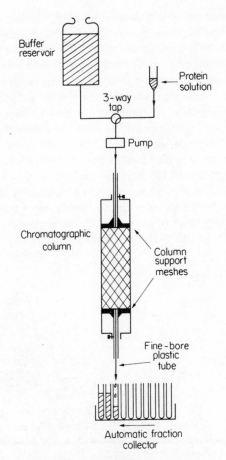

Buffer
reservoir

Protein
solution

3-way
tap

Pump

Chromatographic
column

Column
support
meshes

Fine-bore
plastic
tube

Automatic fraction
collector

Figure 69 A diagrammatic representation of protein chromatography with
the aid of a fraction collector

500 ml had emerged from the column. At least 100 fractions should be collected so as to preserve the separation of components that was effected by the column, so each fraction would be 5 ml in volume. Automatic fraction collectors are designed to move a new test-tube under the dripping plastic tube from the column at preset intervals. Some fraction collectors use photoelectric devices to sense the volume in the tube and in this case would be set to detect 5 ml and then change fractions. Others count the number of drops and change fractions after a preset number of drops have been collected. This type requires foreknowledge of the drop size being delivered and can be unreliable if the drop size alters during an experiment as often happens in protein chromatography because of variation in surface tension with protein concentration. The most reliable method of fraction collection is based upon a preset *time* for each fraction. The buffer must be pumped through the column at a constant rate,

say 200 ml/h, for the hypothetical column just described, and the fraction collector is set to change fractions every 1·5 minutes, thereby ensuring 5 ml in each fraction. The whole experiment would thereby take $2\frac{1}{2}$ h to complete. Columns cannot be pumped at more than about 20 ml/h per square centimetre of cross-sectional area, otherwise the large protein molecules, which diffuse only slowly, do not have time to enter the pores of the gel and separation is not achieved. On the other hand, if pumped too slowly the separated bands of protein will diffuse into each other as they pass down the column, again leading to loss of resolution. After collecting, fractions would be analysed to detect the enzyme of interest. One may also measure the protein content of each fraction so as to obtain some idea of the purification achieved by the column. These two sets of data would be plotted together on a chromatogram such as the one shown in Figure 70. It can be seen that in this example the enzyme has been separated from much protein of lower and higher molecular weight, but is still contaminated by at least two other proteins of similar size.

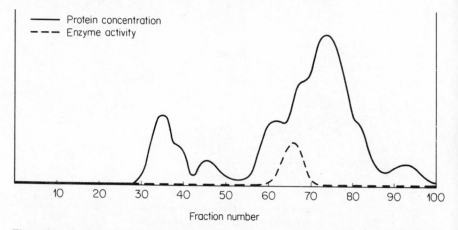

Figure 70 Hypothetical chromatogram showing the partial purification of an enzyme by gel filtration on a column (25 × 5 cm) of Sephadex. Fraction volume 5·0 ml. The first 30 fractions contain no protein because this represents the liquid contained in the column when the sample was applied

2. Ion-exchange chromatography

Ion-exchange methods depend primarily upon differences in the charge carried by protein molecules. At its isoelectric point in a given buffer a protein has zero charge, but at pH s below this value the protein will be positively charged (cationic) while at pH s above the isoelectric point it will carry a negative charge (anionic).

One may use either cation-exchange resins or anion-exchange resins to purify proteins. The former bind cations because they carry covalently attached negatively charged groups such as the sulphonate $(-SO_3^-)$, carboxylate $(-COO^-)$

or phosphate ($-OPO_3^{2-}$) groups. Examples of cation-exchange resins that are suitable for protein chromatography are given in Table 8. In a similar way anion-exchange resins carry positively charged groups such as the protonated amino ($-NH_3^+$), quaternary ammonium ($-^+NR_3$) or guanido

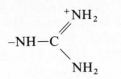

groups covalently bound to a polymeric matrix. Examples of such resins used for proteins are also given in Table 8.

Table 8

Name	Polymer matrix	Charged group
A. Cation-exchange resins:		
Sulphopropyl Sephadex (SP.Sephadex)	crosslinked Dextran	$-CH_2 \cdot CH_2 \cdot CH_2 \cdot SO_3^-$
Carboxymethyl cellulose (CM.cellulose)	Cellulose	$-CH_2 \cdot COO^-$
Carboxymethyl Sephadex (CM.Sephadex)	crosslinked Dextran	$-CH_2 \cdot COO^-$
Phosphocellulose	Cellulose	$-O-PO_3^{2-}$
B. Anion-exchange resins:		
Diethylaminoethyl Sephadex (DEAE.Sephadex)	crosslinked Dextran	$-CH_2 \cdot CH_2 - {}^+NH \overset{\diagup\ CH_2 \cdot CH_3}{\diagdown_{CH_2 \cdot CH_3}}$
Diethylaminoethyl cellulose (DEAE.cellulose)	Cellulose	ditto
Quaternary amino ethyl Sephadex (QAE.Sephadex)	crosslinked Dextran	$-CH_2 \cdot CH_2 \cdot N^+ - CH \cdot CH_3$ with C_2H_5OH, C_2H_5 substituents
Guanidoethyl cellulose	Cellulose	$-CH_2 \cdot CH_2 \cdot NH - C \overset{{}^+NH_2}{\diagdown_{NH_2}}$

One may use these ion-exchange resins in two ways. The pH may be adjusted so that the enzyme in question does not bind to the resin, in which case it passes straight through the column leaving other, oppositely charged, proteins bound to the column. For example, an enzyme whose isoelectric point was at pH 5 could be separated from proteins with isoelectric points above pH 7 by adjusting the solution to pH 6 and passing it through a cation-exchange resin. At pH 6 the enzyme would be negatively charged and would not bind to the negatively charged column, whereas the other proteins would be positively charged and would bind. Conversely the same solution at pH 6 could be passed through an anion-exchange resin, to which the enzyme would bind while the other proteins would pass straight through. The enzyme could then be eluted from the column, either by using a buffer of low pH or, more usually, of high salt concentration which would displace the enzyme.

In practice matters are not as simple as depicted above and one would usually find a number of other proteins bound to the column as well as the desired enzyme. In that case a solution of gradually increasing salt concentration would be used to elute the column. In this way the most weakly bound proteins would be eluted first and separated from the more tightly bound ones which would emerge later in the salt gradient.

3. Adsorption chromatography

In practice adsorption chromatography is very similar to ion-exchange chromatography except that the adsorbent which is packed into the column contains both positive and negative groups, so the interactions between proteins and adsorbent are more complex than in the case of ion-exchange. Hydroxyapatite, often also called hydroxylapatite, is the most common adsorbent, although calcium phosphate deposited on cellulose powder can also be used. Hydroxyapatite $(Ca_{10}(PO_4)_6(OH)_2)$ can be prepared in the laboratory by standard procedures, or obtained commercially as a granular free-running powder. It is usually suspended in a dilute sodium or potassium phosphate buffer at a pH between 6·5 and 7·0 and packed into a column which is then washed with several volumes of the buffer. The protein solution in the same buffer is then introduced into the column and elution is commenced. In general the pH of the eluting buffer is the same as that used in preparing and loading the column, but its concentration is gradually increased so as to displace the bound proteins sequentially. This method will often achieve separation of proteins that cannot be separated by ion-exchange or gel filtration chromatography.

4. Affinity chromatography

A method that has sprung into prominence in recent years makes use of the specific affinity of an enzyme for certain ligands. Molecules such as the substrate, allosteric effector, or substrate analogue are covalently attached to an inert porous polymer such as agarose (a constituent of agar). The polymer–ligand complex is packed into a column in a suitable buffer. When a mixture of proteins is applied to the top of such a column, most proteins will pass straight through, but the enzyme of interest will interact with the bound ligand on the polymeric matrix and will be retarded or even immobilized on the column. After washing the column to remove the unwanted proteins the enzyme may be eluted in one of a number of ways. Perhaps the most elegant of these is to use a solution of free ligand at a high enough concentration to compete effectively with the matrix-bound ligand, thereby weakening the interaction between the enzyme and the column. In other cases the enzyme–column interaction can be weakened by a change in pH or ionic strength so as to bring about elution of the enzyme.

Although the principle of affinity chromatography is delightfully simple and the selectivity inherent in the method is capable of giving 'one-step' purification, nevertheless certain practical difficulties do arise. The first of these concerns the choice of ligand. One which binds tightly to the enzyme is required, but not so tightly that it would be impossible to remove the bound enzyme from the column. Another consideration is how to attach the ligand to the insoluble polymer matrix. Clearly the point of covalent attachment must not be such as to interfere with the enzyme–ligand interaction. A third problem is that, once attached, the ligand must not be sterically inaccessible to the binding site of the enzyme. This last consideration means that it is usually necessary to interpose a 'spacer arm' between the polymer matrix and the ligand, to hold the latter out away from the polymer and into solution where it may interact

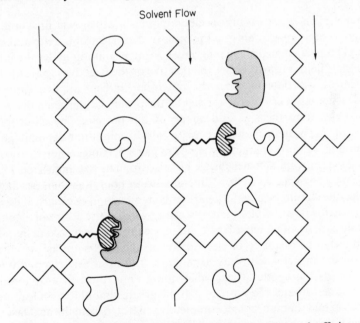

Solvent Flow

Figure 71 Diagrammatic illustration of the principle of affinity chromatography. Three strands of the crosslinked polymer matrix are shown. The spacer arm is shown in bold, attaching the ligand (shaded) to the polymer matrix. Various roughly globular protein molecules are shown diffusing between the strands of the polymer but only the enzyme molecules (shaded) have a binding site for the ligand. One enzyme molecule is shown actually binding to the immobilized ligand. Note that the attachment of ligand to the spacer arm is made at a point where there is no interaction with the enzyme

with the enzyme without interference from the matrix. These factors are illustrated in Figure 71.

The method has been used successfully in the purification of a number of enzymes but of course the effort needed to synthesize a suitable affinity column medium makes it less attractive in the short term than some of the other methods that are available. On the other hand, once a suitable medium has been synthesized, purification can often be achieved with very little effort. A variant on the method described above is to make an antibody to the enzyme by injecting some of the pure enzyme into the blood of an experimental animal. The antibody is then attached covalently to a polymer matrix where it serves the same purpose as the specific ligand discussed above. This method requires a small amount of pure enzyme to have been prepared by some other method, so as to be able to raise specific antibody against it.

ELECTROPHORETIC METHODS

1. Zone electrophoresis

This is a variant of free solution electrophoresis which was first used many years ago to separate proteins on the basis of their electrophoretic mobility. At a given pH a particular protein will move through solution under the influence of an electric field, towards either the positive or the negative electrode. Its rate of movement will depend upon the potential gradient applied between the electrodes, upon the net charge carried by the protein at the given pH and, to a lesser extent, upon the size and shape of the protein. The electrophoretic mobility of the protein is defined as its rate of migration per unit potential gradient and is thus a characteristic of the particular protein at the given pH, depending mainly upon its net charge but also upon its size and shape. Different proteins may be expected to have different electrophoretic mobilities and so, in principle, one should be able to separate them on a kind of electrical racetrack. All the proteins must be placed at a 'starting-gate' in the centre of a column of buffer solution. At either end of the column, electrodes are placed in the solution and connected to a source of electric power (d.c.). As soon as the current is turned on the proteins will start to migrate. The positively charged ones will move towards the anode at rates depending upon their electrophoretic mobilities. Sooner or later the protein with the greatest mobility will lead the field to such an extent that it becomes completely separated from the next fastest, and so on. Negatively charged proteins will migrate towards the cathode in a similar way. When a sufficient degree of separation has been achieved the current may be switched off and the contents of the tube run out into a fraction collector prior to locating and pooling the separated components.

Although this free solution method can be made to work in some cases, a number of practical difficulties arise. Even though the proteins may be introduced in a narrow band of solution at the commencement of electrophoresis, various factors cause the individual bands of protein to spread out during their migration, thus detracting from the optimal separation. Diffusion cannot be avoided although it can be reduced by cooling. The passage of an electric current causes heating effects in the solution and these may actually be greatest in the moving bands of protein, thereby leading to thermal instability and the tendency for the bands to be disrupted by convection currents in the liquid. Overheating in the neighbourhood of the protein bands can also lead to thermal denaturation. Cooling the apparatus and using only moderate potential gradients can overcome some of these effects but, in turn, slow down the process, giving rise to greater diffusional problems.

These problems have led to the use of inert stabilizing media, such as cellulose or Sephadex, being added to the column to reduce the chances of band spreading and force the proteins to move in discrete bands or zones through the apparatus in *zone electrophoresis* as illustrated in Figure 72. A logical extension of this principle is to make the proteins migrate through a continuous gel of controlled

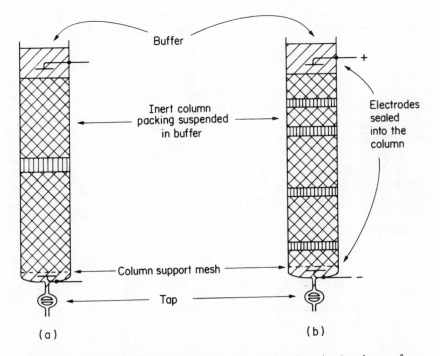

Figure 72 Diagrammatic illustration of zone electrophoresis. A column of an inert granular polymer such as cellulose powder is packed into a glass column in a buffer of suitable pH. A layer of protein solution containing a number of different proteins is introduced onto the top of the packed column and buffer is allowed to flow through the column until the protein band reaches the middle as shown in (a). The tap is then closed and the electrodes are connected to a source of electrical power. After a suitable time has elapsed the protein bands will be separated as shown in (b). The current is turned off, the tap opened and the contents of the column carefully eluted into a fraction collector by washing through with buffer from above

274

pore size, giving rise to *gel electrophoresis*. Extra selectivity may be introduced if the pore size is small enough to hinder the progress of the larger protein molecules but not the smaller, adding a molecular sieving effect in some cases.

2. Isoelectric focusing

In this method a pH gradient is established along the length of a column of solution and proteins are then made to migrate under the influence of an applied electric potential until they reach the pH at which they carry a net charge of zero, that is to say they reach the place in the gradient where the pH is equal to their isoelectric point. Being no longer charged they stop migrating. Any tendency for a protein to diffuse into regions of higher or lower pH on either side is offset to a great extent by the fact that if it does so it immediately acquires a net charge which causes it to migrate back to its isoelectric point in the gradient. The proteins thus focus in tight bands at regions in the column corresponding to their individual isoelectric pH s.

The mixture of proteins can fill the whole column at the start of the experiment—there is no need to introduce a tight band—and there is less need for a stabilizing inert medium to fill the column. It is still necessary to cool the apparatus to prevent convection and thermal denaturation but in principle this is an elegant and sophisticated way of separating proteins on the basis of differences in their isoelectric points. Problems arise in some cases when the method is applied to bulk purification because proteins are least soluble at their isoelectric points. If one attempts to focus a large amount of protein into a tight band at its isoelectric point there is a danger of exceeding its solubility in this region. If the protein precipitates it will fall out of the focused band and disturb the separation process. To some extent this may be overcome by carrying out the procedure in a *horizontal* column and providing a corrugated tray beneath the solution to catch the precipitated protein as shown in Figure 73.

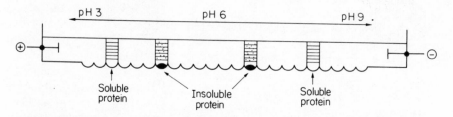

Figure 73 Horizontal isoelectric focusing apparatus. Bands of proteins are shown focusing at points in a pH gradient where the pH is equal to their individual isoelectric points. Two of the proteins have exceeded their solubility at the isoelectric point and are shown precipitating into corrugations in the base of the apparatus. Any tendency to diffuse out of the focused band in a leftward direction causes the protein to acquire a positive charge since it moves into a region of lower pH. The applied potential gradient thus causes it to migrate towards the anode on the right and hence back to its isoelectric point. The opposite applies to rightward diffusion

THE CHOICE AND ORDER OF METHODS OF PURIFICATION

Although in principle the various methods described could be used in any order, practical considerations dictate that the electrophoretic methods should be used towards the end of a purification procedure where the number of components still contaminating the required enzyme is small and the volumes to be handled are small. Similar limitations apply to gel filtration and to some other column chromatographic methods but the restrictions on volume of solution are not as stringent, so such methods tend to be used in the middle of a series of purification steps. The precipitation methods can be used on larger volumes and are usually applied early in a purification, as they also allow a concentration of the desired protein. Affinity chromatography, if it can be applied, might be expected to achieve purification in one step straight from a crude extract, but few examples of such an ideal situation are documented. In practice it is usually necessary to carry out some preliminary purification before applying affinity chromatography.

Apart from considerations such as these, there are no rules about which methods to use. With any given enzyme it is largely a matter of trial and error until a sequence of methods and conditions is found which purify the enzyme from its contaminating proteins with as little trouble and as large a yield of enzyme as possible.

CRITERIA OF PURITY

All of the methods outlined above can be used on an analytical scale to establish the purity of a preparation. When it yields a single peak in various chromatographic systems, when it gives a single band on electrophoresis at more than one pH and a single band on isoelectric focusing, and when it shows only one component in solubility or precipitation tests, there is good reason to believe that it is pure. Chemical tests may provide further evidence of purity. Thus, for a pure protein, an integral number of moles of N-terminal amino acids per mole of protein should be found by quantitative N-terminal analysis, integral ratios of all the amino acids should be found by total amino acid analysis and bound cofactors, if it contains them, should also be present in stoichiometric amounts. The final criterion of purity of course is the demonstration of a unique amino acid sequence, but that is rarely undertaken in order to demonstrate purity.

Lastly, immunological tests may establish the presence of impurities even when they cannot be demonstrated by other tests. The problem of establishing purity is thus tedious and largely based on negative results insofar as a failure to show heterogeneity by a number of different criteria is usually accepted as evidence of homogeneity.

References

REFERENCES FOR CHAPTER 1

1. M. Dixon, 'The history of enzymes and of biological oxidations', in *The Chemistry of Life*, edited by J. Needham, Cambridge University Press, 1970.
2. J. D. Watson, *Molecular Biology of the Gene*, 2nd ed., W. A. Benjamin Inc., 1970.
3. A. G. Loewy and P. Siekevitz, *Cell Structure and Function*, 2nd ed., Holt, Rinehart and Winston Inc., 1969.
   A. B. Novikoff and E. Holtzman, *Cells and Organelles*, Holt, Rinehart and Winston Inc., 1970.

REFERENCES FOR CHAPTER 2

4. A. L. Lehninger, *Bioenergetics*, W. A. Benjamin Inc., 1965.
5. H. Gutfreund, *Enzymes: Physical Principles*, Wiley-Interscience, 1972.
6. V. Massey, *Methods in Enzymology*, **I**, 729 (1955).
7. V. Massey, *Biochemical Journal*, **53**, 67 (1953).
8. M. Dixon and E. C. Webb, *Enzymes*, 2nd ed., Longmans, 1964.

REFERENCES FOR CHAPTER 3

9. T. J. Bowen, *An Introduction to Ultracentrifugation*, Wiley, 1970.
10. W. A. Schroeder, *The Primary Structure of Proteins*, Harper and Row, 1968.
11. W. Kauzman, 'Some factors in interpreting protein denaturation', *Adv. Protein Chemistry*, **14**, 1 (1959).
    H. A. Scheraga, 'Intramolecular bonds in proteins. II Non-covalent bonds' in *The Proteins*, 2nd ed. (H. Neurath, Ed.), Vol. I, p. 478, Academic Press, 1963.
    C. Tanford, 'Protein denaturation. Parts A and B', *Adv. Protein Chemistry*, **23**, 121 (1968).
    C. Tanford, 'Protein denaturation. Part C', *Adv. Protein Chemistry*, **24**, 2 (1970).
12. H. R. Wilson, *Diffraction of X-rays by Proteins, Nucleic Acids and Viruses*, Edward Arnold, 1966.
    D. Eisenberg, 'X-ray crystallography and enzyme structure', in *The Enzymes*, 3rd ed. (P. D. Boyer, Ed.), Vol. I, p. 1, 1970.

REFERENCES FOR CHAPTER 4

13. J. T. Edsall and J. Wyman, *Biophysical Chemistry*, Vol. I, Academic Press, 1958.
14. G. E. Means and R. E. Feeney, *Chemical Modification of Proteins*, Holden-Day, Inc., San Francisco, 1971.
    G. R. Stark, *Advances in Protein Chemistry*, **24**, 261 (1970).
15. D. E. Koshland, Jr., 'Application of a theory of enzyme specificity to protein synthesis', *Proceedings of the National Academy of Sciences, U.S.*, **44**, 98 (1958).

16. J. Monod, J. Wyman and J-P. Changeux, 'On the nature of allosteric transitions: A plausible model', *Journal of Molecular Biology*, **12**, 88 (1965).
17. H. Eyring, R. Lumry and J. D. Spikes, 'Kinetic and thermodynamic aspects of enzyme-catalysed reactions', in *The Mechanism of Enzyme Action* (W. D. McElroy and B. Glass, Eds.), p. 123, 1954.
18. L. Pauling, 'Nature of forces between large molecules of biological interest', *Nature*, **161**, 707 (1948).
19. G. E. Lienhard, 'Enzymatic catalysis and transition-state theory', *Science*, **180**, 149 (1973).
20. M. I. Page and W. P. Jencks, 'Entropic contributions to rate accelerations in enzymic and intramolecular reactions and the chelate effect', *Proceedings of the National Academy of Sciences, U.S.*, **68**, 1678 (1971).
21. D. R. Storm and D. E. Koshland, Jr., 'A source of the special catalytic power of enzymes: Orbital steering', *Proceedings of the National Academy of Sciences, U.S.*, **66**, 445 (1970).
    D. G. Hoare, 'Significance of molecular alignment and orbital steering in mechanisms for enzymatic catalysis', *Nature*, **236**, 437 (1972).

Books on enzyme structure and function

22. W. P. Jencks, *Catalysis in Chemistry and Enzymology*, McGraw-Hill, 1969.
23. S. A. Bernhard, *The Structure and Function of Enzymes*, W. A. Benjamin, 1968.
24. M. L. Bender and L. J. Brubacher, *Catalysis and Enzyme Action*, McGraw-Hill, 1973.
25. E. Zeffren and P. L. Hall, *The Study of Enzyme Mechanisms*, Wiley, 1973.

REFERENCES FOR CHAPTER 5

Books on enzyme kinetics

26. J. M. Reiner, *Behaviour of Enzyme Systems*, 2nd ed. (Van Nostrand Reinhold), 1969.
27. K. J. Laidler and P. S. Bunting, *The Chemical Kinetics of Enzyme Action*, Clarendon Press, Oxford, 1973.
28. I. H. Segel, *Enzyme Kinetics*, Wiley, 1975.
29. J. B. S. Haldane, *Enzymes*, Longmans, 1930.
30. P. D. Boyer (Ed.), *The Enzymes*, 3rd ed., Vol. I, 'Structure and control', Vol. II, Kinetics and mechanisms', Academic Press, 1970. (Also published in a combined paperback.)

Original papers on enzyme kinetics

31. M. Dixon, 'The determination of enzyme inhibitor constants', *Biochem. J.*, **55**, 170 (1953).
32. E. L. King and C. Altman, 'A schematic method of deriving the rate laws for enzyme-catalysed reactions', *J. Phys. Chem.*, **60**, 1375 (1956).
33. J. T. Wong and C. S. Hanes, 'Kinetic formulations for enzymic reactions involving two substrates', *Can. J. Bioch. and Physiol.*, **40**, 763 (1962).
34. L. L. Ingraham and B. Makower, *J. Phys. Chem.*, **58**, 266 (1954).
35. W. Ferdinand, 'The interpretation of non-hyperbolic rate curves for two-substrate enzymes', *Biochem. J.*, **98**, 278 (1966).
36. W. W. Cleland, 'The Kinetics of enzyme-catalysed reactions with two or more substrates or products. I. Nomenclature and rate equations. II. Inhibition: Nomenclature and theory. III. Prediction of initial velocity and inhibition patterns by inspection', *Biochim. Biophys. Acta*, **67**, 104, 173 and 188 (1963). See also the article by Cleland in Reference 30 above.
37. K. Dalziel, 'Initial steady state velocities in the evaluation of enzyme–coenzyme–substrate reaction mechanisms', *Acta Chem. Scand.*, **11**, 1706 (1957).

38. K. Dalziel, 'The interpretation of kinetic data for enzyme-catalysed reactions involving three substrates', *Biochem. J.*, **114**, 547 (1969).

REFERENCES FOR CHAPTER 6

39. J. Monod, J. Wyman and J-P. Changeux, 'On the nature of allosteric transitions: A plausible model', *J. Mol. Biol.*, **12**, 88 (1965).
40. M. M. Rubin and J-P. Changeux, *J. Mol. Biol.*, **21**, 265 (1966).
41. D. E. Koshland, Jr., G. Némethy and D. Filmer, 'Comparison of experimental binding data and theoretical models in proteins containing subunits', *Biochemistry*, **5**, 365 (1966).
42. M. E. Kirtley and D. E. Koshland, Jr., 'Models of cooperative effects in proteins containing subunits. Effects of two interacting ligands', *J. Biol. Chem.*, **242**, 4192 (1967).
43. J. E. Haber and D. E. Koshland, Jr., 'Relation of protein subunit interactions to the molecular species observed during cooperative binding of ligands', *Proc. Nat. Acad. Sci. U.S.A.*, **58**, 2087 (1967).

REFERENCES FOR CHAPTER 7

44. G. N. Cohen, *The Regulation of Cell Metabolism*, Holt, Rinehart & Winston, 1968. E. A. Newsholme and C. Start, *Regulation in Metabolism*, Wiley, 1973.
45. F. S. Rolleston, *Current Topics in Cellular Regulation*, **5**, 47 (1972).
46. H. Kacser and J. A. Burns, in *Rate Control of Biological Processes*, Symposia of Society for Experimental Biology (Ed. D. D. Davies), Number XXVII, p. 65, Cambridge University Press, 1973.
47. E. A. Newsholme and B. Crabtree, *ibid.*, p. 429.
48. H. A. Krebs, *ibid.*, p. 299.

REFERENCES FOR CHAPTER 8

49. F. Jacob and J. Monod, *J. Molecular Biol.*, **3**, 318 (1961).
50. A. L. Goldberger and J. F. Dice, 'Intracellular protein degradation in mammalian and bacterial cells', *Annual Reviews of Biochem.*, **43**, 835 (1974).

REFERENCES FOR APPENDIX 2

51. *Methods in Enzymology*, Vols. I, II, III, V, VI and XXII.
52. H. A. Sober, R. W. Hartley, Jr., W. R. Carroll and E. A. Peterson in *The Proteins*, 2nd ed. (H. Neurath, Ed.), Academic Press, 1965, pp. 2–98.
53. (a) *Sephadex. Gel Filtration in Theory and Practice*, 1970.
    (b) *Sephadex Ion Exchanges*, 1970.
    (c) *Affinity Chromatography. Principles and Methods*, 1974.
    All published by Pharmacia Fine Chemicals, Uppsala, Sweden.
54. A. H. Gordon, 'Electrophoresis of proteins in polyacrylamide and starch gels', in *Laboratory Techniques in Biochemistry and Molecular Biology* (T. S. Work and E. Work, Eds.), North-Holland Publishing Co., 1972.
55. C. R. Lowe and P. D. G. Dean, *Affinity Chromatography*, Wiley, 1974.

Author index

Adair, G. S., 192, 193
Altman, C., 167, 168
Arrhenius, S., 117

Begg, G., 54
Briggs, G. E., 29, 32
Brown, A. J., 26
Buchner, E., 3
Burk, D., 33
Burns, J. A., 228

Changeux, J-P., 194
Cleland, W. W., 174, 176, 185
Crestfeld, A. M., 133

Dalziel, K., 178, 181, 185
Dixon, M., 3, 150

Eadie, G. S., 35
Edman, P., 53, 54, 110
Eyring, H., 120

Fischer, E., 85

Gutfreund, H., 25

Haldane, J. B. S., 29, 32, 144
Hanes, C. S., 167, 168
Hartley, B. S., 76
Henri, V., 26
Hill, A. V., 190
Hoare, D. G., 127
Hofstee, B. H. J., 35

Ingraham, L. L., 168

Jacob, F., 241
Jencks, W. P., 126

Kacser, H., 228
King, E. L., 167, 168
Koshland, D. E., Jr., 104, 107, 127, 202, 207
Krebs, Sir Hans A., 236

Lienhard, G. E., 126
Lineweaver, H., 33
London, F., 74

Makower, B., 168
Massey, V., 37
Mathias, A. P., 134
Menten, M. L., 26
Michaelis, L., 26
Monod, J., 109, 194, 241
Moore, S., 47

Newsholme, E. A., 234
Northrop, J. H., 41

Ogston, A. G., 112

Pauling, L., 121
Payen, A., 3
Persoz, J-F., 3

Rabin, B. W., 134
Richards, F. M., 133
Rolleston, F. S., 227

Scatchard, G., 95, 193
Schleiden, M. J., 1
Schwann, T., 1
Stein, W. H., 47
Sumner, J. B., 41

Willstatter, R., 41
Wong, J. T., 167, 168
Wyman, J., 194

Subject index

Abortive complexes. 183
Absorption spectrum, alteration on ligand binding, 86
Acetone as protein precipitant, 265
Acetylcholine esterase, 257
Acetyl coA, 17, 249, 257
  carboxylase, 18
N-acetyl glucosamine (NAG), 107
N-acetyl glutamic acid, 238
N-acetyl imidazole, 104
N-acetyl muramic acid (NAM), 107
Activated complex, 118
Activation and activators, 115
Activation diagram, 123
Activation energy, 119
Activation free energy, $\Delta F^{\ddagger}$, 119
  relationship to rate constant, 125
Active sites, 85, 109
Activity–pH curve, 159
Adair hypotheses, 192
Adenylate kinase, 236
ADP, 17, 138, 236, 250
Adsorption chromatography, 270
Affinity chromatography, 270
Affinity labelling, 101
Agarose, 266, 270
Ageing phenomena, 84
Alanine, Table 1 (p. 43)
Alcohol dehydrogenase, 250
Aldolase, 235
  ageing of, 84
Aliphatic side chains, 44
Allosteric activator, 194, 208
Allosteric control, 109, 233
Allosteric enzymes, 109, 207, 235
Allosteric inhibitors, 110, 221
Allosteric interaction, 186, 194
Allosteric ligands, 109
Allosteric modifications, 113
Allosteric proteins, repressors, 242
Allosteric sites, 109, 186
Alternative pathways, 166

Amide group in side chains, 42
Amino acid, analyser, 45
  composition, 45, 275
  in proteins, 41
Amino peptidases, 58
Amino terminal analysis, 57, 80, 275
Amino terminus, 51
Ammonium sulphate in protein precipitation, 83, 261
AMP, 235, 250
  cyclic (cAMP), 234, 243
    binding protein, 245
Analytical ultracentrifuge, 47
Anilinothiazolinone (ATZ), 53
Anion exchange resins, 268, Table 8 (p. 269)
Antibiotics, 10
Antibodies, 272
Antimetabolites, 122
Antimicrobial drugs, 10
Apolar side chains, 44
Approximation effect, 126
Arginine, 45, Table 1 (p. 43)
Arginyl residues, chemical modification, 104
Aromatic stacking, 74
Asparagine, 42, Table 1 (p. 42)
  conversion to aspartic acid, 45
  deamidation, 84
Aspartate, 109, 208. 221
  transcarbamoylase, see ATCase
Aspartic acid, 42, Table 1 (p. 42)
Assay of enzymes, 36
ATCase (aspartate transcarbamoylase), 76, 79, 109, 207, 221
  catalytic subunit, 110, 211, 212
  oligomer, 214
  regulatory subunit, 110, 211, 213
ATP, 17, 138, 208, 211, 234, 235, 250, 258
AUG, initiation codon, 245
Autocatalytic effect, 189

Bed volume, in column chromatography, 266

Binding site, 85
  detection of, 86
Binomial theorem, 97
Biotin, 18, Table 2 (p. 88)
Blood clotting, 234
Boltzmann constant, 64, 118
Boltzmann distribution law, 118
'Bottlenecks' in metabolic pathways, 228
Branched kinetic mechanisms, 165
Branched pathways in metabolism, 221
Briggs–Haldane hypothesis, 29
N-bromosuccinimide, 104

Calcium phosphate, protein adsorbent, 270
Cancer, enzymes in, 11
Carbamoyl aspartate, 109, 208
Carbamoyl phosphate, 109, 208, 221, 238
  synthetase (cytoplasmic), 238
  synthetase (mitochondrial), 238
Carboxylases, 257
Carboxy-lyases, 257
Carboxyl groups in side chains, 42
Carboxyl terminal analysis, 59
Carboxyl terminus, 51
Carboxymethyl cysteine, 53
Carboxypeptidases, 59, 60, 107
Cascade process, 234
Catalysis, theory of, 117
Catalytic groups, 86, 109, 128
Catalytic site, 85
Cation exchange, columns, 45, 268
  resins, 268, Table 8 (p. 269)
CD (circular dichroism), 90, 106
CDP, 210
Cellular regulation, 6, 220
Central dogma, 75
Chaotropic reagents, 82
Charge-relay systems, 128
Chemical modification of proteins, 99, 246
Chemotherapy, enzymes in, 10
p-Chloromercuribenzoate (PCMB), 80, 102, 107, 211
Chromatogram, 268
Chromatography, in molecular weight determination, 48
  in protein purification, 265
Chymotrypsin, 55, 70, 101, 112, 113
Chymotrypsinogen, 112
Circular dichroism (CD), 90, 106
cis–trans isomerases, 258
Cistrons, 241, 245
Citrate, 257
  synthase, 257
Classification of enzymes, 12, 248, 250

Cleland nomenclature of enzyme mechanisms, 174
CMP, 210
Coarse control, 8, 240
Cobalamin, vitamin B_{12}, 89
Cobalt as cofactor, 86
Coenzyme A, 88, 257
Coenzymes, 86, Table 2 (p. 87)
Cold labile enzymes, 163
Collision number, 118
Column chromatography, 266
Committed enzymes, 221
Compartmentation, 9, 237
Competitive binding, 99
Competitive inhibition, partial, 152
  pure, 148
Competitive inhibitors, 106, 147, 148
Complementary kinetic mechanisms, 182
Concerted model, 195
Concerted transformation constant, 195
Conformational change, 62, 106, 107, 109, 110, 186, 187, 243, 246
Conformation of proteins, 62
  factors affecting, 82
Conservation equation, 27
Constitutive proteins, 243
Control, coarse and fine, 8, 217, 240
  irreversible and reversible, 113
  of enzyme activity, 7
Controllability (K), 229
Control strength, 229
Cooperative process, 63
Cooperativity, 186, 189
Coordinate induction and repression, 241
Copper as cofactor, 86
Co-repressor, 243
Coupled enzyme assays, 39
Covalent catalysis, 129
Criteria of protein purity, 275
Crystalline enzymes, activity of, 82
C-terminal analysis, 59
CTP, 110, 208, 209, 221
Cyanate, 57, 103
Cyanogen bromide, 55
Cyclic phosphodiester, 131
1,2-Cyclohexanedione, 105
Cyclo-isomerases, 258
Cysteic acid, 53, 76
Cysteine, Table 2 (p. 42), 44, 83
Cystine, 44

DA-cobalamin, 89
Dalziel replotting procedure, 178

Dansyl amino acids, 57
Dansyl chloride, 56, 134
Dansyl–Edman method, 56
Dead end complexes, 183
Deamidation of asparaginyl residues, 84
Deficiency diseases and cofactors, 86
Degree of a rate equation, 173
Dehydrogenases, nomenclature, 256
Denaturation, 62, 263, 265
Deproteinization of biological fluids, 83
De-repression, 243
Desensitization, 194
Diagonal technique, 76
Dialysis, 91
Diazo-1-H-tetrazole, 103
Dielectric constant, 72, 129, 163, 265
Diethyl pyrocarbonate, 103
Difference spectrophotometry, 106
Differentiation, enzymes in, 11
Diffraction pattern, X-ray, 81
Diffusion, coefficient, 47
  studies and protein shape, 81
Dihydroxyacetone phosphate, 122
Dipole, 63
Disequilibrium ratio (ρ), 224, 225, 231
Dissociation constant, K_d, 27
  experimental determination of, 91, 92, 95
Disulphide bond, 44, 53, 75, 76
Dithiothreitol, 84
Dixon plot, 150
DNA, 5, 241, 242
Drugs, as enzyme inhibitors, 10, 122
'Dummy' enzyme, 220

Eadie plot, 35
E. C. Numbers, 250
Edman degradation, 53
Edman subtractive method, 56
Effector, 109, 117, 233
  binding, homotropic, 201
Elastase, 107, 113
Elasticity (ε), 230, 233
Electrodes, specific, 90
Electron delocalization, 73
Electron density map, 81
Electrophoresis, 114, 272
Electrophoretic mobility, 49, 272
Electrostatic interactions, 72
Ellman's reagent, 102
Endergonic reactions, 16
End-product inhibition, 6, 220
Ene-diol anion transition state, 122
Enolase, 82, 257

Enthalpy, change ΔH, 20
  of activation $\Delta H^{\ddagger}$, 125
Entropy and entropy change ΔS, 20
Entropy of activation $\Delta S^{\ddagger}$, 125
Enzyme activity, in crystals, 82
  International Unit of, 36
Enzyme assay, 36, 39
Enzyme concentration, effect on kinetics, 26, 140
  effect on metabolic flux, 229, 233, 240
Enzyme degradation, 245, 246
Enzyme kinetics, effect of activators, 146
  effect of dielectric constant, 163
  effect of enzyme concentration, 26, 140
  effect of inhibitors, 146
  effect of ionic strength, 163
  effect of isomerization, 145
  effect of pH, 159
  effect of product, 141
  effect of temperature, 161
  one-substrate, 25, 140
  two-substrate, 165
Enzyme regulation, 7, 8, 113, 217, 240
Enzyme source, 250
Enzyme–substrate, complex, 26, 27, 85
  compound, 128
Enzyme synthesis, 241
Epimerases, 39, 258
Equilibrium constant, and $\Delta F°$, 20
  definition, 19
  of metabolic reactions, 218
Equilibrium dialysis, 91
Essential groups, 160
Ethanol as protein precipitant, 265
N-Ethyl maleimide (NEM), 102, 107
Exergonic reactions, 16

FAD, 87
Feedback, 6
  inhibition, 6, 208
  principle, 6
  regulation, 220, 236
Feedforward activation, 6
Fibrous proteins, 65
Fine control, 8, 217
Fingerprint of a protein, 80
First order process, 24
Flavin adenine dinucleotide (FAD), 87
Flavin mononucleotide (FMN), 87
Fluorescence of proteins, 90
Fluorodinitrobenzene (FDNB), 134
Flux, analysis, 225
  effect of enzyme concentration on, 229, 233, 240

Flux (*ctd.*)
  factors determining, 219
  glycolytic, 235
  in metabolic pathways, 217
  regulation, 220, 227–229, 233
FMN, 87
Folic acid, 89
N-Formyl methionine, 245
Fractional saturation, 191
Fraction collectors, 266
Free energy, 14
  change ΔF, 15–17
    measurement of, 19
    relation to equilibrium constant, 20
    metabolic, 224, 227
Fructose diphosphatase, 221, 234
Fructose diphosphate, 234
Fructose-6-phosphate, 234
Fumarase, fumarate, 37
Functional groups, 86
Futile cycle, 236

Galactose, 240, 243
  operon, 243
β-Galactosidase, 78, 79
Gel electrophoresis, 114, 274
Gel filtration, 48, 92, 265
General acid base catalysis, 127
General two-substrate kinetic mechanism, 165
Genes, 241
Genetic information and protein synthesis, 5, 241
Glucokinase, 138
Gluconeogenesis, 236
Glucose, 138, 240
  -1-phosphate, 233, 240
  -6-phosphate, 138
Glutamate synthase, 257
Glutamic acid, 42, Table 1 (p. 42)
Glutamine, 41, Table 1 (p. 42), 45, 238
  deamination, 84
  synthetase, 234
Glyceraldehyde-3-phosphate, 100
Glyceraldehyde-phosphate dehydrogenase, 80, 81, 90, 94, 100, 128, 249
Glycine, 42, Table 1 (p. 42)
Glycogen, 233
  phosphorylase, 233
Glycolytic flux, 235
Glycolytic pathway, 220, 221, 233, 240
Governor principle, 6

Gradient elution, 270
Group transfer reactions, 165
Guanidinium chloride, 79, 82

Haemoglobin, O_2 saturation curve, 190
Haldane relationship, 144, 178, 225
α-Helix, 65, 66
Helix–coil transition, 67
Heterotropic effects, 194, 233
Heterotropic interaction, 186
Hexokinase, 138, 240
  product inhibition of, 139
Hill coefficient, 190, 191, 202, 208
Hill plot, 192, 208, 209
Histidine, Table 1 (p. 43), 45, 240
  biosynthesis, 221
  operon, 243
Histidyl residues, chemical modification of, 103
  in chymotrypsin active site, 101
  in RNase active site, 132
Hofstee plot, 35
Homeostasis, 2, 6, 236
Homoserine, 56
Homotropic cooperativity in effector binding, 201
Homotropic effects, 194, 233
Homotropic interaction, 186
Hybrid oligomers, in ATCase, 213
  in LDH, 114
Hydantoin, 58
Hydration shells, 72
Hydrazinolysis, 60
Hydrodynamic studies of protein shape, 81
Hydrogen bonds, in proteins, 65
  in water, 63
Hydrolases, 250, 254, 257
Hydro-lyases, 257
Hydrolysis of proteins, 45
Hydrophilic side chains, 42, 45
Hydrophobic interaction, 62
  effect of ionic strength on, 65
  effect of temperature on, 65
  in quarternary structure, 77
Hydrophobic side chains, 44
Hydroxyapatite, 270
α-Hydroxybutyrate, 39
Hydroxyl groups in side chains, 42
Hydroxyproline, 45
Hyperbolic rate curves, 26, 171, 189, 194, 208

Imino acid, proline, 44
Immobilized enzymes, 12, 45

Inactivation, 115
Independent binding sites, 193
Induced fit hypothesis, 106, 128, 187
Inducer, 243
Inducible enzymes, 241, 243
Industrial use of enzymes, 12
Ingraham–Makower model, 168
Inhibition, 115
  competitive, 147, 148
  end-product, 220
  mixed, 147, 158
  non-competitive, 147, 155
  uncompetitive, 147, 159
Inhibitor constant, K_i, 150
Initial rate, v_0, 23
Initiation of protein synthesis, 245
Interacting sites, 186, 190, 194, 202, 212–216
Interaction, allosteric, 109, 186, 194
  constant, 203
  cooperative, 186, 189, 206
  heterotropic, 186, 199, 207, 208
  homotropic, 186, 196, 206, 208
  intersubunit, 188, 190, 194, 202, 212–216
Intercistronic DNA, 241
International Unit (I.U.) of enzyme activity, 36
Iodoacetamide, 100
Iodoacetic acid, 53, 100, 102, 107, 132
Ion-exchange chromatography, 268
Ion-exchange resins, 269, Table 8 (p. 269)
Ionic strength, 73, 163, 261
Iron, as cofactor, 86
Isoelectric focusing, 274
Isoelectric point, 73, 268, 274
Isoenzymes, 112, 114
Isoionic point, 73
Isoleucine, Table 1 (p. 43)
  biosynthesis, 221
Isomerases, 250, 255, 258
Isomerization in enzyme kinetics, 145
Isomorphous heavy atom derivatives, 81
Iso Ordered Bi Bi mechanism, 175
Iso Theorell–Chance mechanism, 176

K_d, dissociation constant, 27
K_m, Michaelis constant, 26, 140
  measurement of, 32
  for two-substrate mechanisms, 184
  importance in flux regulation, 227
  importance in glucose metabolism, 139
Kacser and Burns theory, 228
Kappa notation, 167
α-Ketobutyrate, 39

Kinetic mechanisms, 141
Kinetic models, 141, 164
King and Altman rules, 168
Koshland's reagent, 104
Koshland's sequential model, 202
K-systems, 200, 209

Lactate dehydrogenase, 39, 107, 114
Lag period in enzyme assays, 39
Leucine, Table 1 (p. 43)
Ligand and ligand binding sites, 85
Ligand binding, experimental measurement
  of, 86
Ligand-linked conformational change, 187
Ligases, 250, 256, 258
Lineweaver–Burk plot, 34, 147, 149, 156–
  158, 173, 174, 179–183
Lock and key hypothesis, 85
London dispersion forces, 74
Lyases, 250, 255, 257
Lysine, Table 1 (p. 43), 45
Lysozyme, 68, 69, 75, 107, 109, 121
Lysyl residue, chemical modification of, 103
  in RNase, 134

Magnesium ion, as cofactor, 86, 238
Malate, 37
Maleate, 209
Maleic anhydride, 103
Malonyl CoA, 17
Manganese, as cofactor, 86
Mass action ratio (Γ), 224
Maximum velocity, V_m, 26, 32, 140, 184, 227, 232
Maxwell–Boltzmann distribution law, 64
Mechanisms, kinetic, 141
Mercaptide ion, 44
Mercaptoethanol, 75, 84, 211
Mercuric ions, in ATCase, 214
Metabolic pathways, map, 220, 222
Metal ions as cofactors, 86
Methionine, Table 1 (p. 42), 44, 83, 245
  sulphoxide, 44, 83
Methionyl residues, cleavage at, 55
O-Methyl isourea, 103
Michaelis constant, K_m, 26, 140
  measurement of, 32
  for two-substrate mechanisms, 184
  importance in flux regulation, 227
  importance in glucose metabolism, 139
Michaelis–Menten hypothesis, 26
Michaelis plateau, 36
Michaelis plot, 33

Mixed inhibition, 147, 158
Modern Rack hypothesis, 109, 128
Modified enzyme, 166
Molecular sieve chromatography, 48, 265
Molybdenum as cofactor, 86
Monod, Wyman and Changeux (M.W.C.) model, 194
Mucopolysaccharide, 107
Multienzyme complexes, 249
Multiple electrophoretic forms, 84, 114
Multiple forms of enzymes, 84, 112, 114, 250
Mutases, 258
Mutations, 6, 112, 243, 246
M.W.C. model, 194

$\bar{N}$, average number of ligand molecules bound per molecule of protein, 95, 97
NAD and NADH, 39, 88, 236, 248
NAG (N-acetyl glucosamine), 107
NAG–NAM polymers, 107
NAM (N-acetyl muramic acid), 107
NBS (N-bromo-succinimide), 104
Near-equilibrium reactions, 224, 227, 236
Negative cooperativity, 202, 206
Negative feedback, 6, 221
NEM, (N-ethyl maleimide), 102, 107
Niacin, 88
Nicotinamide adenine dinucleotide (NAD), 39, 88, 236, 248
Nicotinic acid, 88
Ninhydrin, 45
N.m.r., 90, 106
Nomenclature of enzymes, 12, 248
Non-competitive inhibition, 147, 155
Non-equilibrium reactions, 224, 227, 228
Non-hyperbolic rate curves, 143, 147, 171, 172, 189, 194, 198, 208, 233, 235
Non-polar side chains, 44
Non-sequential mechanisms, 174
N-terminal analysis, 57, 275
Nuclear magnetic resonance (n.m.r.), 90, 106

Obligatory intermediates, 169
Oligomers, oligomeric proteins, 76, 246
Operators, 241
Operon theory, 241
Opsin, 87
Optical activity of amino acids, 42
Optical isomers, 111
Optical rotatory dispersion (ORD), 90, 106
Optimum pH, 160
Optimum temperature, 161
Orbital steering, 127
ORD, 90, 106

Ordered Bi Bi mechanism, 175
Ordered mechanisms, 174
Organic solvents as protein precipitants, 265
Organomercurials, as SH reagents, 102, 107
Orientation in catalysis, 126
Overlaps in sequence determination, 60
Oxaloacetate, 257
Oxidases, 256
Oxidoreductases, 248, 250, 256

Pacemaker enzymes, 228
Pantothenic acid, 88
Papain, 55, 82, 116
Partial specific volume, 47
Pepsin, 55, 76
Peptidases, 55
Peptide bond, 50, 61
Peptide map, 80
Perchloric acid, 82
Performic acid, 53, 77
Peristaltic pump, 266
pH, effect on enzyme kinetics, 159
optimum, 160
Phenylalanine, Table 1 (p. 43), 45
Phenylglyoxal, 105
Phenylisothiocyanate, 53
Phenylthiocarbamyl peptides, 53
Phenylthiohydantoin, 53
Phosphofructokinase, 234
Phosphoglucomutase, 107
Phosphoglycerate mutase, 76, 78
Phosphoglycolohydroxamate, 122
Phosphorylase, 113, 234
b kinase, 234
Phosphotungstic acid, 82
Photooxidation, 132
Picric acid, 82
Ping-pong mechanism, 129, 176
pK_a, environmental effects on, 160
Planck constant, 120
β-Pleated sheet, 65, 66
Polyacrylamide gel, 49
Polycistronic message, 242
Polypeptide chain, 50
Polyvalent regulatory enzymes, 221
Positive feedforward, 6, 221
Post-transcriptional control, 245
Potassium as cofactor, 86
Potassium phosphate, as protein precipitant, 83, 265
Precipitation, in protein purification, 261
Primary structure, 50
Product inhibition, 143, 181

Proelastase, 113
Proenzymes, 112
Proline, Table 1 (p. 43), 44
Prolyl residues, effect on structure, 70
Promoter, 242, 243
Propinquity in catalysis, 126
Prosthetic groups, 86
Protein degradation, 246
Protein–ligand complex, detection of, 86
Protein precipitants, general, 82
Protein purification, 261
Protein structure, non-aqueous core, 62
    primary, 50
    quaternary, 76
    secondary and tertiary, 61, 70
    X-ray crystallography, 81
Protein synthesis, 241
Protein turnover, 246
Proteolytic enzymes, 55, 84, 245
Protomers, 76
Purification of proteins, 261
Purity of proteins, criteria of, 275
Pyridoxal phosphate, 88
Pyridoxine, 88
Pyrimidine biosynthesis, 110, 221, 222, 238
Pyrimidine nucleotides, 208, 222
Pyruvate, 39, 249
    dehydrogenase system, 234, 249
    kinase, 221

Q_{10}, 161
Quaternary structure, 76

Racemases, 258
Random coil, 67
Random mechanisms, 165
Random rapid equilibrium mechanism, 174, 176
Rapid equilibrium assumption, 27
Rapid reaction techniques, 130
Rate constant, k, 23
    first order, 24
    related to $\Delta F^{\ddagger}$, 125
    second order, 24
Rate equation, structure of, 142, 164
Rate of reaction, 22
    effect of temperature, 117
Rates of synthesis and degradation of enzymes, 240, 243, 245
Reaction coordinate, 123
Reaction rate, measurement of, 22
Reaction specificity, 110
Recognition sites on DNA, 242
Recommended names of enzymes, 249

Reductases, nomenclature, 256
Regulatory efficiency, 233
Regulatory enzyme 221, 233, 246
Regulatory steps, identification of, 227
Relaxed form of an oligomer, 195
Replotting procedure of Dalziel, 178
Reporter groups, 106
Repressible enzymes, 241, 243
Repressors, 241, 242, 243
Retinene, 87
Riboflavin, 87
Ribonuclease A, 51, 52, 55, 70, 73, 75, 116, 131–136
    UpcA complex, 136
Ribonuclease S, 133
Ribosomes, 242, 245
mRNA, 241
tRNA, 245
RNA polymerase, 241, 242
Ruheman's Purple, 45

$S_{0.5}$, 200, 208
Salt bridge, 72, 73
Salt gradient, 270
Salting-in, 261
Salting-out, 73, 83, 261
Saturated enzyme reactions, 227
Scatchard plot, 95, 193, 209
SDS, 48, 79
    –gel electrophoresis, 48, 78
Secondary structure, 61, 69
Second order process, 24
Sedimentation, in study of protein shape, 81
    coefficient, 47
    -equilibrium method, 48
    -velocity method, 47
Self-regulation of metabolism, 6, 217
Sensitivity (Z), 229
Sephadex, 48, 266
Sequence determination, 50, 54
Sequential mechanisms, 174
Sequential model of Koshland, 202
Serine, Table 1 (p. 42)
Sigmoid rate curves, 171, 172, 189, 194, 198, 208, 233
β-Sheet, 65, 66
SH groups, 44, 53, 100, 102, 107, 213
Sodium dodecyl sulphate (SDS), 48, 79
Solubility of proteins at isoelectric point, 274
Source, of enzyme, 250
Spacer-arms, 271
Specific affinity, 270

Specific binding sites, 106
Specific cleavage of polypeptides, 55
Specific electrodes, 90
Specificity, 110, 242
Specific rate of reaction, 164
S-peptide, in RNase S, 133
S-protein, in RNase S, 133
Steady state fluxes, 217
Steady state hypothesis, 31
Stereoisomers, 111
Stereospecificity, 107, 111
Strain theory of catalysis, 127
Streptococcal proteinase, 116
Structural rules of Wong and Hanes, 168
Substrate analogue, 39, 106, 132, 209
Substrate distortion, 109
Substrate independent reactions, 227, 228
Substrate protection, 101
Substrate specificity, 111
Subsystems in metabolism, 220, 229
Subtilisin, 55, 133
Subunits, 76, 78
Succinate, 209
Sulphenic acid group ($-$SOH), 83
Sulphosalycilic acid, 82
Sulphydryl group ($-$SH), 44, 53, 100, 102, 107, 213
Surface denaturation, 263
Svedberg equation, 47
Switch mechanisms, 233, 236
Synergistic inhibition, 221
Synthases, 257
Synthetases, 256, 258.
Systematic names of enzymes, 248

Tautomerases, 258
Temperature, effect on enzyme kinetics, 161
    effect on rate of reaction, 117
Template hypothesis, 85
Ternary complex, 166
Tertiary structure, 61, 70
Tetrahydrofolic acid, 89
Tetranitromethane, 104
Theorell–Chance mechanism, 175
Thermodynamic box, 121
Thermolysin, 55
Thiamine, vitamin B_1, 87
Thiamine pyrophosphate, 87
Thioester bond, 129
Thiol–disulphide exchange, 75
Three-dimensional structure of proteins, 61–84
Three-point attachment hypothesis, 112
Threonine, 42

Tight form of an oligomer, 195
Titration of ligand binding sites, 94
Tosyl phenylalanine chloromethyl ketone (TPCK), 101
Tosyl phenylalanine methyl ester, 101
Trace metals, 86
Transcription, 241, 242
Transcriptional control, 245
Transferases, 250, 253, 257
Transfer RNA, 245
Transformation constant, 202
Transition state, 118, 134
    analogues, 121
Translation, 242
Translational control, 245
Trichloracetic acid (TCA), 82
Tri-NAG lactone, 122
Trinitrophenol, 82
Triosephosphate isomerase, 122
Trivial names of enzymes, 249
Trypsin, 55, 80, 112, 113
Trypsinogen, 112
Tryptophan, Table 1 (p. 43), 45, 246
    pyrrolase, 246
Tryptophanyl residues, 104
Turnover of protein, 84, 245, 246
Tyrosine, Table 1 (p. 43), 45
Tyrosyl residues, 104

UDP, 208
UDP-galactose, 39
    epimerase, 39
UDP-glucorunic acid, 39
UDP-glucose, 39
    dehydrogenase, 39
Ultracentrifugation in ligand binding studies, 92
Ultracentrifuge, 47, 265
UMP, 208
Unbranched mechanisms, 174
Uncompetitive inhibition, 147, 158
Unfruitful collisions, 118
UpcA, 132, 136
Urea, 12, 79, 82, 238
    cycle, 238
Urease, 12, 41, 111
Uridine, 210
UTP, 208

V_m or V_{max}, 26, 32, 140, 184, 227, 232
Valine, 43
van der Waal's forces, 74
Velocity of reaction, v, 22
Viruses, enzymes in, 11

Viscosity, and protein shape, 81
Vitamins, 86, Table 2 (p. 87)
V systems, 201

Wasteful cycle, 234
Water, structure of, 63
Wong and Hanes' rules, 168

X-ray diffraction studies, 81, 106, 107, 136, 212, 214

Zinc, as cofactor, 86
in ATCase, 211
Zone electrophoresis, 272
Zymogens, 112